Mathematics,
Matter and Method

Philosophical Papers, Volume 1

D1488446

Mathematics, Matter and Method

Philosophical Papers, Volume 1

SECOND EDITION

HILARY PUTNAM

Professor of Philosophy, Harvard University

CAMBRIDGE
UNIVERSITY PRESS

Published by the Press Syndicate of the University of Cambridge
The Pitt Building, Trumpington Street, Cambridge CB2 1RP
40 West 20th Street, New York, NY 10011-4211, USA
10 Stamford Road, Oakleigh, Melbourne 3166, Australia

First published 1975
Second edition with additional chapter, published 1979
Reprinted 1980, 1985, 1987, 1989, 1993, 1995

ISBN 0-521-29550-5 paperback

Transferred to digital printing 2003

Contents

Volume 1

To Ruth Anna

Introduction: Science as approximation to truth

These essays were written over a fifteen-year period. During that time my views underwent a number of changes, especially on the philosophy of mathematics and on the interpretation of quantum mechanics. Nevertheless they have, I believe, a certain unity.

The major themes running through these essays, as I look at them today, are the following: (1) Realism, not just with respect to material objects, but also with respect to such 'universals' as physical magnitudes and fields, and with respect to mathematical necessity and mathematical possibility (or equivalently with respect to mathematical objects); (2) the rejection of the idea that any truth is absolutely *a priori*; (3) the complementary rejection of the idea that 'factual' statements are all and at all times 'empirical', i.e. subject to experimental or observational test; (4) the idea that mathematics is not an *a priori* science, and an attempt to spell out what its empirical and quasi-empirical aspects really are, historically and methodologically.

Realism

These papers are all written from what is called a *realist* perspective. The statements of science are in my view either true or false (although it is often the case that we don't know which) and their truth or falsity does not consist in their being highly derived ways of describing regularities in human experience. Reality is not a part of the human mind; rather the human mind is a part – and a small part at that – of reality. But no paper in this collection is entirely devoted to the topic of realism, for my interest in the last fifteen years has not been in beating my breast about the correctness of realism, but has rather been in dealing with specific questions in the philosophy of science from a specific realist point of view. (However, the reprinting of these volumes has given me the opportunity to include my essay on 'Philosophy of logic' which does discuss the case for realism.)

To prevent misunderstandings, let me say that by realism I do not mean materialism. Part of the burden of these essays is that a consistent realist has to be realistic not only about the existence of material objects in the customary sense, but also has to be realistic about the objectivity of mathematical necessity and mathematical possibility (or equivalently about the existence of mathematical objects) and about entities which are neither material objects nor mathematical objects – in particular, fields and physical magnitudes. The importance of this latter kind of realism is discussed in the paper 'On properties' and also plays a key role in 'An examination of Grünbaum's philosophy of geometry'. Grünbaum, at the time I criticized him, maintained that the fundamental field in modern cosmology – the gravitational field or metric field – was simply a way of articulating relations between solid bodies; my rejection of this contention is an example of the kind of realism just alluded to.

Rejection of the *a priori*

While realists in the philosophy of science were comparatively scarce in the 1950s, they are today, happily, a flourishing species. My views in the philosophy of mathematics and on quantum logic go beyond what most realists are prepared to defend. The reason that I hold positions which are at least different from, if not incompatible with, the positions held by most realistic philosophers of science is that I subscribe to an epistemological view, which in a certain sense denies the existence of *a priori* truth. That position is briefly stated in the paper 'It ain't necessarily so'; to be properly understood, that paper should really be read together with 'The analytic and the synthetic', a paper in the other volume of these collected papers. (The division of my collected papers into those on the philosophy of science on the one hand, and those on philosophy of language and the philosophy of mind on the other, has necessarily created certain problems; at many places in these papers I appeal to ideas from the philosophy of language.)

The concept of an *a priori* truth that I reject is historically best represented by Descartes' notion of a clear and distinct idea. Descartes' clear and distinct ideas were in a certain sense self-certifying truths. They bore their validity on their face. Not only could it never be rational to give up such a truth; it could never even be rational to doubt one.

That we should give up the idea that there are such truths is the lesson that I believe we should learn from the history of geometry.

Suppose someone says: 'You can return to the place from which you started by travelling in a straight line.' Forget, for the moment, anything you may know about relativity theory or non-Euclidean geometry, and put yourself in the frame of mind that the best human minds were in, say, in 1800. What might one have replied? One might have said: 'Sure, if I travel from New York in a straight line on the surface of the earth, then eventually I will come back to New York.' Suppose the first speaker replies: 'No, I don't mean travelling in a straight line on a curved surface, I mean travelling in a really straight line, a straight line *in space*.' One might have replied: 'You mean travelling in a straight line in one direction for a while and then reversing one's direction and coming back?' Suppose the first speaker says: 'No I don't mean that. I mean that you can come back to the same place by travelling in a straight line in space without ever changing the sense of your motion.' Would not this have been an *evident* impossibility? Wouldn't one have thought that it was an impossibility in 1800 or even in 1840 (unless one knew the works of Lobachevsky, or Gauss or Bolyai)? Would not one have been *entitled* to think that this was an impossibility? Indeed, even after the work of Lobachevsky, might one not have been entitled to think that it would still be an impossibility for this to be the case in actual space until the General Theory of Relativity? Is it not the case that here we have an example of something which seems evidently impossible and which one has the right to believe to be impossible until a certain theory has been produced, in fact a whole group of theories?

Strange to say, this is not the moral that philosophers have drawn from the history of geometry. The received view is that the temptation to think that the statements of Euclidean geometry are necessary truths about actual space just arises from a confusion. One confuses, so the story goes, the statement that one can't come back to the same place by travelling in a straight line (call this statement 'S'), with the statement that S is a theorem of Euclid's geometry. That the axioms of Euclid's geometry imply S is indeed a necessary truth; but there is all the difference in the world between saying that S would be true if space were Euclidean, and saying that S is necessarily true.

To put it bluntly, I find this account of what was going on simply absurd. I think I know why it seemed impossible that one could travel in a straight line and come back to the same place. I think I can easily put myself in a frame of mind in which it still seems impossible, and it is just false that what I am doing is confusing the statement that one can't come back to the same place by travelling in a straight line with the statement that 'Euclid's axioms imply that one can't come back to the same place by travelling in a straight line.' If it was just an illusion that the statements of Euclidean geometry were necessary truths – necessary truths about the space in which bodies move – then philosophers of science owe us a plausible explanation of this illusion. The explanation that they have given us is a non-starter. It will not do to wave about the distinction between Euclidean space as an abstract mathematical object and physical space, i.e. the space in which bodies move. For what seems to be a necessary truth – for what would have been a necessary truth relative to our body of knowledge if it had been a truth – is precisely that physical space is a Euclidean space.

Let me approach this point in another way. In my view physical entities are not literally operationally defined; they are in a very complicated sense 'defined' by systems of laws. (The exact sense is discussed in two papers in the other volume of these collected papers, namely 'The meaning of meaning' and 'Explanation and reference'.) If the term 'straight line' were literally *defined* (say as 'path of a light ray'), then the statement that physical space obeys the laws of Euclid's geometry would never even have seemed an *a priori* truth. What creates the whole problem is the fact that the theory that 'defined' physical space was a theory to which there were no conceptual alternatives prior to some time in the hundred years between 1815 and 1915. In short, Euclidean geometry as a theory of physical space was always a synthetic theory, a theory about the world, but it had the strongest possible kind of paradigm status prior to the elaboration of the alternative paradigm. If this is correct, and I believe it is, then the overthrow of Euclidean geometry is the most important event in the history of science for the epistemologist; and the received account of that overthrow is a philosophical scandal. My account does not deny – indeed it affirms – that there is a distinction between truths which are *a priori* relative to a particular body of knowledge and truths which are empirical relative to a particular body of knowledge. What it denies is that there are truths which are *a priori* relative to the philosopher's favorite context, which is the context of 'all contexts'. The context of 'all contexts' is no context at all.

These ideas, which I had formed in the course of an examination of the history of geometry, were reinforced by the discovery that if the formalism of contemporary quantum mechanics is really representative of the kind of formalism that fundamental physical theories are going to use in the future, then the only way to make sense of this seems to be an acceptance of the idea that we live in a world with a non-classical logic. In no area, not even the area of logic, are 'clear and distinct ideas' immune from overthrow.

Philosophy of mathematics

Not surprisingly, I was led to reexamine the history of classical mathematics. In this enterprise, aided both by some insightful papers by Kurt Gödel, even if I could not agree with all of Gödel's Platonism, and by some discussions with the mathematician Martin Davis, I was led to the conclusion that the differences between mathematics and empirical science have been vastly exaggerated. That in mathematics too there is an interplay of postulation, quasi-empirical testing, and conceptual revolution leading to the formation of contextually *a priori* paradigms. Mathematics is not an experimental science; that is the first thing that every philosopher learns. Yet the adoption of the axiom of choice as a new mathematical paradigm *was* an experiment, even if the experiment was not performed by men in white coats in a laboratory. And similar experiments go all the way back in the history of mathematics. Of course mathematics is much more '*a priori*' than physics. The present account does not deny the obvious difference between the sciences. But seeing the similarities as well as the differences renders the differences for the first time philosophically intelligible, I believe.

In particular, we can recognize that principles of mathematics are relatively *a priori*, without having to conclude that they are conventions or rules of language and therefore say nothing. In particular we do not have to choose between Platonism – the view that mathematics is about objects of which we have *a priori* knowledge – and Nominalism – the view that mathematics is not about real objects, that most of it is just make believe.

Reichenbach was fond of the idea that theories which appear to be incompatible can sometimes be equivalent descriptions of the same states of affairs. Although Reichenbach was led to this view by the verifiability theory of meaning, the view itself is quite independent of that theory. Suppose, for example, that we lived in a Newtonian world. It is easy to show that in such a world physical theory admits of two seemingly different formulations. In the first formulation, space points and point masses are taken to be individuals, and the relation between a point mass and the space point at which it is located, is a relation between individuals. In the second formulation, only point masses are individuals but space points are properties of individuals (logically strongest location properties). Also, there are possible definitions of the primitive terms of each theory in the language of the other theory. It is, moreover, a logical consequence of each theory that the theorems of the other theory are true when the primitive terms of the other theory are appropriately interpreted. So we may say that if Newtonian physics were true, then these two theories would be equivalent descriptions. If Newtonian physics were true, the question 'Are there such objects as space points in addition to mass points?' (i.e. 'Are there such *individuals* as space points in addition to mass points?') would have *no representation-independent answer*. On the other hand, if general relativity theory is

correct, then there are statements about the actual and possible behavior of space–time points, which do not admit of any 'translation'
into statements about the actual or possible behavior of particles. Thus
the question of whether field theory and particle theory are equivalent
descriptions of the world or not is not a question to be settled by meaning
analysis or by any kind of *a priori* analysis; it depends on what the
correct physical theory of the world is. This is what I mean by saying
that the notion of an 'equivalent description' can usefully be taken over
from Reichenbach; but when it is stripped of its positivistic under-
pinnings, it changes in important ways. In particular, whether two
theories are equivalent descriptions or not is now a synthetic or 'factual'
question, not a question to be settled *a priori*. What questions are
representation-relative, not in principle but in practice, depends on what
alternative representations one is scientifically motivated to introduce.
There is no motive for introducing a representation in which the
boiling point of water is 60°F; and if one did introduce such a
representation, it would be correctly said that one had simply trivially
changed the meaning of the expression 60°F, or some other
expression. Thus the question 'What is the boiling point of water?'
is not, in practice, a question with a representation-relative
answer.

I bring this up because it is the burden of two of the papers in
this volume – 'Mathematics without foundations' and 'What is
mathematical truth?' – that the question 'Do mathematical objects
exist over and above ordinary physical objects?' is one which has no
representation-independent answer. The old question 'Would a list
of all the things in the world include chairs *and* numbers, or only
such things as chairs?' is not a good question. Showing that it is not a
good question does not require the assumption of any empiricist
exclusion principle, however disguised.

If I were asked whether I am a 'Platonist' (or better an Aristotelian)
with respect to the existence of mathematical objects, my answer would
have to be a guarded 'yes'. Aristotle eliminated certain absurdities
from Plato's theory (the Third Man Argument). And he moved part
of the way in the direction I have moved here, the direction of making
the knowledge of mathematical truth dependent upon experience.
If we move further in that direction by recognizing that mathematical
methodology involves quasi-empirical elements, and by recognizing
that the existence of mathematical objects is representation-relative,
then I believe we get a better approximation to the truth. If we do this,
then we see that Platonism itself is a *research program*; not something
fixed once and for all but something to be modified and improved by

trial and error. In my view this is something characteristic of all live philosophical positions.

I have included in the collection one paper – 'The thesis that mathematics is logic' – with which I now very largely disagree. I have included this paper, although it may appear to constitute a jarring note, for two reasons: firstly, it does partly anticipate the Modal Logical Picture described in the two papers that follow it; secondly, it was my realization of the literal falsity of one sentence in this paper ('applied mathematics consists of assertions to the effect that *if* anything is a model for certain system of axioms, *then* certain statements "if A then B" hold, where A and B are first order statements *not* about the hypothetical model but about real things') – a sentence which claims, in effect, that mathematics is only used to derive statements in nominalistic language from statements in nominalistic language – that led me to the view espoused in *Philosophy of Logic* and in the two papers mentioned.

So much for the most pervasive themes in these papers. The last papers in this volume are concerned with general issues in the methodology of science and try to suggest, all too inadequately I fear, what scientific methodology looks like from the present perspective. Received philosophy of science suggests that science is highly derived talk about regularities in human observation, or failing that, is myth whose cash value lies in its ability to systematize the prediction of such regularities. Many philosophers who reject this positivistic view in the philosophy of natural science nevertheless seem to feel that mathematics is myth whose cash value lies in its ability to aid natural science. If these papers have a unity, it is perhaps the view that science, including mathematics, is a unified story and that story is not myth but an approximation to the truth. An approximation parts of which may be, at certain times, provisionally '*a priori*', but all of which is subject to modification and improvement.

Since the philosophy of science is, after all, not all of philosophy, it may be well to say a word or two about wider issues. It will be obvious that I take science seriously and that I regard science as an important part of man's knowledge of reality; but there is a tradition with which I would not wish to be identified, which would say that scientific knowledge is all of man's knowledge. I do not believe that ethical statements are expressions of scientific knowledge; but neither do I agree they are not knowledge at all. The idea that the concepts of truth, falsity, explanation, and even understanding are all concepts which belong exclusively to science seems to me to be a perversion. That Adolf Hitler was a monster seems to me to be a true statement (and even a 'description' in any ordinary sense of 'description'), but the term 'monster' is neither reducible to nor eliminable in favor of 'scientific' vocabulary. (This is not something discussed in the present volume. It *is* a subject on which I hope to write in the future.)

If the importance of science does not lie in its constituting the *whole* of human knowledge, even less does it lie, in my view, in its technological applications. Science at the best is a way of coming to know, and hopefully a way of acquiring some reverence for, the wonders of nature. The philosophical study of science, at the best, has always been a way of coming to understand both some of the nature and some of the limitations of human reason. These seem to me to be sufficient grounds for taking science and philosophy of science seriously; they do not justify science worship.

H.P.

Harvard University
September 1974

I

Truth and necessity in mathematics*

I hope that no one will think that there is no connection between the philosophy of the formal sciences and the philosophy of the empirical sciences. The philosophy that had such a great influence upon the empirical sciences in the last thirty years, the so-called 'logical empiricism' of Carnap and his school, was based upon two main principles:

(1) That the traditional questions of philosophy are so-called 'pseudo-questions' (*Scheinprobleme*), i.e. that they are wholly senseless; and

(2) That the theorems of the formal sciences – logic and mathematics – are analytic, not exactly in the Kantian sense, but in the sense that they 'say nothing', and only express our linguistic rules.

Today analytical philosophers are beginning to construct a new philosophy of science, one that also wishes to be unmetaphysical, but that cannot accept the main principles of 'logical empiricism'. The confrontation with the positivistic conception of mathematics is thus no purely technical matter, but has the greatest importance for the whole conception of philosophy of science.

What distinguishes statements which are true for mathematical reasons, or statements whose falsity is mathematically impossible (whether in the vocabulary of 'pure' mathematics, or not), or statements which are mathematically necessary, from other truths? Contrary to a good deal of received opinion, I propose to argue that the answer is *not* 'ontology', *not* vocabulary, indeed nothing 'linguistic' in any *reasonable* sense of linguistic. My strategy will not be to offer a contrary thesis, but rather to call attention to facts to which 'ontological' accounts and 'linguistic' accounts do not do justice. In the process, I hope to indicate just how complex are the facts of mathematical life, in contrast to the stereotypes that we have so often been given by philosophers as well as by mathematicians pontificating on the nature of their subject.

The idea that the 'ontology' (i.e. the domain of the bound variables) in a mathematically true statement is a domain of sets or numbers or

* A German version of this paper, titled 'Wahrheit und Notwendigkeit in der Mathematik', was presented as a public lecture at the University of Vienna on 3 June 1964, under the auspices of the Institut für Höhere Studien und Wissenschaftliche Forschung (Ford Institute).

functions or other 'mathematical objects', and (moreover) that *this* is what distinguishes mathematics from other sciences is a widespread one. On this view, mathematics is distinguished from other sciences by its *objects*, just as botany is distinguished from marine biology by the difference in the objects studied. This idea lives on in a constant tension with the other idea, familiar since Frege and Russell, that there is no sharp separation to be made between *logic* and mathematics. Yet logic, as such, has no 'ontology'! It is precisely the chief characteristic of the principles and inference rules of logic that *any* domain of objects may be selected, and that any expressions may be instantiated for the predicate letters and sentential letters that they contain. (I do not here count set theory, or higher order quantification theory, as 'logic'.) *Prima facie*, then, there is something problematical about one or both of these views.

In point of fact, it is not difficult to find mathematically true statements which quantify only over material objects, or over sensations, or over days of the week, or over whatever 'objects' you like: mathematically true statements about Turing machines, about inscriptions, about maps, etc. Thus, let T be a physically realized Turing machine, and let $P_1, P_2, \ldots, P_n$ be predicates in ordinary 'thing language' which describe its states. (E.g. P_1 might be: 'ratchet G1 of T is pressing against bar T4, etc.') An atomic instruction might be: 'If $P_2(T)$ and T is scanning the letter "y", T will erase the "y", print "z" in its stead, shift one square left on the tape (more tape will be adjoined if T ever reaches the end), and then adjust itself so that $P_6(T)$'. Such an instruction is wholly in 'nominalistic language' (does not quantify over abstract 'entities'). T is completely characterized by a finite set of such instructions, $I_1, I_2, \ldots, I_k$. Then the statement

(1) As long as I_1 and I_2 and ... and I_k, then T does not halt.

could very well be a mathematically true statement, and quantifies only over physical objects.

Again, suppose we use the symbols I, II, III, ... to designate the numbers one, two, three... (i.e. the name of the number n is a string of n 'I's'). In this notation, the sum of two numbers can be obtained by merely concatenating the numerals: nm is always the sum of n and m. Let us write $x = y^*$ to mean 'x equals y cubed', Nx to mean 'x is a number', and ! (read: 'shriek') to indicate absurdity. The following is a rather rudimentary formal system whose axioms can be seen to be correct on this interpretation:

System E.S.

Alphabet I, ., $=$, *, !, N
Axioms

$$N1$$
$$Nx \to Nx1$$
$$Nx \to x = x$$
$$Nx \to x.1 = x$$
$$x.y = z \to x.y1 = zx$$
$$x.x = y, x.y = z \to z = x^*$$
$$z_1 = x_1{}^*, z_2 = x_2{}^*, z_3 = x_3{}^*, z_1 = z_2 z_3 \to !$$

It is easily seen that ! is a theorem of $E.S.$ if and only if some cube is the sum of two cubes. Fermat proved that this is impossible. Thus the following is true:

(2) If X is any finite sequence of inscriptions in the alphabet 1, ., =, *, !, N and each member of X is either an inscription of $N1$, or of a substitution instance of one of the remaining above axioms, or comes from two preceding terms in the sequence by Detachment, then X does *not* contain !.

Now, a finite *sequence* of inscriptions $I_1, \ldots, I_n$ can itself be identified with an inscription – say, with $I_1 \# I_2 \# \ldots \# I_n$, where $\#$ is a symbol not in the alphabet 1, ., =, *, !, N, which we employ as a 'spacer'. (We say that such an inscription has an inscription I in the alphabet 1, ., =, *, ! as a member, or 'contains' I, just in case the string begins with $I\#$, or ends with $\#I$, or has a proper part of the form $\#I\#$.) Thus (2) is once again an example of a mathematically true statement which refers only to physical objects (inscriptions). And such examples could easily be multiplied. Thus we see that, even if no one has yet succeeded in translating *all* of mathematics into 'nominalistic language', still there is no difficulty in expressing a *part* (and, indeed, a significant part) of mathematics in 'nominalistic language'.

Let me now, in the fashion of philosophers, consider an Objection to what I have said.

The Objection is that, even if some mathematically true statements quantify only over physical objects, still the *proofs* of these statements would refer at least to *numbers*, and hence to 'mathematical objects'. The reply to the Objection is that the premise is false! The principle needed to prove (2), for example, is the principle of Mathematical Induction. And this can be stated directly for finite inscriptions, and, moreover, can be perceived to be evidently true when so stated. It is not that one *must* state the principle first for numbers and *derive* the principle for inscriptions *via* goedel numbering. (Indeed, this would assume that every inscription possesses a goedel number, which cannot be proved without assuming the principle for inscriptions.) Rather, the

3

principle can be seen to be evidently correct for arbitrary finite inscriptions, just as one sees intuitively that the principle of Induction in number theory is correct.

Here is the Principle, state as a Rule of proof:

$$P(1)$$
$$P(.)$$
$$P(=)$$
$$P(*)$$
$$P(!)$$
$$P(N)$$
$$P(x) \to P(x1)$$
$$P(x) \to P(x.)$$
$$\dots \dots$$
$$\frac{P(x) \to P(xN)}{P(x)}$$

(In words: If 1, ., =, *, !, N are all P, and if, for every x, if $P(x)$ then $P(x1)$, $P(x.)$, ..., $P(xN)$, then, for every x, $P(x)$.)

It would (I assert) be easy to write down a set of axioms about finite inscriptions (*not* including any axiom of infinity, since we are interested only in proving universal statements, and many of these do not require an assumption concerning the *existence* of inscriptions in their proof), from which by the Rule of Induction just stated and first order logic one could prove (2), and many similar statements besides. Moreover, if one is not content to simply *assume* the Rule of Induction, but wishes to 'derive' it, one can do *that* too! It suffices to assume Goodman's calculus of individuals (parts and wholes), instead of the theory of sets (or higher order quantification theory), as was done by Frege and Russell.† Thus the Objection falls, and I conclude that, whatever may be *essential* to mathematics, reference to abstract 'entities' is *not*. Indeed, one could even imagine a culture in which this portion of mathematics – parts and wholes, inscriptions in a finite alphabet, etc. – might be brought to a

† Quine has shown (Quine, 1964) that the Frege–Russell 'derivation' of mathematical induction can be redone so as to require only finite sets. Quine's methods can also be used, to the same effect, in the Theory of Inscriptions, with finite *wholes* playing the role of *sets*. It should be noted that the principle of induction that Russell claimed to be analytic – that every hereditary property which is possessed by zero is possessed by every number – is *empty* without further 'comprehension axioms' asserting that specific conditions define 'properties' (or sets). What plays the role of an *Aussonderungsaxiom* in the theory of finite wholes is the statement that 'there is a whole, y, which is got by joining all those P's which are parts of x', where x is any whole. If we assume this as an axiom schema, then we can derive the rule of mathematical induction for finite inscriptions in the form given in the text from a definition of 'finite inscription' which is analogous to Quine's definition of number.

high degree of development without anyone ever having mentioned a so-called 'abstract entity'. What seems to characterize mathematics is a certain *style of reasoning*; but that style of reasoning is not essentially connected with an 'ontology'.

The same examples also go to show that, whatever it may be that distinguishes mathematically true statements from other statements, it is not *vocabulary*. Is it anything 'linguistic'? Consider the following statement:

(3) No one will ever draw a (plane) map which requires five or more colors to color. (The restriction being understood that two adjacent regions may not be colored the same.)

Do the Rules of English, in any reasonable sense of 'rule of English', decide whether (3) is the sort of statement that is true (or false) for mathematical rather than for empirical reasons? If we take the 'rules of English' to be the rules of 'generative grammars', such as the ones constructed by Noam Chomsky and his followers, or even the State Regularities and Semantic Rules espoused (respectively) by Paul Ziff and by Katz and Fodor (see the references at the end of this paper), then the answer is clearly 'no'. Some philosophers write as if, in addition to these everyday or 'garden variety' rules of English, which are capable of being discovered by responsible linguistic investigation carried on by trained students of language, there were also 'depth rules' capable of being discovered only by *philosophers*, and as if these in some strange way *accounted for* mathematical truth. Although this is a currently fashionable line, I am inclined to think that all this is a mare's nest. But without allowing myself to be drawn into a discussion of this 'depth grammatical' approach to mathematical truth at this point, there is a point which I would still call to your attention: namely, if one says that the 'rules of the language' *decide* whether (3) is a mathematically true or empirically true (or false) statement, then they must *also* decide the mathematical question *can every map be colored with four colors*? For suppose that (3) is mathematically true, i.e. true because what it says will never be done *could* never be done (in the mathematical sense of 'could'). Then (3) would be analogous to:

No one will every exhibit a cube which is the sum of two cubes.

In this case the answer to the Four Color Problem is 'yes' – every (plane) map *can* be colored with four colors. On the other hand, suppose (3) is an *empirical* statement. Then it must be that it is *possible* (mathematically) to produce a map which requires five or more colors to color. For this is the only case in which it *is* an empirical question whether or

5

not anyone ever *will* produce a map which requires five or more colors to color. Thus, any system of Linguistic Rules which is powerful enough to decide whether (3) is an empirical statement or a mathematically necessary statement will also have to be powerful enough to decide the Four Color Problem. And, in general, if the distinction between statements whose truth (or falsity) is mathematical in character and statements whose truth (or falsity) is empirical in character is drawn by the Rules of the Language, then all mathematical truth (in combinatorial mathematics, anyway) is decided by the Rules of the Language. The apparently modest thesis that purely linguistic considerations can determine what is a question of mathematical truth or falsity and what is a question of empirical (or ethical, or theological, or legal, or anyway other-than-mathematical) truth or falsity, is in reality no more modest than the radical thesis that linguistic considerations can determine not just that 'is it the case that p?' is a mathematical or empirical or whatever question, but also can determine the *answer* to the question, when it is a mathematical one.

Not only do 'linguistic' considerations, in any reasonable sense, seem unable to determine what is a mathematically necessary (or impossible) statement, and what is a contingent statement, but it is doubtful if they can single out even the class of assertions of *pure* mathematics. (Consider assertions like (2), for example, some of which belong to pure mathematics, on any reasonable definition, while some – of the same general form as (2), and referring to systems like *E.S.* – are only contingently true, i.e. true because some mathematically possible finite inscriptions in the alphabet in question do not in fact exist.) However, this difficulty can be overcome by 'cheating': we can specify that an assertion will not *count* as a statement of 'pure' mathematics unless there is some indication either in the wording (e.g. the presence of such an expression as 'mathematically implies'), or in the context (e.g. the assertion appears as the last line of a putative mathematical proof) that the statement is supposed to be true for mathematical reasons.

Once again, I shall succumb to my professional habit of considering Objections. The Objection this time is that (3) has (allegedly) two different *senses*: that (3) means one thing when it is intended as a mathematical statement, and something different when it is intended as an empirical statement. The Reply to the Objection is: what if I simply intend to say that no one will ever do the thing specified? Is it really to be supposed that this statement is *ambiguous*, and that the hearer does not know what I *mean* unless he knows my *grounds*? I believe that Wittgenstein would say that the ambiguity is as follows: if I intend (3) as a *prediction*, then I am not *using* (3) as a mathematical statement; but if I

6

am not prepared to count anything as a counter-example to (3), then I am using (3) as a mathematical statement. It will be the burden of the remainder of this paper to criticize this account of what it is to accept a statement as mathematically necessary.

The revisability of mathematical assertions

According to a currently fashionable view, to accept a statement as mathematically true is in some never-quite-clearly-explained way a matter of accepting a rule of language, or a 'stipulation', or a 'rule of description', etc. Of a certainty, the concepts of possibility and impossibility are of great importance in connection with mathematics. To say that a statement is mathematically true is to say that the negation of the statement is mathematically impossible. To say that a statement is mathematically true is to say that the statement is mathematically necessary. But I cannot agree that Necessity is the same thing as Unrevisability. No one would be so misguided as to urge that 'mathematically necessary' and 'immune from revision' are synonymous expressions; however, it is another part of the current philosophical fashion to say 'the question is not what does X mean, but what does the *act of saying* X signify'; and I fear that the view may be widespread that the act of saying that a proposition p is mathematically necessary, or even of asserting p on clearly mathematical grounds, or in a clearly mathematical context, *signifies* that the speaker would not count anything as a counter-example to p, or even adopts the rule that nothing *is to count* as a counter-example to p. This view is a late refinement of the view, which appears in the writings of Ayer and Carnap, that mathematical statements are consequences of 'semantical rules'. The Ayer–Carnap position is open to the crushing rejoinder that being a *consequence* of a semantical rule is *following mathematically* from a semantical rule; so all that has really been said is that *mathematics* = language plus *mathematics*. This older view went along with the denial that there is any such thing as Discovery in mathematics; or, rather, with the assertion that there is discovery 'only in a psychological sense'. The information given by the conclusion is already contained in the premisses, in a mathematically valid argument, these philosophers said: only we are sometimes 'psychologically' unable to see that it is. On examination, however, it soon appeared that these philosophers were simply *redefining* 'information' so that mathematically equivalent propositions are *said* to give the same information. It is only in this Pickwickian sense of 'information' that the information given in the conclusion of a deductive argument is already given in the premisses; and our inability to see that it is requires

7

no special explanation at all, 'psychological' or otherwise. On the contrary, what *would* require psychological explanation would be the *ability*, if any creature ever possessed it, to infallibly and instantly see that two propositions were mathematically equivalent, regardless of differences in syntactic structure and in semantic interpretation in the *linguistic* sense of 'semantic interpretation'. Far from showing that Discovery is absent in mathematics, Carnap and Ayer only enunciated the uninteresting tautology that in a valid mathematical argument the conclusion is mathematically implied by the premises.

The late refinement described above does avoid the emptiness of the Ayer–Carnap view. To accept a statement as mathematically necessary is then and there to adopt a 'rule'; not to recognize that in some sense which remains to be explained the statement is a *consequence* of a rule. Thus the late refinement is both more radical and more interesting than the older view. We adopt such a rule not *arbitrarily*, but because of our Nature (and the nature of the objects with which we come into contact, including Proofs). That a sequence of symbols exists which impels us to say 'it is mathematically impossible that a cube should be the sum of two cubes' is indeed a *discovery*. But we should not be misled by the picture of the Mathematical Fact waiting to be discovered in a Platonic Heaven, or think that a given object would be a proof if our nature did not impel us to *employ* it as a proof.

What student of the philosophy of mathematics has not encountered this sort of talk lately? And what *intelligent* student has not been irritated by it? *To be sure*, a proof is more than a mere sequence of marks, on paper or in the sand. If the symbols did not have meaning in a language, if human beings did not speak, did not do mathematics, did not follow proofs and so on, the same sequence of marks would *not* be a proof. But it would still be true that no cube is the sum of two cubes (even if no one proved it). Indeed, these philosophers do not deny this later assertion; but they say, to repeat, that we must not be misled by the picture of the Mathematical Fact as Eternally True, independent of Human Nature and human mathematical activity. But I, for one, do not find this picture at all misleading. Nor have I ever been told just *how* and *why* it is supposed to be misleading. On the contrary, I allege that it is the picture according to which accepting a mathematical proposition is accepting a 'rule of description' that is radically misleading, and in a way that is for once definitely specifiable; *this* picture is misleading in that it suggests that once a mathematical assertion has been accepted by me, I will not allow anything to *count* against this assertion. This is an *obviously* silly suggestion; but what is left of the sophisticated view that we have been discussing once this silly suggestion has been repudiated?

8

Indeed, not only can we change our minds concerning the truth or falsity of mathematical assertions, but there are, I believe, different *degrees* of revisability to be discerned even in elementary number theory.

To illustrate this last contention, consider the statement 'no cube is the sum of two cubes'. I accept this statement; I know that it can be proved in First Order Arithmetic. Now, let us suppose that I cube 1769 and add the cube of 269. I discover (let us say) that the resulting number is exactly the cube of 1872. Is it really supposed that I would not allow this empirical event (the calculation, etc., that I do) to *count against* my mathematical assertion? To accept *any* assertion p is to refuse to accept $\bar{p}$ *as long as I still accept p*. (The Principle of Contradiction.) So, to accept $x^3 + y^3 \neq z^3$ is to refuse to accept the calculation *as long as I still accept $x^3 + y^3 \neq z^3$*. But there is all the difference in the world between this and adopting the convention that no experience in the world is to be *allowed to refute $x^3 + y^3 \neq z^3$* – which would be insane.

But what would I in fact do if I discovered that $(1769)^3 + (269)^3 = (1872)^3$? I would first go back and look for a mistake in the proof of $x^3 + y^3 \neq z^3$. If I decided that the latter proof was a correct proof in First Order Arithmetic, then I would have to modify First Order Arithmetic – which would be shattering. But it is clear that in such revision purely *singular* statements ($5 + 7 = 12$, $189 + 12 = 201$, $34 \cdot 2 = 68$, etc.) would take priority over *generalized* statements ($x^3 + y^3 \neq z^3$). A generalized statement can be refuted by a singular statement which is a counter-example.

Gentzen has given a convincing example of a constructive proof (of the consistency of First Order Arithmetic, in fact), which is not formalizable in First Order Arithmetic. This proof has been widely misunderstood because of its use of transfinite induction. The point is not that Gentzen used transfinite methods, but that, on the contrary, transfinite methods can be *justified* constructively in some cases. For ε_0 induction is not taken as primitive in constructive mathematics. On the contrary, for $\alpha < \varepsilon_0$, α-induction can be reduced to ordinary induction, even in Intuitionistic Arithmetic. And if we adjoin to Intuitionistic Arithmetic the constructively self-evident principle: if *it is provable in the system* (Intuitionistic Arithmetic) *that for each numeral n, F(n) is provable, then $(x)F(x)$* – then we can formalize Gentzen's whole proof, even though our bound variables range only over numbers, and not over 'ordinals' (in the classical sense), at all.

I point this out, because I believe that it is important that there are methods of proof which are regarded as completely secure by everyone who understands them, and which are wholly constructive, but which go outside First Order Arithmetic. Now, suppose that we had such a proof that established the ω-inconsistency of First Order Arithmetic. (In fact, this cannot happen – Charles Parsons has given constructive proof of the ω-consistency of First Order Arithmetic.) Imagine that the proof is entirely perspicuous, and that after reading it we are convinced that we could actually construct proofs of $\sim F(o)$, $\sim F(1)$,..., for some formula F such that $(Ex)F(x)$ has been proved in First Order Arithmetic. If F is a decidable property of integers, this amounts to saying that $F(o)$, $F(1)$, $F(2)$,... are all *false* even though $(Ex)Fx$ is *provable*. And *this* amounts to saying that First Order Arithmetic is *incorrect*; since $(Ex)Fx$, although *provable* is clearly *false*.

I see nothing unimaginable about the situation just described. Thus, just as accepting $x^3 + y^3 \neq z^3$ does not incapacitate me from recognizing a counter-example if I ever come across one $((1769)^3 + (269)^3 = (1872)^3$ would be a counter-example if the arithmetic checked): so accepting $(Ex)Fx$ does not incapacitate me from recognizing a disproof if I see one (such a disproof would establish, *by means more secure than those used in the proof of $(Ex)Fx$*, that $\sim F(o)$, $\sim F(1)$, $\sim F(2)$,... would all prove correct on calculation). And, in general, even if a mathematical statement has been *proved*, its falsity may not be literally impossible to imagine; and there is no rule that 'nothing is to count against this'.

Another kind of revision that takes place in mathematics is too obvious to require much attention: we may *discover a mistake in the proof*. It is often suggested that this is not *really* changing our minds about a *mathematical* question, but why not? The question, 'is this object in front of me a proof of S in $E.S.$'? may not be a mathematical question, but am I not *also* changing my mind about the question 'is it true that, S?'? And this latter question *may* be a mathematical question. Indeed, if I previously adopted the stipulation that *nothing was to count against S*, then *how could* my attitude towards the mathematical proposition S be affected by the brutally empirical discovery that a particular sequence of inscriptions did not have the properties I thought it had?

An especially interesting case arises when S is itself a proposition concerning proof in a formal system, say 'F is provable in L'. Such facts are especially recalcitrant to the 'rule of description' account, as are indeed all purely existential combinatorial facts, – e.g. 'there exists an odd perfect number', if that statement is true. We may change our minds about such a statement indefinitely often (if the proof is very long); our attitude towards it depends upon brutally empirical questions of the kind mentioned in the preceding paragraph; there may be no *reason* for such a fact apart from the fact that there just *is* a proof of F in L – i.e. we often have no way of establishing such a statement except to 'give a for-instance'; and, lastly, discovering such 'for-instances' certainly *seems* to be Discovery in every sense of the word.

Yet even here defenders of Rule-of-Language accounts still have a trick up their sleeves: they can say, 'all you have shown is that various Natural Contingencies can lead us to revise our rules indefinitely often. But that doesn't destroy the distinction between revising a description and revising a *rule* of description.' But this trick succeeds *too* well: for *any* statement can be regarded as a 'rule of language' if we are willing to say 'well, they changed their rule' if speakers ever give it up. Why should we not abandon this line once and for all?

In closing, I can only apologize for not having given any *positive* account of either mathematical truth or mathematical necessity. I can only say that I have not given such an account because I think that the search for such an account is a fundamental mistake. It is not that there is nothing special about mathematics; it is that, in my opinion, the investigation of mathematics must *presuppose* and not seek to *account for* the truth of mathematics. But this is the beginning of another paper and not the end of this one.

2

The thesis that mathematics is logic*

'Russell and Whitehead showed that mathematics is reducible to logic.'
This sort of statement is common in the literature of philosophy; but all
too rarely nowadays does one see (1) any discussion of what Russell
meant by 'logic', or (2) any indication of the impact that studies in the
foundations of set theory have had upon sophisticated evaluations of the
statement quoted. In this paper I wish to focus on these two matters,
leaving regretfully aside a host of interesting topics that would and should
be brought into a fuller discussion.

(1) Russell's notion of 'logic'

For Russell,† 'logic' includes not just the elementary rules of deduction
(quantification theory), but also the assumptions Russell wishes to make
concerning certain entities – the so-called 'propositional functions'.
Thus logic has a (many sorted) universe of discourse of its own: the
collection of all individuals (zero level) plus the collection of all propositional functions taking individuals as arguments (level one) plus the
collection of all propositional functions taking level one propositional
functions as arguments (level two) plus... But what is a 'propositional
function'?

Today two simplifications have taken place in Russell's treatment. In
the expression $F(x)$, the symbol F is thought of as standing for an
arbitrary *predicate* (of appropriate level). Thus $F(x)$ means *not* 'the
propositional function $F(\hat{x})$ has the value 'Truth on the argument x'
(or has a true proposition as value on the argument x), but more simply
'x *has* the property F'. Thus Russell's 'propositional functions' came to
be identified simply with *predicates*. Secondly, it came to be assumed
that two predicates are *identical* if they have the same extension (the
so-called 'Axiom of Extensionality'). This assumption, or axiom, is

---

* First published in R. Schoenman (ed.), *Bertrand Russell, Philosopher of the
Century* (London, Allen & Unwin, 1967).

† In the present paper the contribution of Whitehead will be neglected. It should
be noted that all of the foundational ideas in *Principia* were published by Russell *prior*
to his collaboration with Whitehead. This includes the theory of types, the axiom of
reducibility, etc.

simply false on the classical interpretation of 'predicate' as 'universal taken in intension', since Blue and Square, for example, would be *different* predicates, in the classical sense, even if all squares were blue and *only* squares were blue. Thus 'predicate' in turn has come to mean simply 'set'. The net effect of both revisions taken together comes simply to this: $F(x)$ now means 'x belongs to the set F'. Russell's 'propositional functions' have for some time now been taken to be simply sets of individuals, sets of sets of individuals, sets of sets of sets...

This is quite contrary to Russell's explicit statements in *Principia* and elsewhere, however. On the reinterpretation of 'propositional function' just described, Russell's view would have to be that the 'universe of discourse' proper to logic is the system of all *sets* (of arbitrary finite type). 'Logic is set theory.' – But this is a view that Russell explicitly rejects!

Moreover, he surely rejects it with good reason. If we are willing to introduce special mathematical entities, and special axioms governing them (over and above the axioms of quantification theory), then why not just introduce *numbers* outright instead of introducing sets and then reducing numbers to sets? Why not redefine 'logic' to include number theory *by stipulation*? Then 'mathematics' (number theory) would indeed be a part of 'logic' – but only by linguistic stipulation. This would be idle – but it seems equally idle to redefine 'logic' to include set theory, which is what has in effect happened. Thus Russell wished to do something quite different: he wished to reduce number theory to set theory, and set theory *in turn* to 'logic' in his sense – theory of propositional functions.

Whether it is any more justifiable to extend the usage of the word 'logic' to include theory of propositional functions than to extend it *directly* to include set theory or even number theory is not an issue I wish to discuss here. But what is a propositional function?

It seems† that a propositional function is just what the name implies: a function whose values (not whose arguments!) are propositions. Propositional functions are thus one-one correlated to predicates (in the sense of universals taken in intension): corresponding to the predicate Blue we have the propositional function $\hat{x}$ *is blue* – i.e. the function whose value applied to any x is the proposition that x is blue. However, there appear to be some conceptual difficulties. Consider the propositional function *x wrote Waverley*. Suppose x happens to be the individual known as Walter Scott, and also known as 'the author of Waverley'. Then what is the value of the propositional function $\hat{x}$ *wrote Waverley*

† Here I am hazarding an interpretation. Russell's own formulations are not completely clear to me.

on this particular x as argument? is it the proposition 'Walter Scott wrote Waverley' or the proposition 'The author of Waverley wrote Waverley'? We can resolve the difficulty by accepting the first (but not the second) of the simplifications in Russell's treatment mentioned before, and taking the propositional function to be identical with the corresponding predicate. For surely any argument that could be offered for taking 'logic' to be the theory of propositional functions (in Russell's sense) could equally well be offered as an argument for taking logic to be the theory of predicates taken in intension.

(2) Does Russell's view make logic 'theological'?

Let us now consider some more-or-less-stock objections to Russell's view of logic.

(i) *The Cantor diagonal argument*

Bypassing Russell's well-known construction of the integers, let us consider for a moment statements about propositional functions (or, as I shall henceforth say 'predicates') of integers. According to one of the theorems of classical mathematics (which is, along with many others, proved in *Principia*) *there are non-denumerably many sets of integers*. (This was proved by Cantor, using the famous 'Diagonal Argument'.) Since 'sets' are identified with predicates taken in extension† in the system of *Principia*, there are then non-denumerably many pair-wise non-coextensive predicates of integers. However, there are only denumerably many well formed formulas in *Principia*. Hence, *the huge majority of all predicates of integers must be indefinable* (in the system of *Principia*, and indeed in any human language or constructed system). This reveals something about the way in which 'predicate' (or 'propositional function') is being used by Russell: 'consider the class of all predicates of integers', means consider an alleged totality which includes not only

† Actually, the famous 'no-class' theory of *Principia* is formally unsatisfactory. What Russell and Whitehead really needed was an interpretation of set theory, based on the simple theory of types, in *Principia* (or a relative consistency proof of Principia *plus* the Axiom of Extensionality relative to Principia). This has been supplied by Gandy (cf. his papers on the Axiom of Extensionality in the *Journal of Symbolic Logic*). Their own procedure was to define classes 'in use'; but the suggested definition has the annoying consequence that there are two truths which are respectively of the forms P and not-P in *abbreviated* notation. Thus *Principia's definition of class* is unsatisfactory! Gandy's definition of 'class' is more complicated than Russell and Whitehead's, but intuitively (as well as formally) more satisfactory. The definition is inductive: at level one, every propositional function is a 'class'; at level $n+1$, every *extensional* function whose arguments are all 'classes' is a 'class'. An *extensional* function is one which applies to an argument if and only if it applies to every function coextensive with that argument. 'Identity' of classes can then be defined simply as *coextensiveness*.

the predicates of integers that can be defined, but, allegedly, all the predicates of integers that can 'exist' in some absolute sense of 'exist'. Is it any wonder that Hermann Weyl has scoffed that this construction is *theology* rather than mathematics?

(ii) *The Skolem–Löwenheim theorem*

According to this celebrated theorem, every consistent theory has a denumerable model (i.e. a true interpretation over a denumerable universe of discourse). In fact, every consistent theory has a denumerable model which is an *elementary submodel* of any given model – i.e. the entities of the submodel have a property or stand in a relation definable in the theory, when the submodel is used to interpret the theory, if and only if they have that property or stand in that relation when the bigger model is used to interpret the theory. For example, if an entity is 'the successor of zero' in the elementary submodel, that same entity must have been 'the successor of zero' in the bigger model; if an entity is 'the set of all prime numbers' in the elementary submodel, that same entity must have been 'the set of all prime numbers' in the bigger model, etc. In particular, *exactly the same sentences are true* in a model and in any elementary submodel of the given model.

This leads at once to a pseudo-paradox or apparent paradox (it is not a genuine antinomy) that has become known in the literature as the 'Skolem paradox'. Namely, there is a theorem of *Principia* which *says* that there are non-denumerably many sets of integers. Hence there must be non-denumerably many sets of integers (in any model). But this contradicts the Skolem–Löwenheim theorem, which says that *Principia* has a denumerable model (assuming *Principia* is consistent)!

The resolution of the paradox is as follows. The sentence which 'says' that there are non-denumerably many sets of integers runs more exactly as follows:

(S) There does not exist a two-place predicate P such that P is a one-to-one correspondence between the set of all sets of integers and the set† of all integers.

What happens when we reinterpret *Principia* by taking 'predicate' to mean 'predicate in M', where M is some denumerable model? The answer is that the *meaning* of the sentence (S) is affected. What (S) *now* says is:

There does not exist a two-place predicate P *in the model M* (i.e. in the appropriate 'level' of the model M) such that P is a one-to-one

---

† It should be noted in connection with this formula that the 'set of all integers' is reduplicated on every sufficiently high level, in *Principia*.

correspondence between the set (of the model M) which is the set of all sets of integers which exist in the model M and the set of all integers.

And this may well be true although M is denumerable! In short, if M is denumerable, then the set which is 'the set of *all* sets of integers' *from the standpoint of the model M*, is *not* the set of *all* sets of integers 'viewed from outside'. Moreover, the set in question *can* be mapped one-one on to the set of all integers by a certain two-place predicate P, viewing the situation 'from outside'; but the predicate P which is the one-to-one correspondence between the set which is 'the set of all sets of integers' from the standpoint of M, and 'the set of all integers', *is not itself an entity of the model M*.

Let us now re-examine the sentence (S) with all this in mind. What we meant when we said that (S) 'says' that there are non-denumerably many sets of integers is that (S) says this if the model M contains (i) *all* sets of integers, and (ii) *all* two-place predicates (of the appropriate level). But how are we to express these requirements?

The answer is that we cannot, at least not by axiomatic means. Even if, so to speak, God 'gave' us a 'standard' model for *Principia* (one containing *all* sets of integers, *all* sets of sets of integers, etc.), one could find a second model – in fact a submodel of the first – in which the very same sentences would be true! Thus a 'standard' model, if there is such a thing, must be distinguished from a non-standard one by properties *other than the ones expressible in Principia (or in any formal language)*. The only such property that suggests itself is the 'property' of containing *all* sets of integers, *all* sets of sets of integers, etc. But just what does 'all' mean here?

It seems that someone who is convinced that he understands what is meant by '*all* sets of integers' can *explain* what he means to the sceptic only by saying '*you* know, I mean *all*' – but that is no explanation at all! The relevance to Russell's position is immediate. If 'logic' really is the theory of, among other things, the totality of *all* predicates of integers, then just where did we acquire this notion (of all non-denumerably many of them)?

(iii) The continuum hypothesis

Consider the following mathematical assertion, which is known as 'the continuum hypothesis':

(C) If S is any infinite set of sets of integers, and there does not exist a one-to-one correspondence between S and the set of all integers, then there exists a one-to-one correspondence between S and the set of *all* sets of integers.

This asserts that there is no set of real numbers (or, equivalently, of sets of integers) which is properly 'bigger' than N (the set of all integers) but properly 'smaller' than R (the set of all real numbers, or, equivalently, the set of all sets of integers).

Gödel showed in 1939 that given any model M for set theory (or for *Principia*) one can construct a model M' in which this statement holds true. About a month ago (Spring 1963) Paul J. Cohen showed that given any model M one can also construct a model M' in which the *negation* of (C) holds. Moreover, both Gödel's construction and Cohen's construction lead to models which are intuitively 'reasonable'. Thus either (C) or its negation could be adopted as an additional axiom of set theory without forcing a contradiction, or even forcing the models to have intuitively unacceptable properties. Finally, the proofs are relatively 'invariant', in the sense that they would still go through even if further assumptions of an intuitively plausible type (e.g. the existence of inaccessible ordinals) were added to set theory. In short, no axioms for set theory appear even conceivable as of this writing which would close the question of the truth of (C) except by fiat.

I wish to bring this question to bear upon what has been called Russell's 'Platonism'. According to Russell, there is such a well defined totality as the totality of all propositional functions of integers (although we cannot characterize it up to isomorphism by its formal properties, as we know from the Skolem–Löwenheim theorem, and we cannot hope to *list* all the propositional functions in question, even if we were allowed to work for ever, as we know from the non-denumerability proof). Moreover, the other totalities mentioned in (C) are allegedly well defined: the totality of all one-to-one correspondences (of the appropriate level), and the totality of all sets of sets (propositional functions of propositional functions) of integers. *Hence* (C) *must be either true or false as a matter of objective fact.*

It is just at this point that the foregoing arguments against Russell's position seem to me to gain 'bite'. We are now dealing with a *bona fide* mathematical assertion (the first problem on Hilbert's famous list of twenty). Russell's position is that universals ('propositional functions') are just as real as tables and chairs, and that statements about them – even statements about *all* of them – are 'objectively' true or false even when we cannot (because of our human limitations) *check* them, just as it might be true that there were between one million three hundred thousand and one million four hundred thousand people in the area now known as North Vietnam six hundred years ago, but we might be utterly unable ever to verify this fact.

It is a matter of sociological fact that Russell's position on this matter seems absurd to most mathematicians and philosophers. *Why* it seems absurd, whereas the statement about North Vietnam does not seem absurd to most people, notwithstanding the apparently equal uncheckability, is a matter for serious philosophical discussion. The Verificationist answer – that we *might*, after all, find out how many people there were in the area in question six hundred years ago – does not seem adequate (perhaps one could even have a theory of the destruction of information with time from which it would follow that *in principle* it is no longer possible to verify or falsify this statement; yet I, for one, would see no reason to deny that it could be true or false). I would hold that a notion is clear only when it has been adequately explained in terms of notions that we antecedently accept as clear, or when it itself has always been accepted as clear. This is, of course, rough – but it provides at least a rough indication of where the trouble lies: such notions as the notion of a totality of *all* propositional functions of integers, of *all* sets of integers, etc., are relatively novel ones, and they have, as a matter of historical fact, never been explained at all. No one would accept the argument: we know what a language is, in the sense of knowing that English, French, German, are languages: therefore we understand what is meant by the *totality of all languages*, not in the sense of actually spoken languages, but in the sense of an alleged precise and well defined *totality of all possible languages*. Clearly there is no such precise and well defined totality: the notion of a 'language' is much too vague for me to be sure in advance that I can tell under all possible circumstances whether or not some system of sounds (need they be sounds? What about 'dolphinese'? Could there be a language consisting of bee dances?) is or is not, in its actual cultural setting, given the organisms that actually produce it, to be classified as a 'language'. To 'consider the set of all languages' in this 'abstract' sense is to consider no well defined set at all.

Yet just this fallacious argument is what is offered in connection with 'propositional functions' or predicates. Because we know that Blue and Square and Made of Copper are predicates (we have a 'clear notion of predicate' in the sense of *one that is clear in most cases*), we are asked to conclude that there must be a precise and well defined *totality of all predicates*, not in the sense of all actually defined predicates, or in the sense of all predicates which are actually exemplified, or in the sense of all predicates (whether someone has a word for them or not) which *could*, as a matter of physical law, be exemplified, but in the sense of a *totality of all possible predicates.* Clearly there is no such totality. In fact, if we take the 'individuals' of *Principia* to be physical objects, then we boggle at level one: what is meant by the totality of *all possible predicates of physical objects?* Is 'Being Made of Phlogiston' a predicate? And what of 'Being Made of X-on', where 'X-on' is a theoretical term that no scientist ever *did* introduce, but that one *might* have introduced? Indeed, since every *meaningful* theoretical term corresponds to a possible predicate, it would seem that a totality of all possible meaningful theoretical terms (as a precise and well defined totality) is presupposed by the notion of a totality of all predicates of physical objects.

The notion of a totality of all *sets* of physical objects might seem more justifiable than the notion of a totality of all *predicates* (taken in intension), if only because we seem to be sure when a set has been given, and not always sure if a word which has been uttered does or does not correspond to a 'property'. But, if there are infinitely many individuals (and this is postulated by the famous Axiom of Infinity in *Principia*) then this greater sureness is illusory. In the case of infinite sets what we understand is the notion of a *definable* infinite set of things. But the notion of an *undefinable* infinite set – of a *wholly* irregular way of picking out infinitely many things from an infinite collection – has no clear meaning, at least to me, or to many scientists. If it really has a clear meaning, will someone please make it clear to us?†

This request is not simple philosophical obstinacy, like the position of the phenomenalist who refuses to understand 'physical object' until and unless it is made clear in *his* terms. I am willing to allow in the clarification of 'arbitrary set' (or 'arbitrary propositional function') any terms which are generally admitted to be clear. It is a matter of historical fact that this new way of speaking – speaking of a totality of *all* subsets of an infinite set – is quite recent (since Cantor), and that Cantor, who introduced it, did so with inconsistencies (sometimes he says a set is all the things 'falling under a rule', and at other times he speaks of wholly *arbitrary* collections – '*beliebige Mengen*'). Surely it is reasonable in science to ask that new technical terms should *eventually* be explained?

(iv) Remark

One problem in connection with the foregoing is the so-called problem of 'Impredicative definitions'. Such a definition occurs in *Principia* each time that we, for example, define a predicate of level two using a

---

† I no longer (1974) agree that the notion of 'set' presupposes the notion *of definability* or that 'set' is unclear. It is clear *relative to the notion of mathematical possibility* – which I take to be indispensable and irreducible. Cf. the next chapter in this book.

quantifier over all predicates of level two. The fact that impredicative definitions are permitted makes the problem of defining 'all possible predicates' (even of physical objects) all the more hopeless. For suppose that we had an epistemological theory (of a kind that Russell has sometimes accepted) according to which there exists some finite list of basic predicates (of level one), say $P_1, \ldots, P_n$, such that all *possible* predicates of physical objects are allegedly definable in terms of these. Then we could make the notion of a 'predicate of physical objects' precise by taking it to mean 'predicate definable (by specified means) from P_1, $P_2, \ldots, P_n$'. However, if the 'specified means' include *impredicative definition* then some of the predicates included in the totality will presuppose the whole totality. The question whether a given individual has such a predicate will then reduce to a question about the whole totality which may in turn depend on the very question with which we started – whether the individual with which we began has the property with which we began.

Another difficulty is the non-denumerability. The predicates of individuals definable in *Principia* from any finite list of basic predicates $P_1, \ldots, P_n$ are only denumerably infinite in number. This is also related to impredicativity in that the Cantor Diagonal Argument depends upon the fact that 'all sets of integers' includes sets of integers defined by reference to this very totality. If impredicative definitions are excluded from *Principia* (as recommended by Weyl) then it is no longer possible to prove that any set is (absolutely) non-denumerable; but, unfortunately, many sets needed in mathematics – e.g. the set of *all* real numbers – are then not definable! (I omit discussion of the well-known Axiom of Reducibility here, since formally speaking that axiom is equivalent to the allowing of impredicative definitions.)

(3) 'If-thenism' as a philosophy of mathematics

Before he espoused Logicism, Russell advocated a view of mathematics which he somewhat misleadingly expressed by the formula that mathematics consists of 'if-then' assertions. What he meant was not, of course, that all well formed formulas in mathematics have a horseshoe as the main connective! but that mathematicians are in the business of showing that *if* there is any structure which satisfies such-and-such axioms (e.g. the axioms of group theory), *then* that structure satisfies such-and-such further statements (some theorems of group theory or other). In the remainder of this paper I wish to make a modest attempt to rehabilitate this point of view. What are the difficulties?

(i) The study of finite structures

One difficulty is this: when we derive consequences from a set of axioms we are determining the properties *not* just of all *finite* structures which satisfy those axioms, but of *all* structures (including the infinite ones) which satisfy those axioms. But there are certain branches of mathematics which limit their inquiry to finite structures of a certain kind. For example, there is a whole literature dealing with the properties of finite groups. Clearly the theorist here is not just deriving consequences from the axioms of group theory, for in that case he would be proving theorems about *all* groups. He would never prove a statement of the form 'All finite groups have the property *P*' unless indeed the stronger, statement were true: 'All groups have the property *P*'. But the whole object of this branch of mathematics is to obtain properties of all finite groups which are *not* properties of all groups!

This objection may be dealt with in two ways. First of all, one may distinguish the thesis that mathematics studies the properties of all structures of a specified kind (groups, finite groups, fields, etc.), without asserting the *existence* of any such structures (except within a well understood context: e.g. 'existence of an identity', in group theory means *if* there is given any group *G*, *then* there exists an element *e* in *G* with the property of being the identity of *G*), from the thesis that mathematics is the derivation of consequences from axioms by means of logic (in the sense of quantification theory). We may say, for example, that when we wish to prove that all *finite* models for an axiom-set have certain properties, then we have in general to admit *a primitive mode of proof* over and above 'logic'; namely *mathematical induction*. On this view, the use of mathematical induction as a valid way of proving theorems about all structures of interest in a given domain *is* our way of expressing the fact that those structures are all finite, as well as we can.†

---

† Strictly speaking, the situation is this. If we accept the notion of a 'totality of all predicates' (of the structures in question), then we can say that the *finite* structures are *completely characterized* by the fact that mathematical induction holds for an *arbitrary* predicate: i.e. if the empty structure has a property, and the property is preserved under adjoining a single element to a structure, then every finite structure has it. The finite structures form the *largest* class with respect to which this mode of reasoning is valid. As remarked below, however, every formalized theory admits of models in which 'all predicates' does not really mean *all* predicates. In such models, a class may satisfy the formal definition of the 'class of all finite structures', but contain some structures which are infinite 'viewed from outside' (or viewed from a different model). This will happen because *for every predicate which 'exists' from the standpoint of that model*, it is true that if the empty structure has the predicate and the predicate is preserved under adjoining single elements, then every structure in the class *F* has it, where *F* is the class which is taken to be the 'class of all finite structures' in that model. If *F* has a member which is an infinite structure, viewed from outside, then there will be a predicate – say *not being infinite* (in the sense of the 'outside' model) which does *not* obey mathematical

Alternatively, we may imbed the axiom-set in question in a richer axiom set (including a portion of set theory) within which the notion of a 'finite model' for the axiom-set in question is explicitly definable, and within which mathematical induction is formally derivable. Then theorems about all 'finite groups', say, are just a subclass of the theorems about all structures of a larger kind – say all 'models for T', where T is a certain system of set theory. (Of course, T might be left unformalized, in actual practice, or might be a large class of 'acceptable' set theories.)

(ii) The restriction to standard models

A more important difficulty is this: number theory, for example, is usually held to be concerned not with *all* models for, say the Peano Axioms, but solely with all standard models – i.e. all models in which every element is either 'zero' (the unique element with no predecessor), or 'the successor of zero', or 'the successor of the successor of zero', or... In short, we are concerned, as number theorists, with those models in which every element bears a finite power of the successor relation to 'zero'.

This is, in principle, the same difficulty as (i). The view I am suggesting is that *standardness should be viewed as a relation between models*, rather than as an absolute property of a given model.† In other words, when we say that such-and-such (some construction defined in set theory) is a standard model for Peano Arithmetic, this should be interpreted as meaning that *if M is any model for the set theory in question, then* such-and-such entities in M form a model for Peano Arithmetic, *and* that model is standard relative to M. In any usual set theory it can easily be proved that any two standard models for Peano Arithmetic are isomorphic; thus for each model M of the set theory, those entities u of M which are standard models for Peano Arithmetic relative to M will

induction with respect to the whole class F; but this predicate is no predicate at all from the standpoint of the model in which F *is* the class of all finite structures.

There are thus two standpoints one may take: one may say *either* that the notion of a structure being finite is a perfectly clear one, but one that cannot be completely formalized; or one may say that the notion *depends essentially* on some totality of predicates, and thus partakes of the same unclarity as the notion of the 'totality of all predicates' (of a given level). The latter view is the one taken here. I am aware that it is counterintuitive. I believe that it will seem *less* counterintuitive, however, if one fixes it in one's mind that *from the mathematical standpoint* what is *essential* and *all* that is essential about the totality of all finite structures is that this totality obeys mathematical induction *with respect to an arbitrary predicate*, and no larger collection does. Whereas 'intuitively' a finite structure is one that one could 'get through counting' (and thus the intuitive notion of finitude is connected with our experience of space and time), in *mathematics* the finite structures are the ones with respect to which a certain mode of reasoning is valid.

† I no longer (1974) agree with this view. In the next chapter, a definition of standardness in terms of *modal* notions is proposed.

22

THE THESIS THAT MATHEMATICS IS LOGIC

be isomorphic to one another. A 'standard model' is here defined as above: one in which each element bears a *finite* power of the successor relation to 'zero', *where the meaning of 'finite' may vary with the model selected for the set theory.*

To complete this, we introduce the convention that a model which is *non-standard* relative to any model M shall be called 'non-standard' (absolutely) even if it is standard relative to some other model M'. We do not introduce a parallel *absolute* use of 'standard' because we do not wish to presuppose a totality of *all* models (this would put us right back in 'theology'). The point is that if I define a model M' for a set theory as a function of an arbitrary given model M, and then define a model M'' for Peano Arithmetic in terms of M' in such a way that M'' is standard relative to M' but *non*-standard relative to the bigger model M, then we may say simply that 'M'' is a non-standard model for Peano Arithmetic' – meaning that there is some model (really a class of hypothetical models, of course) relative to which M'' has been defined, and relative to which M'' is non-standard. None of this really presupposes the *existence*, in a non-hypothetical sense, of any models at all for anything, of course!

One common objection is as follows: 'Why not restrict models for *set theory* to be such that all "finite" sets (in the sense of the usual definitions) are *really finite*? Then standard models for Peano Arithmetic are models which are standard relative to *such* models for set theory.' My reply, in a nutshell, is that I have no need for the alleged concept of a set being *really* finite. But this needs a little unpacking.

First of all, notice that this added restriction cannot possibly be of *use* in mathematics. The only way we *ever*, in fact, know that a model for number theory is non-standard is by showing that it is non-standard *relative* to some model for set theory relative to which it has been defined. If the alleged concept of a *really standard* model, in an absolute sense, makes sense at all, then it is *still* a concept that we can do completely without in mathematics. In fact, I would not have the faintest conception of how to go about using it!

To make clear what is meant, let us look at a case in which *all* the models of a theory are 'non-standard'. Such a case may arise as follows. There is a term t of the theory (call it T) such that we can *prove* all of the following sentences:

$$t \text{ is a number}$$
$$t \neq 0$$
$$t \neq 1$$
$$t \neq 2$$
$$\vdots$$

Moreover, the fact that all of these infinitely many sentences of T are provable in T is itself provable in, say Peano Arithmetic. Then T has no standard models! But how do we know this?

The answer is *not* by some mysterious act of 'intuition', as some have urged, but rather that this *theorem* – that T has no standard models – is provable in any normal system of set theory – even in T itself (and by just the argument that naturally occurs to one)!

Thus, what we have really proved, when we prove the theorem that 'T has no standard model', is that *if M is any model for T, and M is itself an entity of M', where M' is any model for the set theory in which the proof is being given, then M is non-standard *relative to M'*. What I am contending is that all *uses* of the standard/non-standard distinction in foundational research are uses of this *relational* notion of standardness, in the manner indicated.

Let T_0 be a set theory with some empirical concepts and assertions in which one can express the assertion 'there is a one-to-one correspondence between the sequence of natural numbers and the sequence of successive *minutes* of time, counting from an arbitrary "first minute"'. The intuitive 'definition' of 'finite' is often thought to be this: *a set is finite if one could get through counting it.* There are counter-examples (Zeno's paradox) if one does not require that it should take a minimum time (say, 'one minute') to count out each member of the set: for if I take a half-minute to count 'one', a quarter-minute to count 'two', an eighth-minute...etc., then I could count out an *infinite* collection in a finite time. If the infinite collection has the order $\omega + 1$ (or any other order type, with a last member) then there could even be a *last* element 'counted' (say, after having counted out the first ω elements in one minute, I take one more half-minute and count out the $\omega + 1$st). Thus an infinite set can be 'counted in a finite time', even if we require that there should be a 'last element counted'. But what if we require that there *should* be a minimum time which it takes to 'count' one element? Have we then captured the alleged notion of 'really finite'?

A little reflection shows that what is being proposed is, in effect, this: *a standard model* for Peano Arithmetic (in the alleged 'absolute' sense) should be one which is *standard relative to T_0* (or some such theory). This proposal has some features which are at first blush attractive. For example, we have the attractive-seeming theorem that any finite set can be mapped one-one on to the minutes in a sequence of successive minutes which is 'finite' *in the sense of possessing a first and last minute.* From the point of view of traditional (psychologically oriented) epistemology, this comes close to capturing the 'natural' notion of a finite set. Why, then, should we not accept this?

There are a host of reasons. We have recently learned that the properties which intuitively belong to *space* – infinitude, unboundedness, etc. – are not necessarily properties of space at all (because space may be non-Euclidean). Similarly, the *chronometrical* properties of *time* are an empirical, not a mathematical question. And the proposed definition of 'finite', however 'traditional' and 'epistemologically correct' is riddled with empirical (cosmological) assumptions about time: that there is an infinitude of future time, for example; that the sequence of successive minutes (counting from an arbitrary 'first minute') is *well ordered*; that the sequence has the order type ω. Every one of these assumptions may be false! even if it seems strange to think that, for example, there may be *more than* ω future minutes! No mathematician can consent to let the properties of the *mathematical* notion of finiteness depend upon cosmological hypotheses concerning the true chronometry (= 'time geometry').

This brief discussion, may, however, point to the origin of the widespread belief that we *have* an absolute notion of 'finite'. It is a fact that for a long time the principles of intuitive chronometry, like those of Euclidean geometry, were regarded as synthetic *a priori*. And as long as the usual chronometry (which assigns the order type $\omega^* + \omega$ to the sequence of all *minutes* – from negative to positive infinity) is thought to be *a priori*, then it seems that we *have* (physically) a standard model for Peano Arithmetic – for the sequence of minutes from some arbitrary first minute is itself such a model (assuming 'plus' and 'times' to be inductively defined in the usual way). And we can *fix* the notion of 'standard model' by taking this model to be *the* model. This is in effect what Kant did; but it is erroneous for just the reason that Kant's views on geometry are erroneous: because the cosmological properties of time in the large are no more *a priori* than those of space in the large.

To recapitulate: we cannot fix any precise notion of an 'absolutely standard' model even for Peano Arithmetic, except *relative* to some other model. We might take some empirically given model and define it to be *the* standard model, but such a course is useless in pure mathematics. Thus if I am wrong, and there is indeed some 'absolute' notion of a '*really* standard' model for Peano Arithmetic, I am anxious to know just where in mathematics we could use it. Similar remarks hold with respect to the notion of a 'standard model' in set theory. Such a model is usually taken to be one in which the power set of a given set *really* contains *all* the subsets of the given set. Once again, this makes perfect sense as a *relation between models*; and if it has any further 'absolute' sense, that absolute sense seems totally unusable and irrelevant from the standpoint of mathematics. I conclude, thus, that an 'if-thenist' view of mathematics *can* do justice to talk of 'standard' and 'non-standard' models *in so far as such talk does any work.*

(iii) Gödelian arguments

Suppose I 'accept' a set theory T as 'correct' in the sense that all the number-theoretic statements provable in T are *true* (in *the* standard model, assuming, for the moment, the 'Platonistic' attitude towards standard models which we just criticized). Then the further statement $\mathrm{Con}(T)$ which expresses the consistency of T is one that I should *also* accept, since it must be true if T is consistent. But Gödel showed that Con (T) is not provable in T if T is consistent. Thus we have found a sentence – namely Con (T) – which I have as good grounds for accepting as I have for accepting T itself, but which cannot be proved in T.

It has sometimes been urged that the fact that we can 'see' in this way that we should accept $\mathrm{Con}(T)$, although $\mathrm{Con}(T)$ cannot be formally proved (in T) shows that we *have* and are able to *use* the alleged Platonistic notion of 'truth' – i.e. of truth in *the* standard model for Peano Arithmetic. But this is simply a mistake. If T becomes inconsistent when $\mathrm{Con}(T)$ is adjoined – i.e. if Con (T) is *refutable* in T – then T has no standard model, as is easily proved (even in T itself, in most cases). Thus I have as good grounds for accepting the added axiom $\mathrm{Con}(T)$ as preserving consistency as I have for expecting that it will not turn out to be provable that T has no standard models. If I 'accept T' in the sense of having this very strong expectation (which may easily be unreasonable in the case of some of the extant systems of set theory), then I should also be willing to accept this added axiom; but this argument does *not* turn on the existence of a non-relational notion of 'standardness'.

(iv) The problem of application

We now come to the difficulty that caused Russell himself to abandon 'if-thenism'. This is, in my view, a far more serious one than the ones we have discussed so far (although the ones we have discussed so far tend to worry mathematicians more). In my opinion, the great strength of logicism as compared to other positions in the philosophy of mathematics is its ability to handle the problem of the *application of mathematical methods to empirical subject matters*; and this is just what Russell himself has always stressed.

Let us review briefly this aspect of the logicist account. Consider first the application of logic itself to empirical premises to derive an empirical conclusion. A simple syllogism will do, say

26

> All timid persons fear hungry lions
> All bankers are timid
> _____
>
> (*therefore*) All bankers fear hungry lions

The principle: (for all A, B, C) if all A are B and all B are C then all A are C, is itself a principle of pure logic. But when we assert that the principle is *valid* we are thereby saying that the principle holds even when empirical subject matter terms are 'plugged in' for the capital letters A, B, C; not just when subject matter terms from formal logic and mathematics are 'plugged in'. (In my opinion the Intuitionist account of logic fails to give an adequate account of just this fact; however this falls outside the scope of the present paper.)

Now consider an inference from (applied) arithmetic rather than from logic, say

> There are two apples on the desk
> There are two apples on the table
> The apples on the desk and table are all the ones
>   in this room
> No apple is both on the desk and on the table
> Two plus two equals four
> _____

(*therefore*) There are four apples in this room

The logicist account of such an inference is well known. The logicist definitions of 'there are two As' and 'there are four As' are such that one can *prove* that 'There are two As' is equivalent to a statement of pure quantification theory (with identity) namely: 'There is an x and there is a y such that x is an A and y is an A and $x \neq y$ and such that for every z, if z is an A then either $z = x$ or else $z = y$.' Similarly, 'There are four As' is equivalent to a formula in quantification theory with identity. Thus the entire inference above is equivalent, line by line (except for 'two plus two equals four') to an inference in pure logic, by the narrowest standard – quantification theory with identity. What of the line 'two plus two equals four'? The answer is that the above inference is still valid with that line omitted! Moreover, the logicist translation of 'two plus two equals four' is equivalent, not to a formula of first order logic (quantification theory), but to a formula of second order logic; in fact, to the formula

For every A, B, C, if C is the union of A and B and A and B are disjoint and A has two members and B has two members then C has four members

where the italicized clauses are all expressible in first order logic with identity. And what this formula says is that the above inference is valid!

Thus we see the role of the formula 'two plus two equals four' in the above inference: it is not an added *premiss* (the inference is valid without it); it is rather the *principle* by which the conclusion is derived from the (other) premisses. Moreover, the principle is essentially a first order principle: since the initial universal quantifiers 'for every A, B, C' can be inserted in front of every valid first order principle. Thus the above inference is tantamount to an inference in pure logic even by narrow standards of what constitutes pure logic; and the fact that the principle 'two plus two equals four' can be used to derive empirical conclusions from empirical premisses is simply an instance of the fact that we noted before: the fact that when we assert that a principle of pure logic is 'valid' we thereby assert that the principle is good under all substitutions for the predicate letters A, B, C, etc.; even substitutions of empirical subject matter terms. What has confused people about 'two plus two equals four' is that *unlike* '(for all A, B, C) if all A are B and all B are C then all A are C' it does not *explicitly* contain 'A, B, C' which can have empirical subject matter terms 'plugged in' for them; but it is demonstrably equivalent to a principle which *does* explicitly contain 'A, B, C'.

This discussion contains what is of permanent value in logicism, I think. This account of the application of (discrete) mathematics is neat and intellectually satisfying. Moreover it *does*, I think, show that there is no sharp line (at least) between mathematics and logic; just the principles that Kant took to be 'synthetic *a priori*' (e.g. 'five plus seven equals twelve') turn out to be expressible in the notation of what even Kant would probably have conceded to be logic. Of course, the logicists did *not* show that *all* of mathematics – not even all of number theory – is expressible by means of first order formulas in the manner in which 'two plus two equals four' is. We have already noted that the Continuum Hypothesis, for example, is expressed in *Principia* by a higher order formula which presupposes propositional functions of propositional functions of...of individuals, and that such quantifications over an alleged totality of *all* propositional functions (of a given level) – including the ones defined in terms of the alleged totality itself – pose some real problems of interpretation. But there is an unfortunate tendency in philosophy of mathematics to overlook the fact that the logicist translations of *some* mathematical propositions – in fact, just those propositions about specific integers ('five plus seven equals twelve') which figured in Kant's discussion – do *not* presuppose higher order logic, and thus do not depend on the notion of a totality of 'all' propositional functions except in the harmless way in which even first order logic does: via the idea that we can state principles which contain 'dummy letters' A, B, C for which one can plug in any propositional function one likes. But this use of schematic letters does *not*, I think, require that we suppose the totality of all permissible values of the letters to be a well defined one, as does the explicit definition of a propositional function in terms of the alleged totality in question.

In more formal language, what I am drawing is a distinction between those principles of Second Order Logic whose predicate quantifiers (in prenex form) are all universal (and precede all first level quantifiers), and the remaining principles (which involve existential second level quantifiers). The former principles are simply first order formulas; the function of the second level quantifiers is merely to express the notion that the formula holds for all A, B, C, i.e. the notion that the formula is *valid*. On the other hand, a formula which contains '(EP)' – *there exists a propositional function P* – must be regarded as presupposing that the totality of *all* propositional functions is a well defined one (unless the bound variable P can somehow be restricted to a special class of propositional functions which anyone would admit to be well defined). Since the 'translations' of *some* number theoretic utterances are formulas of the first kind, I am arguing that Russell *has* shown that *some* mathematical truths are part of logic (or that part of logic is identical with part of mathematics), and hence that the line between the two subjects is not sharp, even if one boggles at counting formulas of the second kind (ones which assert that 'there exists a P') as part of 'logic'. To have shown that the 'line' between 'logic' and 'mathematics' is somewhat arbitrary was a great achievement; and it *was* the achievement of *Principia*, I contend.

Consider now such a statement as 'the number of the planets is nine'. What this 'means', according to *Principia*, is that a certain *set*, the set of planets, belong to a certain other set (intuitively, the set of all nine-tuples, although it is definable in pure logical notation, in the fashion illustrated in connection with 'two' – i.e. it is definable without using the term 'nine'). If propositional functions and sets are entities which can be understood only *relative to models* – if a 'set', for example, is any entity in a model which bears a certain relation (the converse of the 'epsilon relation' of the model) to certain other entities of the model – then mathematics, *so understood*, does not assert the *existence* of any sets at all (except in the way in which it asserts the 'existence' of an identity in group theory – that is, it asserts that certain sets exist *in each model*). In particular, then, although it may be a theorem of *Principia* that there is such a set as 'the set of planets', all this means is that *if M* is any model for *Principia*, *then* there is an entity *u* in *M* which bears the converse of the 'epsilon relation' of *M* to each planet and only to the planets. Similarly, 'the number of the planets is nine', if asserted in this 'if-then' spirit, could only mean: *if M* is any model for *Principia*, *then* the entity of *M* which is 'the set of planets' belongs to the entity *v* which is 'the number nine'. So far I have skirted the dangerous question: what does it mean to suppose that a model 'exists'? Does a model have to be a system of *physical objects* (in which case the mathematical theories we have been discussing concern the properties of hypothetical *physically infinite* structures), or can it be a system of *abstract* objects? And what would an 'abstract object' *be* anyway? But one thing is clear: if 'existence of a model' is so interpreted that models for, say, *Principia might* in fact exist, but do not necessarily exist (in whichever sense of 'exist'), then we seem to be in trouble. For, if the sentence '*M* is a model for *Principia*' is *false*, then any 'if-then' sentence with this antecedent is vacuously true. Then for every number *n*, 'The number of the planets is *n*' would be true, on this 'if-then' interpretation!

It is important to notice that this difficulty affects only the problem of understanding *applied* mathematics, not pure mathematics. For, in pure mathematics, the business of the mathematician is not in discovering *materially* true propositions of the form 'If *M* is a model for *T* then so-and-so', but in discovering *logically true* propositions of that form. Even if a proposition of the form in question is true, if it is only 'true by accident' (say, because *there is no M* such that *M* is a model for *T*), then it will not be provable by purely formal means, and hence will not be asserted by the mathematician. But the assertion, 'the number of the planets is nine' cannot be interpreted in the same way – cannot be interpreted as meaning that it is a *logical* truth that, if *M* is any model for, say, *Principia*, then the entity *u* which is the set of planets in *M* belongs to the entity *v* which is the 'number nine' in *M* – because it is *not* a logical truth, but only a material truth that the number of the planets *is* nine. And if we attempt to interpret the assertion 'the number of the planets is nine', as the assertion of the *empirical* truth of a material implication, then the familiar paradoxes of material implication arise to plague us.

For this reason, Russell took the view that since we regard various groups of things as actually having a cardinal number (independently of whether any one ever counts those groups, and ascertains that cardinal number, or not) then the corresponding sets must be accepted by us as actually 'existing' (or rather, the corresponding predicates must be – 'sets' are construed as special kinds of predicates in *Principia*), and so must all the predicates of these predicates (since that is what numbers are, on the *Principia* view), and so on. Indeed, since the truth-value of these applied statements must be well defined, a particular model – *the* standard model – must be fixed once and for all, as the one to be used in interpreting *Principia*.

This argument does not appear to me to be decisive, as it did to Russell forty-five years ago.† Consider, first of all, a single empirical statement involving a number-word, say, 'there are two apples on the desk'. Russell himself has provided us with not one but *two* ways of rendering this statement in *Principia* notation: the one we reviewed before, which requires only quantification theory with identity, and the one involving the existence of sets. The former way of saying 'there are two apples on the desk' can always be used, whether we assume that models for *Principia* exist or not, and it expresses an objective factual claim which is true or false independently of the existence of anything but the apples on the desk (as it should be). Thus we do not *have* to render 'the number of the planets is nine' by an elaborate assertion about a relation between two highly defined sets, which in turn becomes interpreted (on the 'if-thenist' view) as a hypothetical assertion about models for *Principia*. Even if we accept the 'if-thenist' view, the statement in question can be regarded as a simple first order assertion about the planets. The puzzle is how mathematics then ever becomes useful in empirical investigation.

In order to solve this problem, let us abbreviate the statement 'the set of planets belongs to the number nine' as P_1, and the statement 'there is an x and there is a y and...such that x is a planet and y is a planet and...and $x \neq y$ and...and such that for every z if z is a planet then $z = x$ or $z = y$ or...', which expresses 'the number of the planets is nine' in a purely first order way, as P^*. The equivalence, $P \equiv P^*$, is a theorem of *Principia*, and hence holds in all models. Thus, *if we assume Principia has a model*, it does not matter whether we assert P or P^*. Otherwise, as we have just seen, it is necessary to use P^* to express what we wish to say without committing ourselves to sets, models, etc.

In a similar way, let Q be the statement 'the set of all suns (in the solar system) belongs to the number one', and let Q^* be the statement, 'there is an x such that x is a sun (in the solar system) and such that for every z, if z is a sun then $z = x$' [which expresses 'there is exactly one sun (in the solar system)']. Let R be the statement, 'something is a major body in the solar system if and only if it is either a planet or a

† It does now! (1974).

sun', let S be the statement, 'nothing is both a planet and a sun', let T be the statement, 'the set of major bodies in the solar system belongs to the number ten', and let T^* be the corresponding statement in quantification theory with identity ('there is an x and there is a y and... such that x is a major body in the solar system and y is a major body in the solar system and...etc.'). Now, we may visualize the following as happening: someone accepts 'the number of planets is nine', which he renders indifferently by P or P^*; he accepts 'there is exactly one sun (in the solar system)', which he renders indifferently by either Q or Q^*; and he accepts R and S. In *Principia*, he then carries out the deduction 'P & Q & R & S, therefore T', and accordingly accepts T. Finally, since $T \equiv T^*$ (in *Principia*) he treats T and T^* as equivalent, and also accepts T^*. He has been led to the correct conclusion: but virtually all of the 'in between steps' in his reasoning 'make sense' only *given* some model for *Principia*. Or is this true?

Well, let us ask: what would happen if someone some day came to an *incorrect* conclusion in this fashion? Let U be the statement, 'the set of major bodies in the solar system belongs to the number *eleven*', let U^* be the corresponding statement in quantification theory with identity, and suppose that someone some day found a deduction of 'P & Q & R & S, therefore U' in *Principia*. What would this mean? Since a deduction remains valid no matter *what* constants are 'plugged in' for A, B, C, etc., one could replace 'the set of planets belongs to the number nine' by 'the set of natural numbers from one through nine belongs to the number nine', replace 'the set of suns belongs to the number one' by 'the set of natural numbers from ten through ten belongs to the number one', etc. In short, from the premisses 'there are nine numbers from one to nine', 'there is exactly one number ten', 'nothing is both a number from one to nine and the number ten', 'every number from one to ten is either a number from one to nine or the number ten', all of which are theorems of *Principia*, one could deduce 'there are *eleven* numbers from one to ten', which is the *negation* of a theorem of *Principia*. Thus *Principia* would be formally inconsistent, and hence *could have* no models at all!

In short, if I find a proof of, say, 'P & Q & R & S, therefore T' in *Principia*, I can rely on the corresponding first order statement, 'P^* & Q^* & R & S, therefore T^*', *provided* I am confident that *Principia could have* models. Since any discovery of a mistaken inference of this kind would be *ipso facto* a discovery of an inconsistency in *Principia*, I can be as confident in such inferences as I am that no inconsistency will turn up in *Principia*. There is a clear difference between believing in the actual existence of something, and believing in its *possible* existence: and I am contending that the employment of *Principia*, at least in deriving such statements as we have been discussion from each other, does *not* presuppose 'Platonism', i.e. belief in the *actual* existence of sets, predicates, models for *Principia*, etc., but only presuppose a belief that a structure satisfying the axioms of *Principia* is *possible*. What it is to believe that the existence of something is possible, is a philosophical issue which will not be entered into here; however, this much is clear: to think that *there could be* a structure satisfying the axioms of *Principia*, in any sense of 'could', is *at least* to be confident that no contradictions will turn up in *Principia*. And that is all we need to employ *Principia* in the manner which I have been describing.

In sum, I am suggesting that pure mathematics consists of assertions to the effect that *if* anything is a model for a certain system of axioms, *then* it has certain properties, while applied mathematics consists of assertions to the effect that *if* anything is a model for a certain system of axioms, *then* certain statements 'if *A* then *B*' hold, where *A* and *B* are first order statements about real things.† Thus applied mathematics does *not* presuppose that models for our mathematical axiom-sets ('standard' or 'non-standard') actually do exist, but only that they *could* exist.

(v) *Consistency*

On the view just presented, and indeed on *any* view, the application of mathematics to 'the real world' presupposes the *consistency* of the mathematical theory applied. If we do not wish to assume the consistency of as 'strong' a system as *Principia*, then we may try to get by with weaker systems – but the consistency of at least elementary number theory is presupposed by every school in the foundations of mathematics (the relative consistency of Intuitionist and 'classical' elementary number theory is easily demonstrated). On what does this presupposition rest?

This question does not seem to me to pose any *especial* difficulty for the view proposed here. Our confidence in the possible existence of a model for at least Peano Arithmetic reduces to our belief that there is nothing contradictory in the notion of an ordinary infinite sequence. *That* belief is shared by all philosophies of mathematics: the Intuitionists may *say* that they 'only presuppose the potential infinite', but this is only terminologically different from saying that they too reject the idea that a contradiction could ever be found in the idea of a sequence

---

† This assertion is wrong. Cf. chapter 4 in which it is argued that one cannot consistently be a realist in physics and an if-thenist in mathematics.

proceeding for ever. On my view, such framework assumptions† in science neither have nor require any 'justification'; it is only that we require a justification if we ever propose to *revise* them.

With respect to mathematical systems stronger than Peano Arithmetic the situation (*vis-à-vis* consistency proofs) today appears problematical. Hilbert hoped to show the consistency of all the standard systems (including *Principia*) relative to number theory; but Gödel dashed this hope in his Second Incompleteness Theorem. What is still not clear is whether the consistency of, say, *Principia* can be demonstrated *relative to* some system whose consistency is more 'intuitively evident' than that of *Principia*. As long as this is not done, we have to admit that classical mathematics, pure and applied, rests upon foundations which *might* someday turn out to be inconsistent. But so what? That is what the situation is; and we have constantly to put up with much more pressing risks in life and in science. If the 'philosophy of mathematics' proposed here offers no more 'justification' than *our intuitive conviction that certain kinds of structures are possible* for our assuming the consistency of 'strong' systems of mathematics, what 'philosophy of mathematics' *does* offer more justification than this?

(4) Hilbert's 'formalism'

The position just defended, although substantially that of Russell in *The Principles of Mathematics*, has a certain relation to Hilbert's position, as expounded, for example in the famous article on the Infinite.‡ Our first order statements about planets, apples on desks, etc., correspond to the statements that, in Hilbert's view, figure in the ultimate *application* of mathematics in empirical science. It was essential to Hilbert's view that *ultimately* the statements we are concerned with when we *apply* mathematics should be simple assertions of numerical magnitude, such as 'there are two apples on the desk'. The corresponding assertions of pure mathematics – e.g. 'two plus two equals four', which can, as I noted, be expressed either as first order formulas, or (by prefixing predicate quantifiers binding '$A, B, C, ...$') as second order formulas, correspond to Hilbert's 'real statements'. Thus, in the example I gave some paragraphs back, '$P^* \& Q^* \& R \& S$, therefore T^*' was a *real* statement. (In fact, it was equivalent to the real statement 'nine plus one equals ten'.) The remaining statements in mathematics are called

---

† Cf. my paper 'The analytic and the synthetic', chapter 2, volume 2 of these papers.

‡ An English translation of this article appears in *Philosophy and Mathematics*, ed. P. Benacerraf and H. Putnam, 1964.

'ideal statements' by Hilbert. Thus 'P & Q & R & S, therefore T' was an *ideal* statement (from set theory). Hilbert saw just what has been pointed out: that what the passage from the ideal statements to the real statements implied by those ideal statements requires is the *consistency* of the ideal statements – not in the sense that the ideal statements would *not* imply the real statements even if the ideal statements were inconsistent; of course they would, since an inconsistent statement implies every statement; but that the assumption that all real statements implied by the ideal statements of the mathematical theory T are *true* invokes the consistency of T.

The position I have been defending seems superior to Hilbert's position in several respects, however. First of all, the term 'ideal statement' does not convey very much in itself, beyond the suggestive analogy with 'ideals' in algebraic number theory, which suggested the term to Hilbert. Sometimes, in trying to be more precise, Hilbert said that the ideal statements are *meaningless combinations of signs*; but this does extreme violence to our intuitions. The view taken here – that the 'ideal statements' are *meaningful* and indeed *true* assertions about all structures of certain specified kinds, whether such structures actually exist or not – seems far closer to common sense. And this view is perfectly compatible with Hilbert's further view, that even the possible existence of such structures should not be taken for granted, but should be demonstrated, if possible, by elementary number theoretic means – only that further view appears to have been dashed by the metamathematical results of Gödel. Certain mathematicians appear to gain a sense of security from saying that they are only manipulating meaningless concatenations of signs; but this security is illusory. If the theory we are working in is *consistent*, then there is no reason not to say that we are proving (by pure elementary logic) what properties all structures which obey the axioms of the theory *must* have; and if it is inconsistent, then saying, 'it was all a collection of uninterpreted formulas, anyhow' will not render it mathematically interesting.

Under the term 'real statement' Hilbert included, however, not only the statements 'two plus two equals four', etc., so far discussed, but also certain other statements – for example, 'there is always a prime between x and $2x$' – when proved by 'constructive' means, i.e. means acceptable to the Intuitionists. Here I can follow him only in part. The notion of a statement with 'constructive content' – e.g. there is always a prime between x and $2x$ – is clear. 'There is always a prime between x and $2x$', if true (and it is), provides an *effective method* for finding a prime greater than any given prime p. Namely, given p, simply test $p+1, p+2, \ldots$ up to $2p$, for primacy. At least one of the tests must have an affirmative outcome, and then one will have discovered a prime greater than p. Moreover, it is clearly natural to extend the term 'real statement' to cover statements with constructive content. For, sticking with our example, suppose for some p our computing procedure does not yield a prime greater than p – i.e. it is not possible to find a prime greater than p by simply looking between p and $2p$. Then we will have discovered a contradiction in Peano Arithmetic. Thus we have as good reason to rely on our computing procedure in any actual case (provided it does not take too long to actually carry out the computation) as we do rely on the 'real statements' discussed before – the basis for our confidence in both cases is our underlying confidence in the consistency of Arithmetic.

Consider, by way of contrast, the 'non-constructive' statement 'there are infinitely many primes'. *In a sense*, this statement too provides a 'computing procedure' for locating a prime greater than p: namely, *hunt through $p + 1, p + 2, p + 3, \ldots$ until you find a prime*. If the statement is true, this computing procedure (and it is a perfectly feasible one to actually 'program') must always 'terminate'; and when the procedure 'terminates', we have the desired prime greater than p. However, if the procedure never terminates, we cannot conclude that Peano Arithmetic is inconsistent! All we can conclude, if this procedure never terminates in some actual case, is that Peano Arithmetic is at least ω-inconsistent. We can prove (in whatever metatheory we prove that the procedure never terminates) that Peano Arithmetic has no 'standard model'. (The nature of such *relative* 'non-standardness' was discussed above.) But we cannot say: 'if the procedure never terminates, you will obtain an inconsistency in Peano Arithmetic.' For, if the procedure never terminates in a particular case, I will not *find this out* (save by proof in a more powerful theory); all I will find out is that I 'run out of computer time'. And since no bound was given in advance on the length of *this* computation, my 'running out of computer time' does not at all establish that the termination of this computation will never come.

Thus, I can agree with Hilbert that there is a clear difference between relying on theorems of Peano Arithmetic which say that a certain computation will terminate in less than some specified number of steps, and relying on theorems which only say that a computation will 'eventually' terminate, or which don't provide an effective procedure at all. But I see *no* reason to agree with Hilbert that it matters in the least, to a statement's counting as a 'real' statement in the sense we have been discussing, whether it has been proved by Intuitionist means or not! Consider again 'there is always a prime between x and $2x$'. Even if this were proved using the Law of the Excluded Middle, or some other principle that Brouwer rejects, it would *still* provide a way of finding a prime greater than any number x, in a predetermined number of trials (in fact, x trials). Moreover, if this procedure ever failed, then *both* 'classical' arithmetic and Intuitionist arithmetic would be inconsistent.

In sum: *my*, so to speak, 'real statements', 'two plus two equals four', etc., are all 'essentially' first order (since, when written as second order statements in the fashion indicated before, all second order quantifiers are prenex and universal); and the predicate variables A, B, C, etc., correspond to just the empirical subject matter terms that one would naturally 'plug in' in an application. Hilbert's 'real statements' 'there is always a prime between x and $2x$', are not of this simple character. However, the statement that something is a Turing Machine with a specified machine table T is easily written in first order terms, provided we have appropriate predicates to serve as the 'states' (these are now 'empirical subject matter terms', although in the pure theory of the Turing Machine in question they would be replaced by dummy letters A, B, C, etc.); and for *fixed n*, it is easy enough to write down (again in first order terms) the statement that the machine will 'halt' in n (or however many) steps. Thus the statement that a *particular* empirically given Turing Machine will halt in n steps is a first order statement which presupposes nothing but the machine in question and the various predicates used to describe it; and the 'real statement' 'there is always a prime between x and $2x$', or whatever, serves simply as a device for obtaining predictions of this kind. If the statement is *proved* (and thus holds in all models), all the predictions of this kind obtained from it will hold good (or else Arithmetic will turn out to be inconsistent). Thus we see why Hilbert wished to call such statements (when suitably proved) 'real statements'. For me this is no issue. My 'cut' is different than Hilbert's 'cut'; my distinction is not between *statements with real meaning* and *meaningless concatenations of signs*; and hence I do not have to struggle, as Hilbert did, to include as many statements in the former class as possible without being 'metaphysical'. Instead, I have introduced here a distinction (in the case of such theories as *Principia*) between statements which refer to entities which exist independently of any particular 'model' for the mathematical theory at issue (e.g. the planets, the apples on some desk, a particular Turing Machine), and statements which refer to models, and to what I am construing, as entities within models – sets, propositional functions, etc. Even some of Hilbert's 'real' statements fall in the latter class; however, my account of our confidence in the computing procedures based upon them is roughly similar to Hilbert's.

(5) Russell's logical realism

Returning now to Russell's views, we are struck by two differences between the 'if-thenist' position, which he first espoused, and the later position. First, he seems to have come to the conclusion that the application of mathematics presupposes a restriction to *standard* models; and secondly, he seems to have come to the conclusion that it is necessary and possible to select a *particular* standard model (for *Principia*). What led him to the *second* of these views? (The first has already been discussed.) A number of considerations seem to have been operative. Linguistic considerations may have played a role. We do not speak of *a* number two, but of *the* number two. This is in agreement with the idea that there is some one definite model which is presupposed in number theory, and that even the substitution of an *isomorphic* model would be a change of subject matter. On this view it is a perfectly genuine theoretical question: 'is the number two the set of all pairs, or is it the set consisting of the null set and the unit set of the null set (as in von Neumann's theory), or some third set, or not a set at all?' I do not think that any mathematician would or should be impressed by this linguistic point. Secondly, both Frege and Russell sometimes write as if the logicist analysis of 'there are two *As*' (reviewed above) *depended* upon 'two' being defined as 'the set of all pairs'. This is a mistake (as Quine has pointed out). Even if we took the numbers *one, two, three,...* as *primitive* (in direct violation of the Frege–Russell spirit), it would suffice to define '*A* has *n* members' (where *n* is a *variable* over integers) to mean '*A* can be put in one-to-one correspondence with the set of natural numbers less than *n*' (or, alternatively, 'with the set of natural numbers from one *through n*'). Then the equivalences $P \equiv P^*$ discussed before would be forthcoming as theorems. It is these equivalences that underlie the logicist account of the application of mathematics; how exactly the numbers are defined, or whether they are taken as primitive is immaterial as long as these equivalences can be derived.

Let us shift attention, however, to the question: how and why did Russell think it was *possible* to single out a standard model for *Principia*? The answer, in broad outlines, is well known. Russell is a 'realist' with respect to universals; and he believed that these (universals, or predicates, or 'propositional functions') provide a model, and in fact the *intended* model, for *Principia*. His critics reply 'theology' (as we saw above), and the battle rages. But there is something curiously unsatisfying here.

What is unsatisfying is that at least three different issues have been blurred together. There is the hoary issue of 'the existence of universals'. There is the quite different issue of the existence of a hierarchy of propositional functions satisfying the axioms of *Principia*. And, unnoticed since Hume, but connecting the other two issues, there is the question of the relation between the 'natural' and 'philosophical' notions of a predicate or property.

Let us take this third issue to start with. Hume remarked that 'relation' (two place predicate) is used in two senses: in the 'natural' sense, *father of* and *to the left of* are 'relations', but *a thousand miles away from* is not what we ordinarily think of as a relation. This is a 'relation' only in the 'philosophical' sense (i.e. the logician's sense). A similar point can be made about *properties*. *Being either green or a human being or a doughnut* is a 'property' in the logician's sense, but not in the 'natural' sense.

This distinction between the 'natural' and 'philosophical' senses of 'property', 'relation', etc., bears also on the first issue – the 'existence of universals'. Let us review this issue for a moment. Realists insist that universals – colors, shapes, etc. – *exist*. Nominalists insist that they do not. Ordinary language, I note in passing, appears to have a blithe disregard for both positions. Realists should be troubled by the fact that 'the color red *exists*' is such an extraordinary thing to say (but they are not); Nominalists should be equally troubled by the fact that '*there is a shade of green which is as intense as scarlet*' is not at all extraordinary in ordinary English.† The traditional argument of the Nominalist is that assuming the existence of universals is otiose: whatever we want to say by saying 'the color red exists' can just as well be said by saying 'there are red things'. The traditional argument of the Realist proceeds in stages. First a distinction is drawn between classes (like Red) whose members have, intuitively 'something in common', and *arbitrary* collections (whose members need not have anything in common). Then it is argued (correctly, I believe) that the Nominalist has no satisfactory way of analyzing this distinction, i.e. of rendering the simple assertions that red things, for example, have something in common. This cannot be rendered by 'red things are all red' because of the 'for example'. What is wanted is a general analysis of the predicate (as a predicate of a number of things) 'having something in common'. The Nominalist might propose to meet the Realist half way by admitting *extensional* universals (classes), and then taking 'the members of *x*

† These 'idiosyncracies' of ordinary language, as some view them, fit rather well with Carnap's well-known distinction between 'external' existence questions (which purport to question the existence of a whole 'framework', e.g. colors, numbers, physical objects) and 'internal' existence questions, which presuppose some such framework. Ordinary language accepts the latter quite cheerfully, but not the former.

have something in common', where x is a class, as a *primitive* predicate. But then he still would not be able to analyze 'the members of x_1 have something *different* in common from the members of x_2'.

What is interesting about this argument is that it rests the existence of universals upon the intuitive notion of a 'something in common'. Intuitively 'red' and 'square' are universals in *this* sense: all red things and all squares have, intuitively, 'something in common', although what they have in common is different in the two cases. But there is not the slightest reason to believe that 'universals' in *this* sense† are closed under logical operations, or obey any of the axioms of *Principia*.

In particular, consider the familiar proof that the number of propositional functions is non-denumerable. Let us try to adapt this so that it becomes a proof that 'properties', in the intuitive sense, are non-denumerable. Let x_1, x_2, x_3, be some denumerably infinite list of things. (The argument assumes the Axiom of Infinity.) Suppose there were only denumerably many properties, say $P_1, P_2, \ldots$, where each P_i is a 'natural' property, i.e. a 'something in common'. Define a predicate Q as follows: x has Q if and only if x is x_n for some n and x does not have P_n. Then Q is a property and must be different from each property P_i in the list $P_1, P_2, \ldots$ But this was supposed to be an enumeration of *all* properties. Thus we have a contradiction.

This argument is fallacious, on the 'natural' notion of 'property', because we should hardly say that x and y, for example, have 'something in common', *merely* because in some *arbitrary* listing x is the seventeenth element (or x_{17}), y is the onehundredth (or x_{100}) and neither does x have the seventeenth property in some list of properties nor does y have the onehundredth. Even if $P_1, P_2, \ldots$ are all properties in the 'natural' sense, Q would clearly be a 'property' only in the 'philosophical' sense.

I conclude that the traditional debate between Nominalists and Realists, whatever its status, is irrelevant to the philosophy of mathematics. 'Properties' in any 'natural' sense no more form a model for *Principia* than physical objects do. And properties in the 'philosophical' sense are simply *predicates*. I think, myself, that by a 'predicate' what is normally meant is a *predicate in some language*, and that the diagonal argument simply shows that there is no well defined totality of all *languages*. But be that as it may, we have already seen that it is one thing to suppose that we have a *usable* notion of a 'predicate' and quite another to suppose that there is a well defined totality of all predicates (including predicates defined only in terms of that totality and predicates

---

† Obviously the sense in question has not at all been made clear.

which cannot apparently be defined at all). The latter supposition is tantamount to the supposition that there exists a standard model for *Principia* in some absolute sense of 'standard': but no traditional argument to the effect that 'all red things have something in common' will support *that* supposition. And, indeed, what argument could?

I have already contended that the question, whether a model for *Principia* (or any mathematical theory) actually exists lies outside of mathematics altogether. I would now go further and say the same for the question, whether only physical objects exist (in which case a model for *Principia* could only exist if it were physically realized, i.e. each 'propositional function' of the model would have in reality to be a physical object, and the relation of 'applying to' which connects propositional functions with their arguments would have to be a suitable relation among these physical objects), or whether 'universals exist'. If this is a question at all, and not a pseudo-question, it is a question for philosophers; but not, thank God! for philosophers of *mathematics*. Mathematics tells us that if anything is a 'group' then it has certain properties: but it does not matter whether the group is a set of physical objects, or a set of *colors*, or whatever. Similarly, mathematics tell us that if anything is a 'model for *Principia*' then it has certain properties: but it does not matter whether the model is a set of physical objects, or of colors, or whatever. If assuming the 'existence of universals' *ipso facto* guaranteed the existence of models for theories such as *Principia*, then the traditional question of the existence of universals might have some bearing on the philosophy of mathematics; but, as far as I can see, it doesn't.

(6) Mathematics as logic?

In summary, I have rejected the thesis that mathematics is logic in the sense outlined in *Principia*, for the reasons given in the second section of this paper. I have proposed instead to revive Russell's earlier view (the 'if-thenist' view). This view too makes mathematics 'logic', *in a sense*. For the essential business of the pure mathematician may be viewed as deriving logical consequences from sets of axioms. (Although mathematical induction may be viewed *either* as an 'extra-logical' method of proof, or as a 'logical' method in a richer theory, when we are studying *finite* models, as I noted). However, mathematics is not *just* logic. For our intuitive conviction that certain kinds of infinite structures *could* exist plays an essential role in the *application* of mathematics. It is a part, and an important part, of the *total* mathematical picture that certain sets of axioms are taken to describe *presumably possible* structures. It is only *such* sets of axioms that are used in *applied* mathematics. Thus there is a question which remains irreducibly a question in the philosophy of mathematics over and above the 'philosophy of logic': the question of illuminating and clarifying our acceptance of mathematical structures as 'presumably possible', or of mathematical axiom sets as 'presumably consistent'. Today this seems largely arbitrary; but no one can exclude that new results in the foundations of mathematics will radically alter the picture.

3

Mathematics without foundations*

Philosophers and logicians have been so busy trying to provide mathematics with a 'foundation' in the past half-century that only rarely have a few timid voices dared to voice the suggestion that it does not need one. I wish here to urge with some seriousness the view of the timid voices. I don't think mathematics is unclear; I don't think mathematics has a crisis in its foundations; indeed, I do not believe mathematics either has or needs 'foundations'. The much touted problems in the philosophy of mathematics seem to me, without exception, to be problems internal to the thought of various system builders. The systems are doubtless interesting as intellectual exercises; debate between the systems and research within the systems doubtless will and should continue; but I would like to convince you (of course I won't, but one can always hope) that the various systems of mathematical philosophy, without exception, need not be taken seriously.

By way of comparison, it may be salutory to consider the various 'crises' that philosophy has pretended to discover in the past. It is impressive to remember that at the turn of the century there was a large measure of agreement among philosophers – far more than there is now – on certain fundamentals. Virtually all philosophers were idealists of one sort or another. But even the nonidealists were in a large measure of agreement with the idealists. It was generally agreed any property of material objects – say, *redness* or *length* – could be ascribed to the object, if at all, only as a power to produce certain sorts of sensory experiences. When the man on the street thinks of a material object, according to this traditional view, he really thinks of a subjective object, not a real 'external' object. If there are external objects, we cannot really imagine what they are like; we know and can conceive only their powers. Either there are no external objects at all (Berkeley) – i.e. no objects 'external' to minds and their ideas – or there are, but they are *Dinge an sich*. In sum, then, philosophy flattered itself to have discovered not just a crisis, but a fundamental mistake, not in some special science, but in our most common-sense convictions about material objects. To put it

* First published in *The Journal of Philosophy*, LXIV, 1 (19 January 1967).

crudely, philosophy thought itself to have shown that no one has ever really perceived a material object and that, if material objects exist at all (which was thought to be highly problematical), then no one *could* perceive, or even imagine, one.

Anyone maintaining at the turn of the century that the notions 'red' and 'hard' (or, more abstractly 'material object') were reasonably clear notions; that redness and hardness are *non*dispositional properties of material objects; that we see red things and see *that* they are red; and that *of course* we can imagine red objects, know what a red object is, etc., would have seemed unutterably foolish. After all, the most brilliant philosophers in the world all found difficulties with these notions. Clearly, the man is just too stupid to see the difficulties. Yet today this 'stupid' view is the view of many sophisticated philosophers, and the increasingly prevalent opinion is that it was the arguments purporting to show a contradiction in the view, and not the view itself, that were profoundly wrong. Moral: not everything that passes – in philosophy anyway – as a difficulty with a concept is one. And second moral: the fact that philosophers all agree that a notion is 'unclear' doesn't mean that it *is* unclear.

More recently there was a large measure of agreement among philosophers of science – far more than there is now – that, in some sense, talk about theoretical entities and physical magnitudes is 'highly derived talk' which, in the last analysis, reduces to talk about observables. Just a few years ago, we were being told that 'electron' is a 'partially interpreted' term, whereas 'red' is 'completely interpreted'. Today it is becoming increasingly clear that 'electron' is a term that has complete 'meaning' in every sense in which 'red' has 'meaning'; that the 'purpose' of talk about electrons is not simply to make successful predictions in observation language any more than the 'purpose' of talk about red things is to make true deductions about electrons; and that the whole question about how we 'introduce' theoretical terms was a mare's nest. I refrain from drawing another moral.

Today there is a large measure of agreement among philosophers of mathematics that the concept of a 'set' is unclear. I hope the above short review of some history of philosophy will indicate why I am less than overawed by this agreement. When philosophy discovers something wrong with science, sometimes science has to be changed – Russell's paradox comes to mind, as does Berkeley's attack on the actual infinitesimal – but more often it is philosophy that has to be changed. I do not think that the difficulties that philosophy finds with classical mathematics today are genuine difficulties; and I think that the philosophical interpretations of mathematics that we are being offered on every hand are wrong, and that 'philosophical interpretation' is just what mathematics doesn't need. And I include my own past efforts in this direction.

I do not, however, mean to disparage the value of philosophical inquiry. If philosophy got itself into difficulties with the concept of a material object, it also got itself out; and the result is some modest but significant increase in our clarity about perception and knowledge. It is this sort of clarity about mathematical truth, mathematical 'objects', and mathematical necessity that I should like to see us attain; but I do not think the famous 'isms' in the philosophy of mathematics represent the road to that clarity. Let us therefore make a fresh start.

A sketch of my view

I think that the least mystifying way for me to discuss this topic is as follows: first to give a very cursory and superficial sketch of my own views, so that you will at least be able to guess at the positive position that underlies my criticism of others, and then to survey the alleged difficulties in set theory. Of course, any philosopher hates ever to say briefly, let alone superficially, what his own view on any topic is (although he is delighted to give such a statement to the view of any philosopher with whom he disagrees), because a superficial statement may make his view seem naive or even downright stupid. But such a statement is a great help to others, at least in getting an initial orientation, and for that reason I shall accept the risk involved.

In my view the chief characteristic of mathematical propositions is the very wide variety of equivalent formulations that they possess. I don't mean this in the trivial sense of cardinality: of course, every proposition possesses infinitely many equivalent formulations; what I mean is rather that in mathematics the number of ways of expressing what is in some sense the same fact (if the proposition is true) while apparently not talking about the same objects is especially striking.

The same situation does sometimes arise in empirical science, that is, the situation that what is in some sense the same fact can be expressed in two strikingly different ways, the most famous example being wave-particle duality in quantum mechanics. Reichenbach coined the happy expression 'equivalent descriptions' for this situation. The description of the world as a system of particles, not in the classical sense but in the peculiar quantum-mechanical sense, may be associated with a different picture than the description of the world as a system of waves, again not in the classical sense but in the quantum-mechanical sense; but the two theories are thoroughly intertranslatable, and should be viewed as having the same physical content. The same fact can be expressed either by saying that the electron is a wave with a definite wavelength λ or by saying that the electron is a particle with a sharp momentum p and an indeterminate position. What 'same fact' comes to here is, I admit, obscure. Obviously what is *not* being claimed is *synonymy* of *sentences*. It would be absurd to claim that the *sentence* 'there is an electron-wave with the wavelength λ' is *synonymous* with the *sentence* 'there is a particle electron with the momentum h/λ and a totally indeterminate position'. What is rather being claimed is this: that the two theories are compatible, not incompatible, given the way in which the theoretical primitives of each theory are now being understood; that indeed, they are not merely compatible but equivalent: the primitive terms of each admit of definition by means of the primitive terms of the other theory, and then each theory is a deductive consequence of the other. Moreover, there is no particular advantage to taking one of the two theories as fundamental and regarding the other one as *derived*. The two theories are, so to speak, on the same explanatory level. Any fact that can be explained by means of one can equally well be explained by means of the other. And in view of the systematic equivalence of statements in the one theory with statements in the other theory, there is no longer any point to regarding the formulation of a given fact in terms of the notions of one theory as more fundamental than (or even as *significantly* different from) the formulation of the fact in terms of the notions of the other theory. In short, what has happened is that the systematic equivalences between the sentences of the two theories have become so well known that they *function* virtually as synonymies in the actual practice of science.

Of course, the fact that two theories can be related in this way is not by itself either surprising or important. It would not be worth remarking that two theories are related in this way if the pictures associated with the two theories were not apparently incompatible or at least very different. In mathematics, the different equivalent formulations of a given mathematical proposition do not call to mind apparently *incompatible* pictures as do the different equivalent formulations of the quantum theory, but they do sometimes call to mind radically different pictures, and I think that the way in which a given philosopher of mathematics proceeds is often determined by which of these pictures he has in mind, and this in turn is often determined by which of the equivalent formulations of the mathematical propositions with which he deals he takes as primary.

Of the many possible 'equivalent descriptions' of the realm of mathematical facts, there are two which seem to me to have especial importance. I shall refer to these, somewhat misleadingly, I admit, by the titles 'Mathematics as Modal Logic' and 'Mathematics as Set Theory', The second, I take it, needs no explanation. Everyone is today familiar with the conception of mathematics as the description of a 'universe' of 'mathematical objects' – and, in particular, with the conception of mathematics as describing relations among *sets*. However, the picture would not be significantly different if one said 'sets and numbers' – that numbers can themselves be 'identified' with sets seems today a matter of minor importance; the important thing about the picture is that mathematics describes 'objects'. The other conception is less familiar, and I shall say a few words about it.

Consider the assertion that there is a counterexample to Fermat's 'last theorem'; i.e. that there is an nth power which is the sum of two nth powers, $2 < n$, all three numbers positive. Abbreviate the standard formula that expresses this statement in first-order arithmetic as '$\sim$ *Fermat*'. If $\sim$ *Fermat* is provable, then, in fact, $\sim$ *Fermat* is provable already from a certain easily specified finite subset of the theorems of first-order arithmetic. (N.B., this is owing to the fact that it takes only one counterexample to refute a generalization. So the portion of first-order arithmetic in which we can prove all true statements of the form $x^n + y^n \neq z^n$, x, y, z, n *constant* integers, is certainly strong enough to *disprove* Fermat's last theorem if the last theorem be false, notwithstanding the fact that *all* of first-order arithmetic may be too weak to *prove* Fermat's last theorem if the last theorem be true. And the portion of first-order arithmetic just alluded to is known to be finitely axiomatizable.) Let 'AX' abbreviate the conjunction of the axioms of the finitely axiomatizable subtheory of first-order arithmetic just alluded to. Then Fermat's last theorem is *false* just in case '$AX \supset \sim$ *Fermat*' is valid, i.e. just in case

$$\square \, (AX \supset \sim Fermat) \tag{1}$$

Since the truth of (1), in case (1) *is* true, does not depend upon the meaning of the arithmetical primitives, let us suppose these to be replaced by 'dummy letters' (predicate letters). To fix our ideas imagine that the primitives in terms of which AX and $\sim$ *Fermat* are written are the two three-term relations 'x is the sum of y and z' and 'x is the product of y and z' (exponentiation is known to be first-order-definable from these, and so, of course, are *zero* and *successor*). Let $AX(S, T)$ and $\sim$ FERMAT(S, T) be like AX and $\sim$ *Fermat* except for containing the 'dummy' triadic predicate letters S, T, where AX and $\sim$ *Fermat* contain the constant predicates 'x is the sum of y and z' and 'x is the product of y and z'. Then (1) is essentially a truth of pure modal logic (if it is true), since the constant predicates occur 'in-essentially'; and this can be brought out by replacing (1) by the abstract schema:

$$\Box \,[\text{AX}(S,\,T) \supset\, \sim \text{FERMAT}(S,\,T)] \qquad (2)$$

– and this is a schema of pure first-order modal logic.

Now then, the mathematical content of the assertion (2) is certainly the same as that of the assertion that *there exist numbers* x, y, z, n ($2 < n$, $x, y, z \neq 0$) such that $x^n + y^n = z^n$. Even if the expressions involved are not synonymous, the mathematical equivalence is so obvious that they might as well be synonymous, as far as the mathematician is concerned. Yet the pictures in the mind called up by these two ways of formulating what one might as well consider to be the same mathematical assertion can be quite different. When one speaks of the 'existence of numbers' one gets the picture of mathematics as describing eternal objects; while (2) simply says that $\text{AX}(S,\,T)$ entails $\text{FERMAT}(S,\,T)$, no matter how one may interpret the predicate letters 'S' and 'T', and this scarcely seems to be about 'objects' at all. Of course, one can strain after objects if one wants. One can, for example, interpret the dummy letters 'S' and 'T' as quantifiers over 'the totality of all properties', if one wishes. But this is hardly necessary, since one can find a particular substitution instance of (2), even in a nominalistic language (apart from the '$\Box$') which is equivalent to (2) (just choose predicates S^* and T^* to put for S and T such that it is not mathematically impossible that the objects in their field should form an ω-sequence, and such that, if the objects in their field did form an ω-sequence, S^* would be isomorphic to addition of integers, and T^* to multiplication, in the obvious sense). Or one can interpret '$\Box$' as a predicate of statements, rather than as a statement connective, in which case what (2) asserts is that a certain object, namely the statement '$\text{AX}(S,\,T) \supset \sim \text{FERMAT}(S,\,T)$' has a certain property ('being necessary'). But still, the only 'object' this commits us to is the statement '$\text{AX}(S,\,T) \supset \sim \text{FERMAT}(S,\,T)$', and one has to be pretty compulsive about one's nominalistic cleanliness to scruple about *this*. In short, if one fastens on the first picture (the 'object' picture), then mathematics is wholly extensional, but presupposes a vast totality of eternal objects; while if one fastens on the second picture (the 'modal' picture), then mathematics has *no* special objects of its own, but simply tells us what follows from what. If 'Platonism' has appeared to be *the* issue in the philosophy of mathematics of recent years, I suggest that it is because we have been too much in the grip of the first picture.

So far I have only indicated how one very special mathematical proposition can be treated as a statement involving modalities, but not special objects. I believe that, by making a more complex and iterated use of modal notions, one can analyze the notion of *a standard model for set theory*, and thus extend the objects–modalities duality that I am discussing to the whole of classical mathematics. I shall not show this now; but, needless to say, I would not deal at such length with this one special example if I did not believe it to represent, in some sense, the general situation. For the moment, I shall ask you to accept it on faith that this extension to the general case can be carried out.

What follows, I believe, is that each of these two ways of looking at mathematics can be used to clarify the other. If one is puzzled by the modalities (and I am concerned here with necessity in Quine's narrower sense of logical validity, excluding necessities that depend on alleged synonymy relations in natural languages), then one can be helped by the set-theoretic notion of a *model* (necessity = truth in all models; possibility = truth in some model). On the other hand, if one is puzzled by the question recently raised by Benacerraf: how numbers can be 'objects' if they have *no* properties except order in a particular ω-sequence, then, I believe, one can be helped by the answer: call them 'objects' if you like (they *are* objects in the sense of being things one can quantify over); but remember that these objects have the special property that each fact about them is, in an equivalent formulation, simply a fact about *any* ω-sequence. 'Numbers exist'; but all this comes to, for mathematics anyway, is that (1) ω-sequences are *possible* (mathematically speaking); and (2) there are *necessary* truths of the form 'if α is an ω-sequence, then...' (whether any *concrete* example of an ω-sequence exists or not). Similarly, there is not, from a mathematical point of view, any significant difference between the assertion that *there exists a set of integers* satisfying an arithmetical condition and the assertion that *it is possible to select* integers so as to satisfy the condition. Sets, if you will forgive me for parodying John Stuart Mill, are permanent possibilities of selection.

The question of decidability

The sense that there is a 'crisis in the foundations' of mathematics has many sources. Morris Kline cites the development of non-Euclidean geometry (which shook the idea that the axioms of a mathematical discipline must be *truths*), the lack of a consistency proof for mathematics, and the lack of a universally acceptable solution to the antinomies. In addition to these, one might mention Gödel's theorem (Kline does mention it, in fact, in connection with the consistency problem). For Gödel's theorem suggests that the truth or falsity of some mathematical statements might be impossible in principle to ascertain, and this has led some to wonder if we even know what we mean by 'truth' and 'falsity' in such a context.

Now, the example of non-Euclidean geometry does show, I believe, that our notions of what is 'self-evident' have to be subject to revision, not just in the light of new observations, but in the light of new *theories*. The intuitive evidence for the proposition that two lines cannot be a constant distance apart for half their length (i.e. in one half-plane) and then start to approach each other (as geodesics can in General Relativity, e.g. light rays which come in from infinity parallel and then approach each other as they pass on opposite sides of the sun) is as great as the intuitive evidence for the axioms of number theory. I believe that under certain circumstances revisions in the axioms of arithmetic, or even of propositional calculus (e.g. the adoption of a modular logic as a way out of the difficulties in quantum mechanics), is fully conceivable. The philosophical ploy which consists in saying 'then terms would have changed meaning' is uninteresting – except as a question in the philosophy of linguistics, of course – unless one can show that in their 'old meaning' the sentences of the theory in question can still (after the transition to non-Euclidean geometry, or non-Archimedean arithmetic, or modular logic) be admitted to have formerly expressed propositions that are clear and true. If in some sense there are 'Euclidean straight lines' in our space, then the transition to, say, Riemannian geometry *could* (not necessarily *should*) be regarded as a mere 'change of meaning'. But (1) there are *no* curves in space (if the world is Riemannian) that satisfy Euclid's theorems about straight lines; and (2) even if the world is Lobatchevskian, there are no *unique* such curves – to choose any particular remetricization which leads to Euclidean geometry and say '*this* is what "distance", "straight line", etc., *used* to mean' would be arbitrary. In short, the price one pays for the adoption of non-Euclidean geometry is to deny that there are *any* propositions which might *plausibly* have been in the minds of the people who believed in Euclidean geometry and which are simultaneously clear and true. Similarly, if one accepts the interpretation of quantum mechanics that is based on modular logic, then one has to deny that there has been a change in the meaning of the relevant sentences, or else deny that there are any unique propositions which might have been in the minds of those who formerly used those sentences and which were both clear and true. You can't have your conceptual revolution and minimize it too!

Yet all this does not, I think, mean that there is a crisis in the foundations of mathematics. It does not even mean that mathematics becomes an empirical science in the ordinary sense of that term. For the chief characteristic of empirical science is that for each theory there are usually alternatives in the field, or at least alternatives struggling to be born. As long as the major parts of classical logic and number theory and analysis have no alternatives in the field – alternatives which require a change in the axioms and which effect the simplicity of total science, including empirical science, so that a choice has to be made – the situation will be what it has always been. We will be justified in accepting classical propositional calculus or Peano number theory not because the relevant statements are 'unrevisable in principle' but because a great deal of science presupposes these statements and because no real alternative is in the field. Mathematics, on this view, does become 'empirical' in the sense that one is allowed to try to *put* alternatives into the field. Mathematics can be wrong, and not just in the sense that the proofs might be fallacious or that the axioms might not (if we reflected more deeply) be really self-evident. Mathematics (or rather, some mathematical theory) might be wrong in the sense that the 'self-evident' axioms might be false, and the axioms that are true might not be 'evident' at all. But this does not make the pursuit of truth impossible in mathematics any more than it has in empirical science, nor does it mean that we should not trust our intuitions when we have nothing better to go on. After all, a mathematical theory that has become the basis of a successful and powerful scientific system, including many important empirical applications, is not being accepted *merely* because it is 'intuitive', and if someone objects to it we have the right to say 'propose something better!' What this does do, rather, is make the 'foundational' view of mathematical knowledge as suspect as the 'foundational' view of empirical knowledge (if one cares to retain the 'mathematical-empirical' distinction at all).

Again, I cannot weep bitter tears about the lack of a consistency proof for classical mathematics. Even if such a proof were possible, it would only be a development within mathematics and not a foundation for mathematics. Not only would it be possible to raise philosophical questions about the branch of mathematics that was used for the consistency proof; but, in any case, science demands much more of a mathematical theory than that it should merely be *consistent*, as the example of the various alternative systems of geometry already dramatizes.

The question of the significance of the antinomies, and of what to do about the existence of several different approaches to overcoming them, is far more difficult. I propose to defer this question for a moment and to consider first the significance of Gödel's theorem and, more generally, of the existence of mathematically undecidable propositions.

Strictly speaking, all Gödel's theorem shows is that, in any particular consistent axiomatizable extension of certain finitely axiomatizable sub-theories of Peano arithmetic, there are propositions of number theory that can neither be proved nor disproved. (I think it is fair to call this 'Gödel's theorem', even though this statement of it incorporates strengthenings due to Rosser and Tarski, Mostowski, Robinson.) It does not follow that any proposition of number theory is, in some sense, absolutely undecidable. However, it may well be the case that some proposition of elementary number theory is neither provable nor refutable in any system whose axioms rational beings will ever have any good reason to accept. This has caused some to doubt whether every mathematical proposition, or even every proposition of the elementary theory of numbers, can be thought of as having a truth value.

A similar consideration is raised by Paul Cohen's recent work in set theory, when that work is taken together with Gödel's classical relative consistency proof of the axiom $V = L$ (which implies the axiom of choice and the generalized continuum hypothesis). Together these results of Gödel and Cohen establish the full independence of the continuum hypothesis (for example) from the other axioms of set theory, assuming those other axioms to be consistent. A striking feature of both proofs is their invariance under small (or even moderately large) perturbations of the axioms. It appears quite possible today that no decisive consideration will ever appear (such as a set-theoretic axiom we have 'overlooked') which will reveal that a system in which the continuum hypothesis is provable is the correct one, and that no consideration will ever appear which will reveal that a system in which the continuum hypothesis is refutable is the correct one. In short, the truth value of the continuum hypothesis – assuming it has a truth value – may be undiscoverable by rational beings, or at least by the 'rational beings' that actually do exist, or ever will exist. Then, what reason is there to think that it has a truth value?

This 'argument' is sometimes taken to show that the notion of a set is unclear. For, since the argument 'shows' (sic!) that the continuum hypothesis has no truth value and the continuum hypothesis involves the concept of a set, the only plausible explanation of the truth-value failure is some unclarity in the notion of a set. (It would be an interesting exercise to find *all* the faults in this particular bit of reasoning. It is horrible, isn't it?)

The first point to notice is that the existence of propositions whose truth value we have no way of discovering is not at all peculiar to mathematics. Consider the assertion that there are infinitely many binary stars (considering the entire space–time universe, i.e. counting binary stars past, present, and future). It is not at all clear that we can discover the truth value of this assertion. Sometimes it is argued that such an assertion is 'verifiable (or at least confirmable) in principle', because it may *follow from a theory*. It is true that in one case we can discover the truth value of this proposition. Namely, if either it or its negation is derivable from laws of nature that we can confirm, then its truth value can be discovered. But it could just happen that there are infinitely many binary stars, without this being required by any law. Moreover, the distribution might be quite irregular, so that ordinary statistical inference could not discover it. Indeed, at some point I cease to understand the question 'Is it always possible *in principle* to discover the truth value of this proposition?' – for the methods of inquiry permitted ('inductive' methods) are just too ill defined a set. But I suspect that, given any *formalizable* inductive logic, one could describe a logically possible world in which (1) there were infinitely many binary stars; and (2) one could never discover this fact using that inductive logic. (Of course, the argument that the proposition is 'confirmable in principle' because it could follow from a theory does not even purport to show that in every possible world the truth or falsity of this statement could be induced from a finite amount of observational material using some inductive method; rather it shows that in *some* possible world the truth of this statement (or its falsity) could be induced from a finite amount of observational material.) Yet I, for one, see no reason – not even a prima facie one – to suspect that this proposition does not have a truth value. Why *should* all truths, even all empirical truths, be discoverable by probabilistic automata (which is what I suspect we are) using a finite amount of observational material? Why does the fact that the truth value of a proposition may be undiscoverable by us suggest to some philosophers – indeed, why does it count as a *proof* for some philosophers – that the proposition in question doesn't *have* a truth value? Surely, some kind of idealistic metaphysics must be lurking in the underbrush!

What is even more startling is that philosophers who would agree with me with respect to propositions about material objects should feel differently about propositions of mathematics. (Perhaps this is due to the pernicious tendency to think of mathematics solely in terms of the mathematical–objects picture. If one doesn't understand the nature of these objects – i.e. that they don't have a 'nature', that talk about them is equivalent to talk about what is impossible – then talk about them may seem like a form of theology, and if one is anti-theological, that may be a reason for rejecting mathematics as a make-believe.) Surely, the *mere* fact that we may never know whether the continuum hypothesis is true or false is by itself just *no* reason to think that it doesn't have a truth value!

'But what does it *mean* to say that the continuum hypothesis is true?' someone will ask. It means that if S is a set of real numbers, and S is not finite and not denumerably infinite, then S can be put in one-to-one correspondence with the unit interval. Or, equivalently, it means that the sentence I have just written holds in any standard model for fourth-order number theory (actually, it can be expressed in third-order number theory). 'But what is a *standard* model?' It is one with the properties that (1) the 'integers' of the model form an ω-sequence under the $<$ of the model – i.e. it is not *possible* to select positive 'integers' $a_1, a_2, a_3, \ldots$ from the model so that, for all i, $a_{i+1} < a_i$ – and (2) the model is maximal with this property – i.e. it is not *possible* to add more 'sets' of 'integers' or 'sets of sets' of 'integers' or 'sets of sets of sets' of 'integers' to the model. (This last explanation contains the germ of the idea which is used in expressing the notion of a 'standard model' in modal-logical, as opposed to set-theoretic, language.)

I think that one can see what is going on more clearly if we imagine, for a moment, that physics has discovered that the physical universe is finite in both space and time and that all physical magnitudes are discrete (finiteness 'in the small'). That this is a possibility we must take into account was already emphasized by Hilbert in his famous article on the infinite – it may well be, Hilbert pointed out, that we cannot argue for the consistency of any theory whose models are all infinite by arguing that physical space, or physical time, or anything else physical, provides a model for the theory, since physics is increasingly tending to replace infinities and continuities by finites and discretes.

If the whole physical universe is thoroughly finite, both in the large and in the small, then the statement '$10^{100} + 1$ is a prime number' may be one whose truth value we can never know. For, if the statement is true (and even intuitionist mathematicians regard this decidable statement as possessing a truth value), then to verify that it is true by using any sieve method might well be physically impossible. And, if the shortest proof from axioms that rational beings will ever have any reason to accept is too long to be physically written out, then it might be physically impossible for beings to whom only those things are 'evident' that are in fact 'evident' (or ever will be 'evident' or that we will ever in fact have good reason to believe) to know that the statement is true.

Now, although many people doubt that the continuum hypothesis has a truth value, everyone believes that the statement '$10^{100} + 1$ is a prime number' has a truth value. Why? 'Because the statement is decidable.' But what does that mean, 'the statement is decidable'? It means that it is *possible* to try out all the pairs of possible factors and see if any of them 'work'. It means that it is *possible* to decide the statement. Thus, the man who asserts that this statement is decidable, is simply making an assertion of mathematical possibility. Moreover, he believes that just one of the two statements:

If all pairs n, m $(n, m < 10^{100} + 1)$ were 'tried' by actually computing the product nm, then in some case the product would be found to equal $10^{100} + 1$. (3)

If all pairs $n, m \ldots$ [same as in (3)], then in no case would the product be found to equal $10^{100} + 1$. (4)

expresses a *necessary* truth, although it may be *physically* impossible to discover which one. Yet this same mathematician or philosopher, who is quite happy in this context with the notion of mathematical possibility (and who does not ask for any nominalistic reduction) and who treats mathematical necessity as well defined in this case, for a reason which is essentially circular, regards it as 'platonistic' to suppose that the continuum hypothesis has a truth value.† I realize that this is an ad hominem argument, but still – if there is such an intellectual sin as 'platonism' (and it is remarkably unclear what this supposed sin consists of), why is it not already to commit it, if one supposes that '$10^{100} + 1$ is a prime number' has a truth value, even if no nominalistic reduction of this statement can be offered? (When one is defending a commonsense position, very often the only argument is ad hominem – for one has to keep throwing the burden of the argument back to the other side, by asking to be told *precisely* what is 'unclear' about the notions being attacked, or why a 'reduction' of the kind being demanded is necessary, or why a 'foundation' for the science in question is needed.)

In passing, I should like to remark that the following two principles, which many people seem to accept, can be shown to be inconsistent, by applying the Gödel theorem:

(1) That, even if some arithmetical (or set-theoretical) statements have no truth value, still, to say of any arithmetical (or set-theoretical)

---

† Incidentally, it may also be 'platonism' to treat statements of physical possibility or counterfactual conditionals as well defined. For (1) 'physical possibility' is *compatibility* with the laws of nature. But the relation of compatibility is interdefinable with the modal notions of possibility and necessity, and, of course, the laws of nature themselves require many mathematical notions for their statement. (2) A counterfactual conditional is true just in case the consequent *follows* from the antecedent, together with certain other statements that hold both in the actual and in the hypothetical world under consideration. And, of course, no nominalistic reduction has ever succeeded, either for the notion of physical possibility or for the subjunctive conditional.

statement that it has (or lacks) a truth value is itself always either true or false (i.e. the statement either has a truth value or it doesn't).
(II) All and only the decidable statements have a truth value.

For the statement that a mathematical statement S is decidable may itself be undecidable. Then, by (II), it has no truth value to say 'S is decidable'. But, by (I), it has a truth value to say 'S has a truth value' (in fact, *falsity*; since if S has a truth value, then S is decidable, by (II), and, if S is decidable, then 'S is decidable' is also decidable). Since it is false (by the previous parenthetical remark) to say 'S has a truth value' and since we accept the equivalence of 'S has a truth value' and 'S is decidable', then it must also be *false* to say 'S is decidable'. But it has no truth value to say 'S is decidable'. Contradiction.

The significance of the antinomies

The most difficult question in the philosophy of mathematics is, perhaps, the question raised by the antinomies and by the plurality of conflicting set theories. Part of the paradox is this: the antinomies do not at all seem to affect the notion 'set of material objects', or the notion 'set of integers', or the notion 'set of sets of integers', etc. Yet they *do* seem to affect the notion '*all* sets'. How are we to understand this situation?

One way out might be this: to conclude that we understand the notion 'set' in some contexts (e.g. 'set of integers', 'set of sets of integers'), but to conclude that we do not understand it in the context 'all sets'. But we do seem to understand *some* statements about all sets, e.g. 'for every set x and every set y, there is a set z which is the union of x and y'. So must we really abandon hope of making sense of the locution 'all sets'?

It is at this point that I urge we attend to the objects–modalities duality that I pointed out a few pages ago. The notion of a set has been used by a number of authors to clarify the notions of mathematical possibility and necessity. For example, if we identify the notion of a 'possible world' with the notion of a model (or, more correctly, with the notion of a structure of the appropriate type), then the rationale of the modal system S5 can easily be explained (as, for instance, by Carnap in *Meaning and Necessity*), and this explanation can be extended to the case of quantified modal logic by methods due to Kripke, Hintikka, and others. Here, however, I wish to go in the reverse direction, and assuming that the notions of mathematical possibility and necessity are clear (and there is no paradox associated with the notion of necessity as long as we take the '$\square$' as a statement connective (in the degenerate sense of 'unary connective') and not – in spite of Quine's urging – as a predicate of sentences), I wish to employ these notions to try to give a clear sense to talk about 'all sets'.

My purpose is not to start a *new* school in the foundations of mathematics (say, 'modalism'). Even if in some contexts the modal-logic picture is more helpful than the mathematical-objects picture, in other contexts the reverse is the case. Sometimes we have a clearer notion of what 'possible' means than of what 'set' means; in other cases the reverse is true; and in many, many cases both notions seem as clear as notions ever get in science. Looking at things from the standpoint of many different 'equivalent descriptions', considering what is suggested by *all* the pictures, is both a healthy antidote to foundationalism and of real heuristic value in the study of scientific questions.

Now, the natural way to interpret set-theoretic statements in the model-logical language is to interpret them as statements of what would necessarily be the case if there were standard models for the set theories in question. Since the models for von Neumann–Bernays set theory and its strengthenings (e.g. the system recently proposed by Bernays) are also models for Zermelo set theory, let me concentrate on Zermelo set theory. In order to 'concretize' the notion of a model, let us think of a model as a graph. The 'sets' of the model will then be pencil points (or some higher-dimensional analogue of pencil points, in the case of models of large cardinality), and the relation of membership will be indicated by 'arrows'. (I assume that there is nothing inconceivable about the idea of a physical space of arbitrarily high cardinality; so models of this kind need not necessarily be denumerable, and may even be standard.) Such a model will be called a 'concrete model' (or a 'standard concrete model', if it be standard) for Zermelo set theory. The model will be called standard if (1) there are no infinite-descending 'arrow' paths; and (2) it is not possible to extend the model by adding more 'sets' without adding to the number of 'ranks' in the model. (A 'rank'consists of all the sets of a given – possibly transfinite – type. 'Ranks' are cumulative types; i.e. every set of a given rank is also a set of every higher rank. It is a theorem of set theory that every set belongs to some rank.) A statement that refers only to sets of less than some given rank – say, to sets of rank less than $\omega \times 2$ – will be called a statement of 'bounded rank'. I ask the reader to accept it on faith that the statement that a certain graph G is a *standard* model for Zermelo set theory can be expressed using no 'non-nominalistic' notions except the '$\square$'.

If S is a statement of bounded rank and if we can characterize the 'given rank' in question in some invariant way (invariant with respect to standard models of Zermelo set theory), then the statement S can easily be translated into modal-logical language. The translation is just the statement that if G is any standard model for Zermelo set theory – i.e. any standard concrete model – and G contains the invariantly characterized rank in question, then necessarily S holds in G. (It is trivial to express 'S holds in G' for any *particular* S without employing the set-theoretic notion of 'holding'.) Our problem, then, is how to translate statements of *un*bounded rank into modal-logical language.

The method is best indicated by means of an example. If the statement has the form $(x)(\exists y)(z)Mxyz$, where M is quantifier-free, then the translation is this:

If G is any standard concrete model for Zermelo set theory and if P is any point in G, then it is possible that there is a graph G' that extends G (i.e. G is a subgraph of G') and a point y in G' such that G' is a standard concrete model for Zermelo set theory and such that

(if G'' is any graph that extends G' and such that G'' is a standard concrete model for Zermelo set theory and if z is any point in G'', then $Mxyz$ holds in G'').

Obviously this method can be extended to an arbitrary set-theoretic statement.

So much for technical matters. I apologize for this brief lapse into technicality, but actually this was only the merest sketch of the technical development, and this much detail is necessary for my discussion. The real question is this: what, if any, is the philosophical significance of such translations?

If there be any philosophical significance to such translations – and I don't claim a great deal – it lies in this: I did not assume that any standard concrete model for Zermelo set theory is maximal. Indeed, I would be inclined to say that no concrete model could be maximal – nor any *non*concrete model either, as far as that goes. Even God could not make a model for Zermelo set theory that it would be *mathematically* impossible to extend, and no matter what 'stuff' He might use. Yet I succeeded in giving a clear sense to statements about 'all sets' (clear relative to the notions I assumed to start with) *without* assuming a maximal model. In metaphysical language, it is not necessary to think of sets as one system of objects in some one possible world in order to follow assertions about all sets.

Furthermore, in construing statements about sets as statements about standard concrete models for set theory, I did not introduce possible concrete models (or even possible worlds) as objects. Introducing the modal connectives '$\square$', '$\lozenge$', '$\dashv$' is not introducing new kinds of objects, but rather extending the kinds of things we can say about ordinary objects and sorts of objects. (Of course, one *can* construe the statement that it is possible that there is a graph G satisfying a condition C as meaning that *there exists a possible graph G* satisfying the condition C; that is one way of smoothing the transition from the modal-logic picture to the mathematical-objects picture.)

The importance of Zermelo set theory and of the other set theories based upon the notion of 'rank' lies in this: we have a strong intuitive conviction that whenever As are possible, so is a structure that we might call 'the family of all sets of As'. Zermelo set theory assumes only this intuition and the intuition that the process of unioning such structures can be extended into the transfinite. Of course, this intuitive conviction *may* be mistaken; it could turn out that Zermelo set theory has no standard models (even if Zermelo set theory is consistent – e.g. the discovery of an ω-inconsistency would show that there are no standard models). But so could the intuitive conviction upon which number theory is based be mistaken. If we wish to be cautious, we can assume only predicative set theory up to some 'low' transfinite type. (It is necessary to extend predicative type theory 'just past' the constructive ordinals if we wish to be able to define *validity* of schemata that contain the quantifiers 'there are infinitely many x such that' and 'there are at most a finite number of x such that', for example.) Such a weak set theory may well give us all the sets we need for physics, and also the basic notions of validity and satisfiability that we need for logic, as well as arithmetic and a weak version of classical analysis. But the fact that we do have an intuitive conviction that standard models of Zermelo set theory, or of other set theories based upon the notion of 'rank' are *mathematically possible structures* is a perfectly good reason for asking what statements necessarily hold in such structures – e.g. for asking whether the continuum hypothesis necessarily holds in such structures.

The real significance of the Russell paradox, from the standpoint of the modal-logic picture, is this: it shows that *no* concrete structure can be a standard model for the naive conception of the totality of all sets; for any concrete structure has a possible extension that contains more 'sets'. (If we identify sets with the points that represent them in the various possible concrete structures, we might say: it is not possible for all *possible* sets to exist in any one world!) Yet set theory does not become impossible. Rather, set theory becomes the study of what must hold in, e.g. any standard model for Zermelo set theory.

4
What is mathematical truth?

In this paper I argue that mathematics should be interpreted realistically – that is, that mathematics makes assertions that are objectively true or false, independently of the human mind, and that *something* answers to such mathematical notions as 'set' and 'function'. This is not to say that reality is somehow bifurcated – that there is one reality of material things, and then, over and above it, a second reality of 'mathematical things'. A set of objects, for example, depends for its existence on those objects: if they are destroyed, then there is no longer such a set.† (Of course, we may say that the set exists 'tenselessly', but we may also say the objects exist 'tenselessly': this is just to say that in pure mathematics we can sometimes ignore the important difference between 'exists now' and 'did exist, exists now, or will exist'.) Not only are the 'objects' of pure mathematics conditional upon material objects; they are, in a sense, merely abstract possibilities. Studying how mathematical objects behave might better be described as studying what structures are abstractly possible and what structures are not abstractly possible.

The important thing is that the mathematician is studying something objective, even if he is not studying an unconditional 'reality' of non-material things, and that the physicist who states a law of nature with the aid of a mathematical formula is abstracting a real feature of a real material world, even if he has to speak of numbers, vectors, tensors, state-functions, or whatever to make the abstraction.

Unfortunately, belief in the objectivity of mathematics has generally gone along with belief in 'mathematical objects' as an unconditional and nonphysical reality, and with the idea that the kind of knowledge that we have in mathematics is strictly *a priori* – in fact, mathematical knowledge has always been the paradigm of *a priori* knowledge. The

† The null set is an exception to this statement, of course; but set theory is relatively interpretable in the theory of *non-empty* sets, provided we are willing to assume that at least one object (other than a set) exists. Thus, 'unconditional' sets are not in any way necessary in mathematics (either in pure mathematics, or in mathematics as part of total science), except as constructions out of 'conditional' sets. One might, nonetheless, insist on the *a priori* existence of the null set *an sich*; but this seems a little strained, even for a metaphysician.

present paper will argue that, on the contrary, mathematical knowledge resembles *empirical* knowledge – that is, that the criterion of truth in mathematics just as much as in physics is success of our ideas in practice, and that mathematical knowledge is corrigible and not absolute.

The method of mathematical proof

The first apparent stumbling block that confronts us if we wish to argue against the *a priori* character of mathematical knowledge is the method of mathematical proof. It does seem *at first blush* as if the sole method that mathematicians do use or *can* use is the method of mathematical proof, and as if that method consists simply in deriving conclusions from axioms which have been fixed once and for all by rules of derivation which have been fixed once and for all. In order to start our investigation, let us, therefore, first ask whether this is really the only conceivable method in mathematics. And, since the axioms are most clear and most 'immutable' in elementary number theory,† let us restrict our attention to elementary number theory: if we can make the case that even the elementary theory of non-negative integers is not *a priori*, then we shall not have much trouble with, say, set theory.

Martian mathematics

Let us now imagine that we have come in contact with an advanced civilization on the planet Mars. We succeed in learning the language of the Martians without too much difficulty, and we begin to read their newspapers, magazines, works of literature, scientific books and journals, etc. When we come to their mathematical literature, we are in for some surprises.

What first surprises us is the profundity of the results they claim to have obtained. Many statements that *our* best mathematicians have tried without success to prove – e.g. that every map can be colored with four colors, that the zeroes of the Riemann zeta functions in the strip above the unit interval all lie on the line $\frac{1}{2}$ – appear as assertions in their mathematical textbooks. Eagerly we start reading these textbooks in order to learn the proofs of these marvelous results. Then comes our

---

† Actually, they are not 'immutable' at all; only the *consequences* (the set of theorems) is – more-or-less – immutable. Elementary number theory was not axiomatized until the end of the nineteenth century ('Peano's axioms'). And how 'immutable' is the set of theorems? Was mathematical induction in its seventeenth century form completely grasped by the ancients? Did even the great seventeenth century number theorists go beyond *recursive* induction?

biggest surprise: the Martians rely on quasi-empirical methods in mathematics!

By 'quasi-empirical' methods I mean methods that are analogous to the methods of the physical sciences except that the singular statements which are 'generalized by induction', used to test 'theories', etc., are themselves the product of proof or calculation rather than being 'observation reports' in the usual sense. For example, if we decided to accept the Riemann Hypothesis (the statement about the zeroes of the Riemann zeta function mentioned a moment ago) because extensive searches with electronic computers have failed to find a counter-example – many 'theorems' have been proved with its aid, and none of these has been disproved, the consequences of the hypothesis (it has, in fact, important consequences in the theory of prime numbers and in other branches of ordinary number theory and algebraic number theory) are plausible and of far-reaching significance, etc. – then we could say, not that we had *proved* the Riemann Hypothesis, but that we had 'verified' it by a quasi-empirical method. Like empirical verification, quasi-empirical verification is relative and not absolute: what has been 'verified' at a given time may later turn out to be false. But is there any reason, other than a sociological one, why quasi-empirical methods should not be used in mathematics? If it turned out the Martians do use quasi-empirical methods, and their mathematical practice is highly successful, could we say that they are irrational?

One standard response ('standard' for a philosopher of recent vintage, anyway) might be to argue that the Martians would be conceptually confused because they 'don't know what a proof is'. And one might go on to argue that if one doesn't know what a proof is, then one doesn't know what *mathematics* is, and (more dubiously) that if one doesn't know what mathematical proof is, then one doesn't understand the assertions in question (the Riemann Hypothesis, or whatever) *as* mathematical assertions.

But before we allow this line of argument to spin itself out too far, we may as well ask: what makes you say that they don't know what a proof is? Suppose the Martians say something like this when queried on this point:

Mathematics is much like any other science in this respect: some assertions appear self-evident (e.g. $F = ma$ in physics, or, perhaps, some of the conservation principles) and others don't (the Law of Gravitation). Moreover, again as in other sciences, some assertions that don't *look* self-evident turn out to be consequences of evident principles (e.g. in Newtonian physics the Third Law – action equals reaction – is a consequence of the other laws) – and others are not. What you call 'proof' is simply deduction from principles that are (more or less) self evident. We recognize proof, and we value proof as highly as you do – when we can get it. What we don't understand is why you restrict yourself to *proof* – why you refuse to accept *confirmation*. After all, there are true mathematical statements that are neither immediately nor demonstratively necessary – epistemologically contingent mathematical truths. Not recognizing confirmation as well as proof debars you from ever discovering these truths.

If the Martians make *this* reply, then we cannot say they don't have the concept of proof. If anything, it's *we* who lack a concept – the concept of *mathematical confirmation*. The Martians know what a proof is; they use both methods – mathematical proof and confirmation; they are highly successful (so might we be if we developed the knack of making quasi-empirical mathematical inferences).

Finally, it might be objected that such methods are not necessary in principle; that mathematical statements just have the property that if they are true then they can be proved. But Gödel's theorem shows the contrary. Even if all statements that can be proved are epistemologically *a priori* and conversely,† the statements that can be proved from axioms which are evident to us can only be a recursively enumerable set (unless an infinite number of irreducibly different principles are at least potentially evident to the human mind, a supposition I find quite incredible). And Gödel's theorem can (in a version due, fundamentally, to Tarski) be expressed by the statement that the class of truths of just elementary number theory is not recursively enumerable.

In particular, then, even if it were the case that all the axioms we use in mathematics are 'analytic', as some philosophers have claimed, and that deduction preserves 'analyticity' (which is never shown), it would not follow that all truths of mathematics are analytic. Indeed, if the analytic sentences are all consequences of some finite list of Meaning Postulates (in the first order logic sense of 'consequences'), then it is a consequence of the theorem just cited that there must be synthetic truths in mathematics. Worse, it is a consequence of this view that all the statements we can prove are analytic; that, although there are synthetic truths in mathematics, our refusal to use quasi-empirical methods debars us from ever discovering a single one of them. Since philosophers who

---

† I will argue later in this paper that some of the axioms of mathematics – in particular, the assumption of a one-to-one correspondence between points in space and triples of reals (or points on a line and reals), and the axiom of choice – are quasi-empirical; thus I do not myself accept the claim that *proved* statements (e.g. consequences of these assumptions) are epistemologically *a priori*. (In fact, I don't think there *is* any such thing as an *a priori* statement, unless 'a priori' just means unrevisable within a particular theoretical frame, characterized both by positive assumptions and a 'space' of theoretical alternatives.)

favor this jargon generally hold that analytic truths have 'no content' and that synthetic truths have 'factual content', one wonders why these philosophers do not insist that we *must* use quasi-empirical methods!

Why have we not used quasi-empirical methods?

We are, then, faced with the following puzzle: if the use of quasi-empirical methods (not to say, *empirical* methods) is, in principle, justified in mathematics, then why have we not used them? Our answer to this puzzle is that the foregoing science fiction story about the Martians was a deliberate hoax: the fact is that *we* have been using quasi-empirical and even empirical methods in mathematics all along – we, us humans, right here on earth!

Thus, consider the basic postulate upon which the subject of analytical geometry is founded (and with it the whole study of space in modern mathematics, including the topological theory of manifolds). This is the postulate that there is a one-to-one order preserving correspondence between the points on the line and the real numbers. Consider the real numbers themselves. Were the real numbers and the correspondence postulate introduced in a rigorous mathematical fashion with a rigorous mathematical justification? They certainly were not. The fact is that the ancient Greeks lacked the mathematical experience, and hence lacked also the mathematical sophistication, to generalize the notion of 'number' to the extent required for the correspondence to exist. Thus, when they ran into the existence of incommensurables, they could only abandon the correspondence postulate, and with it the possibility of an algebraic treatment of geometry. Descartes, on the other hand, was willing to simply *postulate* the existence of a number – a 'real' number, as we now would say – corresponding to each distance.† He did not identify these numbers with sets of rationals or with sequences of rationals. But once he had shown how great the 'pay off' of the correspondence postulate was, not only in pure mathematics but also in

† It may be argued that this postulate – due to Fermat as well as Descartes – is *no longer* assumed in mathematics. For, one can argue, we now distinguish between *physical* space and abstract Euclidean space. Mathematics is concerned with the latter (and with other abstract spaces) not the former. But the latter can simply be identified with the set of triples of reals; thus the Correspondence Postulate is true by *definition*.

Against this we would argue that geometry as the theory of physical space (the space in which *objects* are located and moved about) *was* part of *mathematics* from Euclid until (approximately) the time of Riemann's Inaugural Dissertation. Without the Correspondence Postulate there would have been no *motivation* for calling the set of triples of reals an abstract 'space', or for identifying anything as a *metric* or a *line* or a *curve*. Indeed, talk of sets and functions itself became accepted only after talk of 'curves' had paved the way.

mechanics, there was not the slightest question of abandoning either the correspondence postulate or these generalized numbers, the 'real' numbers. In particular it would be a mistake to argue that Descartes was only 'justified' *because* it was possible (even if he did not know it) to 'identify' real numbers with sets or sequences of rationals. Suppose it were *not* possible to identify real numbers with sets or sequences (i.e. to 'construct' them out of rationals – i.e. suppose these constructions had not been discovered). Would we have *given up* analytical geometry and mechanics? Or would we not rather have come simply to regard real numbers as *primitive* entities, much as most mathematicians regard the natural numbers (*pace* Frege, *pace* Russell!) or as Frege regarded *concepts*, or Zermelo regarded *sets*, or some mathematicians today regard *categories* and *functors*? And suppose we had a consistent axiomatizable mathematics of this kind, mathematics taking real numbers as primitive. Would it be unjustified? It doubtless increases the security of the system to find a way to introduce real numbers by definition (although the degree of security is hard to measure, since part of the price one has to pay is to take *sets* as primitive, and it seems weird today to regard sets as 'safer' than real numbers). But it is not, contrary to the logicists, *essential* to identify real numbers with logical constructions out of rationals.

The fact is that once the assumption of real numbers and of the correspondence between points and reals had shown their fertility in both physics and mathematics, there was no question, barring the discovery of mathematical contradiction (and possibly not even then – we would certainly have tried to circumvent any contradiction by means less drastic than abandoning the real number system, and doubtless we would have succeeded), there was, repeat, no question of abandoning the real number system. The existence of real numbers and the correspondence between real numbers and points on the line were discovered in part quasi-empirically, in part empirically. This is as much an example of the use of hypothetico-deductive methods as anything in physics is.

The same story repeats itself with the introduction of the methods of the differential and integral calculus by Newton and Leibnitz. If the epsilon-delta methods had not been discovered, then infinitesimals would have been postulated entities (just as 'imaginary' numbers were for a long time). Indeed, this approach to the calculus – enlarging the real number system – is just as consistent as the standard approach, as we know today from the work of Abraham Robinson.

The remarks we made about the introduction of the methods of analytical geometry apply with full force to this case too. If the calculus had not been 'justified' Weierstrass style, it would have been 'justified' anyway.† The point is that the real justification of the calculus is its *success* – its success in mathematics, and its success in physical science.

A very recent example of the fully conscious and explicit use of quasi-empirical argument to justify enlarging the axiomatic foundations of mathematics is the introduction of the axiom of choice by Zermelo. In his 1908 paper,‡ Zermelo defends his axiom against the critics of his 1904 paper. Peano, in particular, had pointed out that the axiom appeared to be independent of the axioms in Peano's *Formulaire*, and had gone on to suggest that Zermelo's proof of the proposition that every set can be well ordered was, therefore, no proof at all, since it rests on the 'unproved' assertion of the axiom of choice. Here is Zermelo's reply:§

First, how does Peano arrive at his own fundamental principles and how does he justify their inclusion in the *Formulaire*, since, after all, he cannot prove them either? Evidently by analyzing the modes of inference that in the course of history have come to be recognized as valid and by pointing out that the principles are intuitively evident and necessary for science – considerations that can all be urged equally well in favor of the disputed principle. That this axiom, even though it was never formulated in textbook system, has frequently been used, and successfully at that, in the most diverse fields of mathematics, especially in set theory, by Dedekind, Cantor, F. Bernstein, Schoenflies, J. König, and others is an indisputable fact, which is only corroborated by the opposition that, at one time or another, some logical purists directed against it. Such an extensive use of a principle can be explained only by its self-evidence, which, of course, must not be confused with its provability. No matter if this self-evidence is to a certain degree subjective – it is surely a necessary source of mathematical principles, even if it is not a tool of mathematical proofs, and Peano's assertion‖ that it has nothing to do with mathematics fails to do justice to manifest facts. But the question that can be objectivei decided, whether the principle is *necessary for science*, I should now like to submit to judgment by presenting a number of elementary and fundamental theorems and problems that, in my opinion, could not be dealt with at all without the principle of choice. [Here follows a list of theorems that need the axiom of choice.]

In my opinion, Zermelo is right on two counts. First of all, he is right that 'self evidence' is somewhat subjective, but nonetheless counts

---

† I *don't* mean to deny the importance of removing contradictions from our theories. I mean that there is no unique way of removing contradictions from a somewhat useful theory, and in particular reductive definition is not the unique way, ever.

‡ 'A New Proof of the Possibility of a Well Ordering', reprinted in Heijenoort (1967), pp. 183–98.

§ *Ibid.* p. 187.

‖ 'Additione', *Revista de mathematica* 8, pp. 143–57; reprinted in Peano (1957) Vol. 1. The assertion Zermelo refers to is on p. 147.

for *something*. In empirical science too, it is wrong to think that intuition plays no role at all. Intuition is a *fallible* guide – that is what Francis Bacon taught us – but a fallible guide is still better than no guide at all. If our intuition were totally untrustworthy, we would never think of a correct or approximately correct theory to test in the first place. In mathematics, the desire that our axioms should be intuitively necessary is a legitimate one, especially when combined with the desideratum that Zermelo mentions – that they should formalize the actual practice of mathematicians. But it is noteworthy that what Zermelo characterizes as 'objective' is not the 'self evidence' of the axiom of choice but its *necessity for science*. Today it is not just the axiom of choice but the whole edifice of modern set theory whose entrenchment rests on great success in mathematical application – in other words, on 'necessity for science'. What argument, other than a quasi-empirical one, can we offer for the axiom of Replacement? And the current rumblings in Category theory are evidence that the hypothetico-deductive evolution and testing of new mathematical existence statements (new 'objects') and axioms and methods is still going on.

The use of quasi-empirical methods in mathematics is not by any means confined to the testing of new axioms or new 'ontological commitments'. Although it is rare that either mathematicians or philosophers discuss it in public, quasi-empirical methods are constantly used to discover truths or putative truths that one then tries to prove rigorously. Moreover, some of the quasi-empirical arguments by which one discovers a mathematical proposition to be true in the first place are totally convincing to mathematicians. Consider, for example, how Euler discovered that the sum of the series $1/n^2$ is $\pi^2/6$. Euler proceeded in analogy with the factorization

$$P(x) = c_0 \left(1 - \frac{x}{\varepsilon_1}\right) \left(1 - \frac{x}{\varepsilon_2}\right) \left(1 - \frac{x}{\varepsilon_3}\right) \dots \left(1 - \frac{x}{\varepsilon_n}\right)$$

where $P(x)$ is a polynomial with roots ($\neq 0$) $\varepsilon_1, \dots, \varepsilon_n$. He 'factored' $\sin \pi x$ by considering the 'roots' to be the values for which $\sin \pi x = 0$, i.e. $x = 0$, $x = \pm 1$, $x = \pm 2, \dots$. Thus

$$\sin \pi x = c_0 x \left(1 - \frac{x}{1}\right) \left(1 + \frac{x}{1}\right) \left(1 - \frac{x}{2}\right) \left(1 + \frac{x}{2}\right) \dots$$

(The factor 'x' is present because 0 is one of the 'roots'.) To evaluate the 'constant term' c_0 he used

$$\lim_{x \to 0} \frac{\sin \pi x}{x} = \pi = c_0$$

Thus:

$$\sin \pi x =_{(?!)} \pi x \left(1 - \frac{x^2}{1}\right)\left(1 - \frac{x^2}{4}\right)\left(1 - \frac{x^2}{9}\right) \ldots \tag{1}$$

But by Taylor's theorem:

$$\sin \pi x = \frac{\pi x}{1!} - \frac{1}{3!}\pi^3 x^3 + \frac{1}{5!}\pi^5 x^5 \ldots \tag{2}$$

Equating the coefficients of x^3 in (1) and (2) gives:

$$-\frac{\pi^3}{3!} = \pi(-\tfrac{1}{1} - \tfrac{1}{4} - \tfrac{1}{9} \cdots) \tag{3}$$

or

$$-\frac{\pi^2}{6} = -\sum \frac{1}{n^2} \tag{4}$$

so

$$\sum \frac{1}{n^2} = \frac{\pi^2}{6}$$

Euler, of course, was perfectly well aware that this was not a proof. But by the time one had calculated the sum of $1/n^2$ to thirty or so decimal places and it agreed with $\pi^2/6$, no mathematician doubted that the sum of $1/n^2$ was $\pi^2/6$, even though it was another twenty years before Euler had a proof. The similarity of this kind of argument to a hypothetico-deductive argument in empirical science should be apparent: intuitively plausible though not certain analogies lead to results which are then checked 'empirically'. Successful outcomes of these checks then reinforce one's confidence in the analogy in question.†

Let me give another example of this kind, this time from present-day mathematics. Many mathematicians are quite convinced that there are infinitely many 'twin primes' (i.e. infinitely many pairs n, $n+2$, both prime, such as 5, 7, or 11, 13) even though there is no mathematical proof of this assertion. The argument they find convincing goes as follows: it seems plausible (and agrees with 'empirical' data) that the 'events' *n is a prime* and *n+2 is a prime* are *independent* events in the statistical sense. But the frequency of primes less than n is approximately $1/\log n$. Hence the frequency of twin primes less than n must be (asymptotically) like $1/(\log n)^2$, which implies that the number of twin primes is infinite.

Bas van Frassen has asserted that it is a consequence of my view that the following is a good quasi-empirical inference in mathematics:

† The foregoing example comes from Polya, a great exponent of the importance of plausible reasoning in mathematics.

computers have failed to turn up a counter-example to the Goldbach conjecture, *therefore* the Goldbach conjection is true. Of course, this is not a good quasi-empirical inference. And I do not pretend to be able to give rules by means of which we can tell which are and which are not good quasi-empirical inferences. After all, the analogous problem in philosophy of empirical science – the problem of inductive logic – has resisted solution for centuries, but people have not abandoned empirical science on that account. But I can say what is wrong with this simple 'induction' that the Goldbach conjecture is true. The fact is that neither in mathematics nor in empirical science do we trust the conclusion of a simple 'Baconian' induction to be exactly and precisely correct. A universal generalization – a statement that can be overthrown by a single 'for instance' – cannot be verified by mere Baconian induction in any science. But just contrast the 'inductive' argument we gave for the existence of infinitely many twin primes with the bad argument for the Goldbach conjecture. Even if the events *n is a prime* and *n + 2 is a prime* are not strictly statistically independent, the conclusion will still be correct. In other words, the deduction that there are infinitely many twin primes is 'stable under small perturbations of the assumptions'. One confirms inductively a statistical statement,† not an exceptionless generalization, and then deduces from even the approximate truth of the statistical statement that there will be infinitely many twin primes. My impression is that there are very few mathematicians who are *not* convinced by this argument, even though it is not a proof.

Since we do use quasi-empirical methods a great deal in mathematics (and we aren't even Martians!) I believe that it would be of great value to attempt to systematize and study these methods. Perhaps such an enterprise is premature in the present state of our knowledge. However, a mathematical friend has suggested that model theoretic methods might be used, for example, to try to convert 'probability' arguments like the one for the existence of infinitely many twin primes, into proofs.

Realism in the philosophy of mathematics

I am indebted to Michael Dummett for the following very simple and elegant formulation of realism: A realist (with respect to a given theory or discourse) holds that (1) the sentences of that theory or discourse are

---

† In fact, more careful reasoning shows that the events in question cannot be *strictly* independent, and therefore 'the only reasonable conjecture' – the words are those of a world famous number theorist – is that the number of twin primes less than x 'must' be $1.23 \ldots 1/(\log x)^2$. Another world famous mathematician described this argument as 'totally convincing' – that is, the argument that there must be infinitely many twin primes.

true or false; and (2) that what makes them true or false is something *external* – that is to say, it is not (in general) our sense data, actual or potential, or the structure of our minds, or our language, etc. Notice that, on this formulation, it is possible to be a realist with respect to mathematical discourse without committing oneself to the existence of 'mathematical objects'. The question of realism, as Kreisel long ago put it, is the question of the objectivity of mathematics and not the question of the existence of mathematical objects.

One way to spell this out is the following. Mathematics has, since Frege and Russell and Zermelo and Bourbaki been thought of as describing a realm of mathematical objects. In principle, all these objects can be identified with *sets*, in fact. The language in which these objects are described is highly asceptic – no modal notions, no intensional notions (e.g. 'proof'), indeed, in the by now standard case, no notions except those of the first order theory of 'epsilon' (set-membership). Mathematics has, roughly speaking, got rid of *possibility* by simply assuming that, up to isomorphism anyway, all possibilities are simultaneously *actual* – actual, that is, in the universe of 'sets'.

There is another possible way of doing mathematics, however, or at any rate, of viewing it. This way, which is probably much older than the modern way, has suffered from never being explicitly described and defended. It is to take the standpoint that mathematics has *no* objects of its own at all. You can prove theorems about anything you want – rainy days, or marks on paper, or graphs, or lines, or spheres – but the mathematician, on this view, makes no existence assertions at all. What he asserts is that certain things are *possible* and certain things are *impossible* – in a strong and uniquely mathematical sense of 'possible' and 'impossible'. In short, mathematics is essentially *modal* rather than existential, on this view, which I have elsewhere termed 'mathematics as modal logic'.†

Let me say a few things about this standpoint here.

(1) This standpoint is not intended to satisfy the nominalist. The nominalist, good man that he is, cannot accept modal notions any more than he can accept the existence of sets. We leave the nominalist to satisfy himself.

(2) We do have to say something about Hume's problem. It was Hume more than any other philosopher who drove the notions of possibility and necessity into disrepute. What bothered Hume was the following argument: *we only observe what is actual*. Since the only generalizations we can make on the basis of the observation of actual

† In 'Mathematics without foundations', chapter 3 in this volume.

things are to the effect that all *A*s are *B*s – not that all *possible A*s are *B*s, or that all *A*s are *necessarily B*s, Hume concluded that necessity must necessarily be a subjective matter.

It seems to us that this argument rests on much too simple a view of the structure of scientific knowledge. Physical theory, for example, has not for a long time been a mere collection of statements of the form $(x)(Fx \to Gx)$. From classical mechanics through quantum mechanics and general relativity theory, what the physicist does is to provide mathematical devices for representing all the *possible* – not just the physically possible, but the mathematically possible – configurations of a system. Many of the physicist's methods (variational methods, Lagrangian formulations of physics) depend on describing the actual path of a system as that path of all the *possible* ones for which a certain quantity is a minimum or maximum. Equilibrium methods in economics use the same approach. It seems to us that 'possible' has long been a theoretical notion of full legitimacy in the most successful branches of science. To mimic Zermelo's argument for the axiom of choice, we may argue that the notion of possibility is intuitively evident and necessary for science. And we may go on to argue, as he did, that the intuitive evidence is somewhat subjective, but the necessity for science is objective. It seems to us that those philosophers who object to the notion of possibility may, in some cases at least, simply be ill-acquainted with physical theory, and not appreciate the extent to which an apparatus has been developed for *describing* 'possible worlds'. That we cannot directly *observe* the possible (unless it happens to be actual) should not count as an argument against the notion of possibility in this day and age.

(3) The notion of possibility does not have to be taken as a *primitive* notion in science. We can, of course, define a structure to be *possible* (mathematically speaking) just in case a model exists for a certain theory, where the notion of a model is the standard set theoretic one. That is to say, we *can* take the existence of sets as basic and treat possibility as a derived notion. What is often overlooked is that we can perfectly well go in the reverse direction: we can treat the notion of possibility as basic and the notion of set existence as the derived one. Sets, to parody John Stuart Mill, are permanent possibilities of selection.

It is clear that number theoretic statements, with however many quantifiers, can be translated into possibility statements. Thus a statement to the effect that for every number x there exists a number y such that $F(x, y)$, where $F(x, y)$ is a recursive binary relation, can be paraphrased as saying that it is not *possible* to produce a tape with a numeral written on it which is such that if one *were* to produce a Turing machine of a certain description and start it scanning that tape, the machine would never halt. In a previous paper, I showed that an arbitrary statement† of set theory – even one that quantifies over sets of unbounded rank – can be paraphrased by a possibility statement.

(4) The main question we must speak to is simply, *what is the point*? Given that one can either take modal notions as primitive and regard talk of mathematical existence as derived, or the other way around, what is the advantage to taking the modal notions as the basic ones? It seems to us that there are two advantages to starting with the modal concepts. One advantage is purely mathematical. Construing set talk, etc., as talk about possible or impossible structures puts problems in a different focus. In particular, different axioms are evident. It is not my intention to discuss these purely mathematical advantages here. The other advantage is philosophical. Traditionally, realism in the philosophy of mathematics has gone along with Platonism, as we remarked at the outset, where 'Platonism' connotes simultaneously an epistemological theory and an ontology. The main burden of this paper is that one does not have to 'buy' Platonist epistemology to be a realist in the philosophy of mathematics. The modal logical picture shows that one doesn't have to 'buy' Platonist ontology either. The theory of mathematics as the study of special *objects* has a certain implausibility which, in my view, the theory of mathematics as the study of ordinary objects with the aid of a special concept does not. While the two views of mathematics – as set theory and as 'modal logic' – are intertranslatable, so that there is not here any question of one being true and the other being false, the modal logical view has advantages that seem to me to go beyond mere provision of psychological comfort to those distressed by Platonism. There are real puzzles, especially if one holds a causal theory of reference in some form, as to how one can refer to mathematical objects at all. I think that these puzzles can be clarified with the aid of modal notions. But again, this goes beyond the burden of this paper.

Let us return now to the topic of realism. Realism with respect to empirical science rests on two main kinds of arguments, which we may classify loosely as negative arguments and positive arguments. Negative arguments are to the effect that various reductive or operationalist philosophies are just unsuccessful. One tries to show that various attempts to reinterpret scientific statements as highly derived statements about sense data or measurement operations or whatever are unsuccessful, or hopelessly vague, or require the redescription of much ordinary scientific discovery as 'meaning stipulation' in an implausible way, or

† *Ibid.*

72

something of that kind, with the aim of rendering it plausible that most scientific statements are best not philosophically reinterpreted at all. The positive argument for realism is that it is the only philosophy that doesn't make the success of science a miracle. That terms in mature scientific theories typically refer (this formulation is due to Richard Boyd), that the theories accepted in a mature science are typically approximately true, that the same term can refer to the same thing even when it occurs in different theories – these statements are viewed by the scientific realist not as necessary truths but as part of the only scientific explanation of the success of science, and hence as part of any adequate scientific description of science and its relations to its objects.

I believe that the positive argument for realism has an analogue in the case of mathematical realism. Here too, I believe, realism is the only philosophy that doesn't make the success of the science a *miracle*.

In my view, there are *two* supports for realism in the philosophy of mathematics: *mathematical experience* and *physical experience*. The construction of a highly articulated body of mathematical knowledge with a long tradition of successful problem solving is a truly remarkable *social* achievement. Of course, one might say: 'well, in the middle ages they would have said "the construction of a highly articulated body of theological knowledge with a long tradition of successful problem solving is..."'. But 'Theological knowledge' was in fact highly *inconsistent*. Moreover, if one 'fixed it up' so as to restore consistency, the consistency would be a trivial result – doubtless it would follow from the existence of some kind of finite model. In mathematics we have (we think) a *consistent* structure – consistent notwithstanding the fact that no science other than mathematics deals with such *long* and rigorous deductive chains as mathematics does (so that the risk of *discovering* an inconsistency, if one is present is immeasurably higher in mathematics than in any other science) and notwithstanding the fact that mathematics deals with such complex infinite structures that, as we know from Gödel's work, no hope of a finitistic consistency proof exists. If there is *no* interpretation under which most of mathematics is *true*, if we are really just writing down strings of symbols at random, or even by trial and error, what are the chances that our theory would be consistent, let alone mathematically fertile?

Let us be careful, however. If this argument has force and I believe it does, it is not quite an argument for mathematical realism. The argument says that the consistency and fertility of classical mathematics is evidence that it – or most of it – *is true under some interpretation*. But the interpretation might not be a *realist* interpretation. Thus Bishop might say, 'indeed, most of classical mathematics is true under some interpretation; it is true under an intuitionist *reinterpretation!*' Thus our argument has to stand on two legs: the other leg is *physical experience.* The interpretation under which mathematics is true has to square with the application of mathematics *outside* of mathematics.

In a little book I published not long ago (Putnam, 1971), I argued in detail that mathematics and physics are integrated in such a way that it is not possible to be a realist with respect to physical theory and a nominalist with respect to mathematical theory. In a sense, this means that our intuitions are inconsistent. For I believe that the position most people find intuitive – the one that I certainly found intuitive – *is* realism with respect to the physical world and some kind of nominalism or if-thenism with respect to mathematics. But consider a physical law, e.g. Newton's Law of Universal Gravitation. To say that this Law is true – to even say that it is approximately true at nonrelativistic distances and velocities – one has to quantify over such non-nominalistic entities as forces, masses, distances. Moreover, as I tried to show in my book, to account for what is usually called 'measurement' – that is, for the numericaliza-tion of forces, masses and distances – one has to quantify not just over forces, masses, and distances construed as physical properties (think of a particular mass as a property that any given thing may or may not have, where the notion of a property is such that the property does not have any intrinsic connection with one particular *number* rather than another), but also over *functions from* masses, distances, etc. *to* real numbers, or at any rate to rational numbers. In short – and this is an insight that, in essence, Frege and Russell already had – a reasonable interpretation of the *application* of mathematics to the physical world *requires* a realistic interpretation of mathematics. Mathematical ex-perience says that mathematics is true under some interpretation; physical experience says that that interpretation is a realistic one.

To sketch the argument in a nutshell: if one is a realist about the physical world, then one wants to say that the Law of Universal Gravitation makes an objective statement about bodies – not just about sense data or meter readings. What is the statement? It is just that bodies behave in such a way that the quotient of two numbers *associated* with the bodies is equal to a third number *associated* with the bodies. But how can such a statement have any objective content at all if numbers and 'associations' (i.e. functions) are alike mere fictions? It is like trying to maintain that God does not exist and angels do not exist while main-taining at the very same time that it is an objective fact that God has put an angel in charge of each star and the angels in charge of each of a pair of binary stars were always created at the same time! If talk of numbers and 'associations' between masses, etc. and numbers is 'theology' (in the pejorative sense), then the Law of Universal Gravitation is likewise theology.

A digression on intuitionism

It seems to me that the argument against nominalism just sketched also goes through against intuitionism. Let me be more precise. Intuitionism has two parts: intuitionism gives us a set of *notions* to use in doing mathematics (an *ideology*, in Quine's sense of the term) and it gives us a set of objects to quantify over (an ontology). The two questions: is intuitionist *ideology* adequate for mathematics/physics? and is intuitionist *ontology* adequate for mathematics/physics? are almost never separated when people discuss these questions (the work of Georg Kreisel is a happy exception to this sad state of affairs), but it is essential that they should be. It is my claim that even if the ideology and ontology of intuitionism prove adequate to derive *all* of classical mathematics, the *ideology* of intuitionism is wholly inadequate for physics. The Law of Universal Gravitation, for example, has the form

$$(x)[\theta(x) = \psi(x)] \qquad (1)$$

where θ and ψ are empirically given sequences – 'lawless' sequences. On the intuitionist interpretation of the logical connectives, what (1) *means* is that there is an integer n such that given the first n decimal places of both θ and ψ one can *prove* that (1) is true. Since this is absurd for lawless sequences, and since a proof of the absurdity of a proof of (1) counts as a proof of the intuitionist negation of (1), it is actually a theorem of intuitionist mathematics that

$$\sim (x)[\theta(x) = \psi(x)]$$

– i.e. the Law of Universal Gravitation is intuitionistically false! The reason Brouwer does not notice this is that he treats the empirical world as a 'decidable case' that is, as a finite system. But this requires him to be a thorough-going fictionalist. Indeed, in his Dissertation he not only takes the point of view that physical objects are fictions, but also asserts that other selves and even future states of his own mind are 'fictions'!

Physical application and nondenumerability

I have argued that the hypothesis that classical mathematics is largely *true* accounts for the success of the physical applications of classical mathematics (given that the empirical premisses are largely approximately true and that the rules of logic preserve *truth*). It is worth while pausing to remark just how much of classical mathematics has been developed *for* physical application (the calculus, variational methods, the current intensive work on nonlinear differential equations, just for a start), and what a surprising amount has *found* physical application. Descartes' assumption of a correspondence between the points on a line and the reals was a daring application of what we now recognize to be nondenumerable mathematics to physical space. Since space is connected with physical experience, it is perhaps not surprising that *this* found physical application. Likewise, the calculus was explicitly developed to study *motion*, so perhaps it is not surprising that this too found physical application; but who would have expected *spectral measure*, of all things, to have physical significance? Yet quantum mechanical probabilities are all computed from spectral measures. (In a sense, nothing has *more* physical significance than spectral measure!)

This raises a question which is extremely interesting in its own right, if somewhat tangential to our main concern: do we have evidence for the nondenumerability of physical space, or is this merely a physically meaningless, albeit useful 'idealization', as is so often asserted by philosophers of science?

The reason that I regard this question as tangential to the main question of this paper is that even if physical space turns out to be discrete, even if it only behaves as a nondenumerable space *would* behave (up to a certain approximation), still the explanation of the behavior of space presupposes a correct understanding of how a nondenumerable space *would* behave, and the claim we are making for classical mathematics is that it provides *this*.

The importance of proof

In this paper, I have stressed the importance of quasi-empirical and even downright empirical methods in mathematics. These methods are the source of new axioms, of new 'objects', and of new theorems, that we often know to be true *before* we succeed in finding a proof. Quasi-empirical/empirical inferences support the claim that mathematics is (largely) true, and place constraints on the interpretation under which it can be *called* 'true', but a word of caution is in order. None of this is meant to downgrade the notion of proof. Rather, Proof and Quasi-empirical inference are to be viewed as complementary. Proof has the great advantage of not increasing the risk of contradiction, where the introduction of new axioms or new objects does increase the risk of contradiction, at least until a relative interpretation of the new theory in some already accepted theory is found. For this reason, proof will continue to be the primary method of mathematical verification. But given that formal deductive proof is likely to remain the primary method of mathematical verification, and that it is developed to an astounding extent in the science of mathematics, it is surprising how little we really know about it. In part this is because proof theory developed as an ideological rather than a scientific weapon. Proof theory was burdened with the constraint that only finitist methods must be used – a constraint with no mathematical justification whatsoever. Only recently have workers like Georg Kreisel, Takeuti, Prawitz, and others begun to view proof theory as a non 'ideological' branch of mathematics which simply seeks to give us information about what proof really does.

I should like to conjecture that the modal logical interpretation (or, rather, family of interpretations) of classical mathematics may help in this enterprise. Modal logical interpretations sometimes bear a formal similarity to intuitionist reinterpretations while being fully realistic. Thus they may play a role in the study of proofs similar to the role that has been played by intuitionist and allied interpretations, while giving more or less different 'information'.

Physics and the future of mathematics

In this paper, I have not argued that mathematics is, in the full sense, an *empirical* science, although I have argued that it relies on empirical as well as quasi-empirical inference. The reader will not be surprised to learn that my expectation is that as physical science develops, the impact on mathematical axioms is going to be greater rather than less, and that we will have to face the fact that 'empirical' versus 'mathematical' is only a relative distinction; in a looser and more indirect way than the ordinary 'empirical' statement, much of mathematics too is 'empirical'.

In a sense, this final collapse of the notion of the *a priori* has already begun. After all, geometry was a part of mathematics – not just un-interpreted geometry, but the theory of physical space. And if space were Euclidean, doubtless the distinction between 'mathematical' and 'physical' geometry would be regarded as silly. When Euclidean geometry was dethroned, the argument was advanced that 'straight line' only means 'light ray' and 'any fool can plainly see' that interpreted geometry is empirical. It was kind of an oversight, in this view, that the theory of physical space was ever regarded as *a priori*. In the last few years the standard interpretation of quantum mechanics – viz. that it no longer makes sense to separate epistemology and physics, that henceforth we can only talk about physical magnitudes as they are measured by particular experimental arrangements – has begun to be challenged by the upstart view that quantum mechanics is a complete realistic theory, that there is nothing special about measurement, and that we just happen to live in a world that does not obey the laws of Boolean logic.† Just as those who defended non-Euclidean geometry sought to minimize the impact of their proposals (or, rather, to make them more palatable) by adopting an extreme operationist style of presentation, so the main advocates of quantum logic – Finkelstein, Jauch, Mackey, Kochen – also adopt an extreme operationist style of presentation. They only claim that quantum logic is true given the precisely specified operational meaning of the logical connectives. Mackey and Jauch go so far as to suggest that there is some other study, called 'logic' (with, of course, no operational meaning at all) which they are not challenging. In my opinion, whatever their intentions, they *are* challenging logic. And just as the almost unimaginable fact that Euclidean geometry is false – false of *paths in space*, not just false of 'light rays' – has an epistemological significance that philosophy must some day come to terms with, however long it continues to postpone the reckoning, so the fact that Boolean logic is false – false of *the logical relations between states of affairs* – has a significance that philosophy and physics and mathematics must come to terms with.

The fact is that, if quantum logic is right, then not only the propositional calculus used in physics is affected, but also set theory itself. Just what the effects are is just beginning to be investigated. But it may well be that the answer to fundamental questions about, say, the continuum will come in the future not from new 'intuitions' alone, but from physical/mathematical discovery.

† See my 'The logic of quantum mechanics', chapter 10 in this volume.

5
Philosophy of physics*

The philosophy of physics is continuous with physics itself. Just as certain issues in the Foundations of Mathematics have been discussed by both mathematicians and by philosophers of mathematics, so certain issues in the philosophy of physics have been discussed by both physicists and by philosophers of physics. And just as there are issues of a more epistemological kind that tend to concern philosophers of mathematics more than they do working mathematicians, so there are issues that concern philosophers of physics more than they do working physicists. In this brief report I shall try to give an account of the present state of the discussion in America of both kinds of issues, starting with the problems of quantum mechanics, which concern both physicists and philosophers, and ending with general questions about necessary truth and the analytic–synthetic distinction which concern only philosophers.

I. The problem of 'measurement' in quantum mechanics

Quantum mechanics asserts that if A and B are any two possible 'states' of a physical system, then there exists at least one state (and in fact a continuous infinity of states) which can be described as 'superpositions of A and B'. If A and B are the sorts of states talked about in classical physics – definite states of position, or momentum, or kinetic energy, etc. – then their superpositions may not correspond to classically thinkable states. For example, let A be the 'state' of 'going through slit 1', and let B be the 'state' of 'going through slit 2'. How are we to think of a state C described as $\frac{1}{2}A + \frac{1}{2}B$? We might say 'the particle is *either* going through slit 1 or through slit 2', but this interpretation is excluded by the phenomenon known as *interference*. A particle in $\frac{1}{2}A + \frac{1}{2}B$ cannot be said to be *either* in state A (going through slit 1) *or* in state B (going through slit 2) nor can it be going through both slits at once (although that is in some ways how it behaves). We can only say the *superposition of the states A and B* is a primitive notion of quantum mechanics, answering to nothing at all in pre-quantum mechanical

* First published in Franklin H. Donnell, Jr. (ed.), *Aspects of Contemporary American Philosophy* (Würzburg, Physica-Verlag, Rudolf Liebing K.G. 1965).

physics. We cannot hope to 'visualize' or in any way 'intuit' what it is for a particle (or a whole system) to be in a superposition of states, but we know that it happens, and the equations of quantum mechanics enable us to predict what we will *observe* if we perform any given measurement upon a particle (or a whole system) in a superposition of states, at least with probability. This fundamental feature of quantum mechanics – the existence of superpositions – enables the states of a physical system to be represented as forming a linear vector space. Each vector represents a maximally specified state – i.e. a state concerning which no more information could be specified without violating the Uncertainty Principle. From the standpoint of quantum mechanics, each 'state vector' represents in a certain sense a *complete description* of the state of a physical system – even if the description is incomplete from the standpoint of classical physics. Any possible knowledge concerning the condition of a physical system can then be represented as either a single state vector or a statistical mixture of state vectors, each associated with an appropriate probability.

Now, let X be a macro-object, say a cat in an isolated rocket ship in interstellar space. Let A be a state in which X is in one macro-condition (say, the cat is alive) and B be a state in which X is in a different macro-condition (say, the cat is dead). What would $\frac{1}{2}A + \frac{1}{2}B$ correspond to in this case? If X is alive and also in $\frac{1}{2}A + \frac{1}{2}B$, then X must be alive and also in a state which is incompatible with being alive (for, as already remarked, it is incompatible with quantum mechanics to maintain that a system in a superposition is 'really' in one of the components of the superposition only we don't know which). Thus X cannot be alive. Neither can X be dead, or in any other thinkable macro-condition. However, it is one of the assumptions of orthodox (Copenhagen) quantum mechanics that ordinary macroscopic realism is tenable. That is, although a micro-system (e.g. an individual particle) cannot in general be thought of as possessing any classically intelligible properties when we are not observing it, we may think of, say, the city of Paris as continuing to exist in some classical condition even if all its inhabitants leave it for a month. Similarly, we may think of the cat in the rocket ship as being either alive or dead, even though we cannot observe it as long as the rocket ship is in interstellar space. In short, if X is a *macro*-object, and A and B are different *macro*-conditions, then X *cannot* be in such a state as $\frac{1}{2}A + \frac{1}{2}B$. (The scalar '$\frac{1}{2}$' is only chosen for illustration; states in quantum mechanics form a vector space over the *complex* numbers, in point of fact.)

The writer views these two assumptions of conventional quantum mechanics as constituting a contradiction. On the one hand, we are told that states form a linear vector space, i.e. any two states can be superimposed. On the other hand, we are told that macro-conditions cannot be superimposed. Conclusion: something is wrong with the theory. Similar views, critical of the received interpretation of quantum mechanics, have been expressed by Paul Feyerabend (1957), Abner Shimony (1965) and (in a milder form) by Henry Mehlberg (1958).

The orthodox form of the theory does not lack distinguished defenders, however. In a recent book (Hanson, 1963) Hanson tries to meet the difficulty alluded to above by denying that macro-conditions and macro-objects fall within the scope of quantum mechanics at all. This defence would be regarded as unacceptable by most quantum physicists, who would regard macro-systems as falling in principle within the scope of their theory. Thus in a recent article (see bibliography) Margenau and Wigner defend the adequacy of the received theory along other lines.

Margenau and Wigner restate some of the leading ideas of the so-called Copenhagen Interpretation (which is the standard interpretation of quantum mechanics). According to them quantum mechanics presupposes *a cut between the observer and the object.* Any system whatsoever can be taken as the object; however the observer himself cannot be included. (In particular, the entire universe is *not* a system in the sense of quantum mechanics, i.e. a possible 'object'.) The observer always treats himself as possessing definite states which are known to him. Here Margenau and Wigner deviate slightly from the Copenhagen Interpretation. According to Bohr and Heisenberg, the observer must treat himself as a *classical* object, i.e. everything on the observer side of the 'cut' (including measuring apparatus) is treated as obeying the laws of *classical* physics. Margenau and Wigner do not mention this. What they rather say is that the observer must include a 'consciousness'. Thus they deviate from the Copenhagen Interpretation in a *subjectivistic* direction. Whereas the fact that we do not get superpositions on the observer side of the 'cut' is explained on the Bohr–Heisenberg story by the fact that we use classical physics on this side, it is explained on the Margenau–Wigner story by the fact that we have a faculty of 'introspection' (cf. London and Bauer (1939) for the source of this interpretation) which enables us to perform 'reductions of the wave packet' upon ourselves.

Another resolution of the difficulty we described has been proposed by Margenau in a series of publications (separate from and seemingly incompatible with his publication with Wigner). Margenau proposes, first, that we assign pure states only to systems upon which a precise measurement can be performed (this would exclude assigning a pure state to the whole universe; however, our knowledge of the state of the entire universe can be represented by a statistical mixture, on Margenau's account). He argues, secondly, that it can be derived at once that measurement, considered as an act which 'opens' a previously closed system, results in a state of affairs which can *only* be described by a statistical mixture. For after the measurement the object is in interaction with the rest of the universe, i.e. in interaction with an environment whose state cannot in principle be known exactly, and is thus itself in a state which cannot be known exactly.

To the writer it seems correct that we can show, in the manner sketched by Margenau, that after the measurement (say, opening the rocket ship after its return to earth, in the case of the cat example) that the state of the object X as known by us can only be represented by a mixture. But that mixture need not be a mixture of macro-conditions. In the cat example, for instance, it cannot be derived from the Schrödinger equation that after we look the cat is in fact dead or alive. If we treat ourselves as part of the 'rest of the universe', and represent the state of the rest of the universe *including ourselves* by a statistical mixture (Margenau allows this, although in his publication with Wigner he insists, contrarily, on the 'cut between the object and the observer'), then we can show that if the cat is in the pure state $\frac{1}{2}A + \frac{1}{2}B$ prior to the measurement, the whole universe, including the cat, will be in a certain mixture after the measurement. But it will be the *wrong* mixture: i.e. we get a mixture of superpositions of 'live cat' and 'dead cat' and not a mixture of states of the entire universe in each of which the cat is either alive or dead.

This point may be a little unclear, so permit me to elaborate it somewhat. Suppose that I open the rocket ship and look at the cat. There are (at least) three possibilities: (i) I see the cat alive (ii) I see the cat dead (iii) I am thrown into a new state by the interaction (looking at the cat), namely into the state $\frac{1}{2}$ (Hilary Putnam seeing live cat) $+\frac{1}{2}$ (Hilary Putnam seeing dead cat). The problem is to account for the exclusion of the third 'possibility'. Margenau does not account for it at all. He only accounts for the fact that after the measurement we have a mixture and not a pure case. But some mixtures can be represented as mixtures of possibilities (i) and (ii) and some include subcases like (iii). So Margenau's answer is insufficient.

Sometimes it is said that my *looking* 'throws the cat into a definite state' (this is von Neumann's 'Projection Postulate'). This is just to assert that (iii) is excluded without *explaining* that fact.

Another attempt to overcome the difficulties in present-day quantum mechanics involves the use of *nonstandard logics*. Some proposals again only provide ways of formalizing the fact that (iii) is excluded without explaining it. In Reichenbach's approach, for example, it is simply assumed that statements about macro-observables have the conventional two truth values while statements about micro-observables may have a third truth value; but this radical dichotomy between macro- and micro-observables is not derived from anything, but simply built into the theory *ad hoc*.†

II. The problem of geometry

There is no doubt that the examination into the structure of physical geometry and relativity theory initiated by Reichenbach and continued by Grünbaum has called our attention to matters of philosophical importance. Unfortunately, I find myself unable to agree with Grünbaum on a number of central points. I do agree that the ordinary standard of congruence in physical geometry is the solid rod. I do not agree that one can *define* 'congruent' in terms of solid rods in such a way as to make possible an empirical determination of the metric, especially if the need for correcting for perturbational or differential forces in accordance with some not exactly known system of physical laws is to be taken into account. What appears rather to be the case is that the metric is implicitly specified by the whole system of physical and geometrical laws. No very small subset by itself fully determines the metric. Even if we specify that the laws are to be formulated in such a way that it is said that a solid rod would not change its length at all in transport if it were not for the action of differential forces, it is still possible to reformulate the laws of physics in many equivalent ways which all conform to this condition, and which lead to different geometries.

Reichenbach calls forces which are permanently associated with a spatial region, and which produce the same deformations in all bodies independently of the chemical composition of those bodies, 'universal forces'. What I have just said is that the system of physical laws can be reformulated in many different ways, all consistent with the principle 'there are no universal forces', but leading to *different* geometries. In other words, the principle 'there are no universal forces' does not

---

† [Added 1974] The interpretation proposed by David Finkelstein and discussed in chapter 10 of this volume appears to overcome this difficulty. Assuming the universality of superposition is correct, we are driven to a non-classical logic.

even *implicitly* determine the metric. The contrary statement has often been made, for example, by Reichenbach and Grünbaum; but it rests on a fallacious argument.

Secondly, I believe it is wrong to say that physical geometry is about the coincidence behavior of solid rods, as many authors do. Space–time geometry is not about 'bodies and transported solid rods' except in a derivative sense, but rather about the *metrical field*: that is to say, the universal space–time field whose tensor is the g_{ik} tensor. This is a physical field in the sense of 'physical' relevant to scientific inquiry; we can detect its presence in a variety of ways (detecting the bending of light rays as they pass the sun, for example), and its presence enables us to explain and not just describe the behavior of solid bodies and of clocks. The major open question in the general theory of relativity today is whether the metrical field may not possess enough degrees of freedom to account for all the effects that have hitherto been explained by postulating the electromagnetic field. The character of such a discovery would be fundamentally distorted by regarding the metrical field itself as merely a descriptive convenience which enables us to systematize the relations holding between solid bodies and clocks.

Grünbaum's position is that any continuum, i.e. any dense non-denumerable space, is *intrinsically metrically amorphous*. Exactly what he means is far from clear. However, his discussion indicates the following. Grünbaum supposes that there is an objective topology of physical space. We assign a Riemannian metric, i.e. one of the form $ds^2 = \sum g_{ik}d^i d^k$. The choice of any *such* metric is *a matter of pure convention*, a continuum cannot intrinsically have one metric as opposed to another; but the choice of a non-Riemannian metric would represent a change in the meaning of 'congruent' and would not be the sort of thing Grünbaum has in mind.

It has to be emphasized that the choice of what Grünbaum regards as the intrinsic topology is justified by the fact that it leads to simpler laws, e.g. causal chains do not have to be described as discontinuous. It is also to be emphasized that Grünbaum is not saying that all the admissible metrics lead to equally simple laws. On the contrary, Grünbaum admits that if we choose the metric in the General Theory of Relativity given by the usual g_{ik} tensor, then we obtain much simpler laws than if we choose any other g_{ik} tensor. (The issue is complicated by the fact that Grünbaum does not think of what is chosen as a tensor, but rather holds that we choose a *definition of congruence*; but as already pointed out, 'congruent' is a term which does not admit of explicit definition at all, apart from purely verbal definitions, e.g. 'two intervals are congruent if they are the same length', which assume that a metric is already available.) But even the choice of a g_{ik} tensor on which the description of physical facts becomes incredibly complex (or of a 'congruence definition' which is based on such a tensor) represents a permissible convention for Grünbaum. In short, space has an intrinsic topology, but it does not have an intrinsic metric even though the consequences of changing the metric would be essentially the same as the consequences of changing the topology, namely that physical laws would become incredibly complex. This position seems to me to verge on contradiction.

III. The special theory of relativity

According to Grünbaum, what Einstein realized was that (a) in any universe in which there exists a limit to the velocity with which causal signals can travel (and this can happen in a quasi-Newtonian world),† we are free to define 'simultaneity' as we wish subject to one important restriction, (b) the only important restriction is that a cause may not be said to be simultaneous with or later than its effect. It is *logically* irrelevant that some 'definition' would lead to vastly more complicated physical laws, although it is required that the system of physical laws should not become more complicated in the *one* respect prohibited by (b), i.e. that causes should not be said to be simultaneous with or later than their effects.

This interpretation of the Special Theory of Relativity (STR) makes the relation between causal order and temporal order quite mysterious. In Reichenbach's and Grünbaum's writings it is asserted that in some

† By a 'quasi-Newtonian' world is meant one which obeys the laws of Newtonian physics except that gravitational attraction propagates with a *finite* velocity (say, the speed of light) relative to a privileged reference system (the 'ether frame'). If we suppose that the total energy of the world is finite, then there will also be an upper limit to the speed with which any particle can be accelerated (assuming all particles have at least a certain minimum mass). Thus there will be a finite limit to the velocity with which an arbitrary causal signal can be propagated, although there will *not* be time-dilation, Lorentz-contraction, etc. In fact, in a quasi-Newtonian world, one could determine one's absolute velocity (relative to the ether frame), by performing the Michaelson–Morley experiment.

Reichenbach and Grünbaum have claimed that the principle of relativity is a *consequence* of the existence of a finite limit to the velocity with which causal signals can be propagated. The example of the quasi-Newtonian world shows that this is not so. For in a quasi-Newtonian world one would insist on the requirement of invariance of the laws of nature under the Galilean group of transformations, not under the Lorentz group, and this would lead to an absolute simultaneity. In short, establishing the existence or non-existence of an absolute simultaneity requires considering the overall *simplicity of laws*, and not just the feasibility of some experimental or 'operational' procedure involving the transmission of causal signals.

sense spatio-temporal order *really is* causal order ('time is causality'), and this view has led to a series of attempts to *translate* all statements about spatio-temporal order into statements about causal order. In fact, the relation between the two is quite simple. One of the *laws of nature* in our customary reference system is that *a cause always precedes its effects*. Since this law holds in one admissible reference system, it must hold in all. (The admissible reference systems in the STR are singled out by the two conditions that (a) the laws of nature must assume a simplest form relative to them, and (b) they must assume the same form relative to all of them.) Hence, the admissible reference systems are restricted to be such that a cause always does, indeed, precede its effect.

If, as Grünbaum claims, Einstein's 'insight' was that we are 'free' to define the to and fro velocities of light to be equal, even relative to a system in accelerated motion relative to the fixed stars, then an important part of the General Theory of Relativity was already included in the STR. In fact, however, Einstein allowed the use of systems in accelerated motion relative to the fixed stars as 'fixed bodies of reference' only *after* he became convinced that *as a matter of physical fact* no complication in the form of the laws of nature results thereby.

Given this conceptual setting, it is completely clear that if the world had been discovered to be quasi-Newtonian, but with a limit to the velocity with which causal signals can propagate, Einstein would certainly have insisted on the requirement of invariance of the laws of nature under the Galilean group of transformations as the appropriate requirement, and would not have stressed this 'freedom' to adopt non-customary definitions of congruence which so impresses Grünbaum. We also see that the requirement imposed by Grünbaum, that all the admissible metrics should agree on the topology (the topology is 'intrinsic') makes sense only within the framework of a desire to single out a class of permissible reference systems which will lead to the same laws of nature. If we are willing to change our definitions in a way which involves a change of the laws of nature, then no justification whatsoever can be given for insisting that the principle must be retained that causal signals should not be discontinuous.

I conclude that the alleged insight that we are 'free' to adopt a non-customary 'definition' of simultaneity whenever there is a limit to the speed with which causal signals propagate is correct only in the trivial sense that one may press a noncustomary sense upon *any* word by arbitrary semantic convention, and that Grünbaum's interpretation of Einstein therefore *trivializes* Einstein's profound logical–physical insight.

IV. The direction of time

An important event in the philosophy of physics was the posthumous publication of Reichenbach's book *The Direction of Time*. The problem with which it deals is the problem of the physical basis for the existence of irreversible processes. The orthodox answer to the question is that a process is irreversible only when its being so follows from the Second Law of Thermodynamics.

Reichenbach's book has two notable achievements. First, it is the most careful consideration of all the difficulties presently available. This is a book from which a philosopher can 'educate himself' in a good deal of physical theory, while at the same time exploring the ins and outs of a fascinating conceptual question. Secondly, it elaborates and reformulates the orthodox view so as to meet the difficulties, as well as they can be met on the basis of present knowledge.

The first difficulty that Reichenbach discusses is the problem of stating the Second Law without assuming that a 'positive direction for the time-axis' has already been singled out by some 'non-entropic clock'. (The Law says that the entropy of a closed system is always *greater* at a *later* time.)

The Reichenbach idea is as follows. Consider the universe as a great collection of 'branch systems' – that is, systems that interact with their environment for a time, are then isolated for a time, and then interact with their environment again. The Second Law may then be stated in two parts. (1) There is a direction of entropy increase in each branch system. (2) The direction is the *same* in *all* branch systems.

Part 1 is made precise using the concept 'between with respect to time' which is available from particle mechanics, and does not presuppose a direction of time has already been defined. The problem of making part 2 precise involves the notion of simultaneity at a distance, and thus involves relativity theory. Suffice it to say that, in my opinion, Reichenbach shows that this program can be carried out.

Reichenbach then turns to a more fundamental problem: the physical and/or mathematical basis for the truth of the Second Law. The usual answer is that Boltzmann's H-theorem provides a satisfactory 'micro-reduction' of the Second Law. Reichenbach establishes that this is not so. The H-theorem does indeed imply that after a 'low' entropy will increase -- but it also implies that *before* a 'low', entropy will *decrease*. In fact, for a very large class of universes, the assumptions underlying the classical proof of the H-theorem are *incompatible* with the always-truth of the Second Law. We are thus left with 'unfinished business'

87

on our hands: the one great law of irreversibility (the Second Law) cannot be explained from the reversible laws of elementary particle mechanics – how then is it to be explained?

V. 'Necessary truths' in physics

It has often been urged that there are *no* 'necessary truths' in physical theory. If this means that no statement – or virtually no statement – is immune from revision, well and good. But today we have to worry about the distortions of contemporary empiricism much more than the distortions of Kantianism. And if it was a distortion on the part of Kantianism to hold that such statements as 'every event has a cause', 'space has three dimensions', the principles of geometry, etc., are immune from revision, it is equally a mistake to assimilate them, as some empiricist philosophers of science have, to ordinary empirical generalizations. One might indeed say, if it were not for the appearance of straining after smartness, that there *are* necessary truths in physics, but they can be revised if necessary! The important point, however one expresses it, is that there may not be 'necessary truths' in the sense of unrevisable 'apodictic' judgments, but there are such things as *framework principles* in science, and the revision of framework principles is (a) possible (this is often overlooked by philosophers who mistakenly classify them as 'analytic', just as much as by philosophers of a more old fashioned kind who called them 'synthetic a priori'), but (b) quite a different matter from the revision of an ordinary empirical generalization.

The principles of geometry are a case in point. Before the development of non-Euclidean geometry by Riemann and Lobachevski, the best philosophic minds regarded them as virtually analytic. The human mind could not conceive their falsity. Hume would certainly not have been impressed by the claim that 'straight line' means 'path of a light ray', and that the meeting of two light rays mutually perpendicular to a third light ray could show, if it ever occurred, that the Parallels Postulate of Euclidean geometry is false. He would have contended that it rather showed that light does not travel in straight lines. Thus he would not have admitted that, since we can visualize light rays behaving in the manner just described, it follows that we can 'visualize non-Euclidean space'.

Ever since, philosophers have divided into (broadly speaking) two camps on this question. Those in the one camp – the 'analyticity theorists', as we might call them – have contended that, at least given the pre-relativity meaning of the crucial geometrical terms ('straight line', 'point', etc.), the laws of Euclidean geometry are analytic. On this view, if we today say that space is Riemannian in the large (or may be), then we are simply *changing the meaning* of these crucial geometrical terms. Those in the other camp – the self-styled 'empiricists' – have contended that the laws of geometry are and always were empirical, and that the question which geometry holds in the physical world is and always was a purely empirical question, to be decided by experimental tests. Both positions run into grave difficulties, however. Consider the statement 'there are only finitely many distinct disjoint places (of over a certain minimum size) to get to in the entire physical universe', where this is interpreted as meaning *not* that the 'entire physical universe' occupies only a finite portion of space, but that space itself is finite (although absolutely unbounded – i.e. there are no 'walls'). This state-ment is true if the world is Riemannian and false if it is Euclidean or Lobachevskian. And the only geometrical notion it contains is 'place'. However, this is *not* a notion that appears to have changed its meaning in post-Relativity physics (we are not counting 'places' of microscopic dimensions as places). Anything the layman would call a 'place' (of, say, over a thousand cubic feet size) the physicist would accept as a 'place', and vice versa.

Now, let us suppose that the world is Riemannian, and of very high curvature (so that the number of distinct disjoint 'places' of over a thousand cubic feet volume, in any way of counting, does not exceed a certain not too large number N), and that we have succeeded in visiting each and every one of these 'places', and also in verifying that travel as we may, we never succeed in visiting any *other* places. If someone maintains that all we have done is 'change the meaning of words', then he must maintain that in the old sense of 'place' these are *not* all the places there are, and that there are in addition to these 'physically accessible' places that we can visit in our rocket ships, still other, un-visited, physically inaccessible places and paths, which, together with the 'places' counted by the Riemannian physicist, fill out a Euclidean world. It is clear that this is no longer a mere insistence on retaining an old way of speaking, or an old 'meaning of words', but a metaphysical hypothesis which commits the holder to countenancing a host of 'ghost entities' – the physically inaccessible places and paths.

On the other hand, if 'line' and 'point' and 'congruent' have a fixed meaning which is independent of the geometry chosen (so that the question which geometry is 'right' can be an empirical question in the ordinary sense), what is that meaning? If we say that these are *theoretical* terms which can only be defined in terms of each other (and related notions, e.g. 'distance'), then this seems to be correct, but then we have to recognize that the use of these notions rested for a very long time on a particular framework of assumptions (Euclidean geometry) and that the upheaval of the very framework upon which the concepts depended should not be assimilated to a *mere* empirical discovery. Unfortunately, it is just *this* assimilation that the 'empiricists' have wanted to make.

In order to make the assimilation, they have had recourse to one of two approaches. The Operationists, like Bridgeman, have tended to say that 'straight line' simply *means* 'path of a light ray' (or something of that kind). Unfortunately this is not true, as we can see by observing that there are many quite foreseeable situations in which we would cheerfully give up the principle that 'light travels in straight lines' (*in vacuo*). The more sophisticated approach is the one we mentioned before due to Reichenbach and Grünbaum: it is to contend that *the following principle is an implicit definition of 'congruent'*: 'a transported solid rod remains congruent with itself after correction for the effect upon it of disturbing forces'. This is quite a sophisticated view, because at first blush the 'escape clause' 'after correction for the effect upon it of disturbing forces' appears to make the entire 'definition' operationally useless. However, Reichenbach and Grünbaum contend, if we restrict the *allowable* kinds of 'disturbing forces' in a certain way (ruling out so-called 'universal forces') then we do get an *implicit* unique specification of a metric.

Sophisticated or not, the view fails, however. It fails first because the assertion upon which it rests – that there is only *one* metric (apart from the choice of unit, which is not at issue) which leads to a world system (i.e. a geometry-plus-physics) which is compatible with 'universal forces do not exist' is false.† Secondly, and more simply, it assumes (1) that a body cannot change its shape without accelerating, which is not true in quantum physics (a body might simply 'jump' into a different shape without being acted upon by *any* forces whatsoever); and (2) that all accelerations must be caused by forces – i.e. that Newton's Second Law must take precedence over Euclidean geometry in the event of any conflict. But in the actual practice of Newtonian science there is no evidence that any such priority ordering of fundamental principles actually existed, and the decision to give up one rather than another was dictated by considerations of overall simplicity, and not by a semantical fact (the meaning of 'congruent').

In short, even if Reichenbach and Grünbaum had been right in their mathematics,‡ all they would have shown is that the decision to go over

† This is shown in the Appendix to chapter 6 in this volume, 'An Examination of Grünbaum's Philosophy of Geometry'.

‡ The mathematical mistakes in Reichenbach are at bottom two. Firstly, Reichenbach's 'Theoren θ' (Reichenbach, p. 33) is incorrect. Reichenbach asserts that a fixed

to non-Euclidean geometry in the General Theory of Relativity was prompted by a desire to *retain* certain other laws. This indeed is right – even if more laws are involved than just the one law that 'universal forces' (in Reichenbach's sense) do not exist. These other laws are not, however, merely analytic statements reflecting some arbitrarily chosen definition of 'congruent', and so this is miles apart from the claim that it was the desire to retain *the definition of congruence* that forced the abandonment of Euclidean geometry (given the new experimental evidence).

Do not try, then, to classify the laws of geometry as analytic or 'empirical'. There is an important difference between statements (in a given theoretical context, characterized by a given space of alternatives) which can be confirmed or disconfirmed by specifiable experimental results, and statements which enjoy the kind of conceptual 'self evidence', 'necessity', etc., that the laws of Euclidean geometry once enjoyed. But this does not mean that the traditional empiricist dichotomy between statements necessary in an eternal sense ('necessary' = '*a priori* = 'analytic') and statements contingent in an eternal sense ('empirical' statements) can be excepted. As I argue in chapter 15, the various rescue moves that have been proposed to save the necessary/contingent distinction do not work. It does not work to say that 'necessary' truths only *appear* to be given up, because lords have changed their meanings. This philosopher's ploy cannot explain away the conceptual revolutions of Relativity and indeterministic physics. And to say that the principles of Euclidean geometry or the principle that every event has a cause (and

force, independent of the composition of the body it acts upon, can produce the *same* deformations in all measuring rods. This ignores the fact that the *resistance of the body to deformation* depends upon its chemical composition. Moreover, Reichenbach's 'proof' of Theorem θ is mathematical nonsense (Reichenbach simply asserts that the difference of two g_{ik} tensors is a tensor F_{ik} which gives 'the potentials of the force' acting upon the measuring rod — i.e. of a force which will produce the appropriate deformations in all measuring rods). Secondly, Reichenbach reasons somewhat as follows: suppose that I go over to a different metric from the standard one (call the standard metric g_{ik}, and the new metric g'_{ik}). Then, if the usual metric is compatible with the assumption that a measuring rod does not change its length when transported, except as accounted for by the internal and external forces acting upon it (this assumes a system of physics, as well as a given metric g_{ik}), when I go over to the new description g'_{ik}, I will be forced to say that the measuring rod changes its length when transported (here Reichenbach fails to add: *relative to the description of its length afforded by g_{ik} plus the original physics*). Reichenbach assumes that this universal change of length (which exists only *relative to the original description*) must be accounted for by keeping the original physics and adding appropriate 'universal forces'; if so every other 'geometry plus physics' than the one we started with will involve 'universal forces'. However, the assumption that we must (or even can) proceed in this way is quite gratuitous. We can perfectly well revise the laws obeyed by the differential forces, instead of introducing 'universal forces', and we then arrive at a situation which is quite symmetrical as between g_{ik} and g'_{ik}.

'same cause, same effect') were only mistaken for necessary truths does not work either. For scientists were perfectly correct to assign a special status to these statements. Holding them immune from revision (until conceptual alternatives were elaborated) was good methodology – in fact, it could not lead to any errors until one reached the stage of cosmic physics and micro physics. It is the task of the methodologist to explain this special status, not to explain it away.

6

An examination of Grünbaum's philosophy
of geometry*

There is no doubt that the examination into the structure of physical geometry and relativity theory initiated by Reichenbach and continued by Grünbaum has extended our understanding and called our attention to matters of philosophical importance.† Unfortunately, I find myself unable to agree with Grünbaum on a number of central points. I do agree that the ordinary standard of congruence in physical geometry is the solid rod. I do not agree that one can *define* 'congruent' in terms of solid rods in such a way as to make possible an empirical determination of the metric, especially if the need for correcting for perturbational or differential forces in accordance with some not exactly known system of physical laws is to be taken into account. The prevalent assumption to the contrary is due to an error committed by Reichenbach and frequently found in the literature of the past thirty years (the error of supposing that there must be 'universal forces' in any non-standard metric).‡ What appears rather to be the case is that the metric is implicitly specified by the whole system of physical and geometrical laws and 'correspondence rules'.§ No very small subset by itself fully

* First published in Baumrin (ed.) in *Philosophy of Science, The Delaware Seminar*, 2 (New York, Interscience Publishers 1963). Reprinted by permission of John Wiley & Sons, Inc.

† See bibliography. Quotations from Grünbaum in this paper are from Grünbaum (1962). I wish to acknowledge my indebtedness to Grünbaum for many stimulating conversations, and to our fellow participants in discussion sponsored by the Minnesota Center for the Philosophy of Science.

The reader who is not acquainted with the notion 'g_{ik} tensor' will be able to follow this paper without difficulty by replacing 'g_{ik} tensor' throughout by 'd-function', where by the d-function associated with a metric he may understand the function $d(P_1, P_2)$ which associates with any two 'labeled' points P_1, P_2 (labeled by means of a suitable coordinate system) a total space–time distance $d(P_1, P_2)$. Given a coordinate system, the g_{ik} tensor and the d-function are completely interdefinable. For a popular exposition of the notion of a 'space–time distance' see Gamow's delightful *One, Two, Three, Infinity* (Mentor, 1953).

‡ See the Appendix to this chapter for a detailed discussion.

§ By a 'correspondence rule' I shall mean such a statement as 'temperature as measured by a thermometer is usually approximately correct', 'congruence as established by transporting a solid rod is usually correct to within such-and-such accuracy', etc. As will be clear in the sequel, I do not regard such assertions as *analytic* nor as *stipulations*. Thus the term 'correspondence rule' in this sense should be distinguished from Grünbaum's 'coordinative definitions', e.g. 'the length of a solid rod corrected

93

determines the metric; and certainly nothing that one could call a 'definition' does this.

Secondly, I believe that it is misleading to say, as Grünbaum does, that physical geometry provides 'the articulation of the system of relations obtaining between bodies and transported solid rods quite apart from their substance–specific distortions' (p. 510). What Grünbaum means is not really 'quite *apart* from their substance–specific distortions', but *after proper correction has been made for* their substance–specific distortions. What it is to 'articulate' relationships which exist only after *appropriate* corrections have been made, is far from clear. In my opinion, this error is analogous to the error we would be making if we described electromagnetic theory as 'the articulation of the relations holding among voltmeters quite apart from perturbing influences'.†

Space–time geometry is not about 'bodies and transported solid rods' except in a derivative sense, but rather about the *metrical field*: that is

---

for differential forces does not change'. This last statement is an intratheoretic statement, and does not, Grünbaum to the contrary, restrict the choice of a metric at all unless considerations of *descriptive* as well as *inductive* simplicity are allowed to operate. However, to say 'considerations of descriptive simplicity partially determine the metric' is to say in another way that the whole system of laws is involved.

† It might be objected that the above is unfair on the grounds that *solid body* is a concept of far greater generality and physical importance than voltmeter. Against this it should be noted that even solid bodies do not provide a universal standard of congruence; to define congruence in dimensions which get close to 10^{-13} centimeters, it is necessary to introduce radiative processes, as Reichenbach does. Even this, however, is not the fundamental issue. The fundamental issue is whether the laws of modern *dynamical geometry* (i.e. laws describing a geometry of variable curvature changing with time) explain or only describe the behavior of solid bodies, clocks, etc. I maintain that independently of which of the currently fashionable models of scientific explanation one accepts, one must say that geometro-dynamics explains the behavior of clocks, light rays, etc. For instance, one can explain why it is that a clock which is accelerated in an inertial system appears to slow down in that inertial system by pointing out that, according to the laws governing the elementary physical processes, the rate at which the processes in the clock go on depends upon the proper time elapsed along the clock's world line and not upon the projection of the proper time on the t axis of the reference system. In other words, the clock *measures proper time along its own world line* – and it continues to do this whether it is at rest in the reference frame, moving with a constant velocity in the reference frame, or being accelerated in the reference frame. This explanation of the behaviour of the clock is quite different from saying that the time dilation phenomenon is necessary since otherwise we would not have Lorentz invariance. Note that in this case 'proper time' has direct physical significance in much the way that 'voltage' does. We can construct a meter to measure (voltage)² instead of voltage; but this is most naturally accomplished by first measuring voltage and then including a computing device which squares the result. Similarly, we can construct a clock which measures (proper time)², or even time in a particular reference system rather than time along its own world line; but such a device is most naturally constructed by using an ordinary clock, i.e. one which measures proper time along its own world line, and then including a computing device to perform the appropriate transformations. Proper time is thus something 'physical' in the sense that physical processes depend upon it in a simple way.

to say, the universal space–time field whose tensor is the g_{ik} tensor. This is a physical field in the sense of 'physical' relevant to scientific inquiry; we can detect its presence in a variety of ways (detecting the bending of light rays as they pass the sun, for example), and its presence enables us to explain, and not just describe, the behavior of solid bodies and of clocks. The major open question in the general theory of relativity today is whether the metrical field may not possess enough degrees of freedom to account for all the effects which have hitherto been explained by postulating the electromagnetic field. The character of such a discovery would be fundamentally distorted by regarding the metrical field itself as merely a descriptive convenience which enables us to systematize the relations holding between solid bodies and clocks.

In order to see this point more clearly, let us consider for a moment the difference between articulating a set of relationships and explaining them. We may articulate the relationships which obtain among a set of charged bodies in a given electromagnetic field by specifying the totality of possible space–time trajectories of the charged bodies in question. Such a specification need not be simple. (However, the 'descriptive simplicity' of an articulation does not affect its *correctness*.) On the other hand, we may *explain* the behavior of these charged bodies by specifying the electromagnetic field tensor and giving Maxwell's laws. The explanation differs from the mere *systematization of the relations explained* partly in that the relations explained are derived from higher level laws and partly in that these higher level laws have 'surplus meaning' – i.e. they may be used in conjunction with suitable auxiliary hypotheses to explain yet other phenomena. (Of course, this is not meant to be an exhaustive characterization of the differences between explanation and mere 'articulation'.) In the case of geometry the features of explanation just alluded to are present. Given the metrical properties of space–time as summarized in a g_{ik} tensor, and given the laws – e.g. *light-ray paths are extremal paths, world lines of freely falling bodies are geodesics*, etc – we can explain a wide range of phenomena, and this explanation will have both of the features we just mentioned: derivation of the phenomena from laws, and surplus meaning. The coincidence behavior of solid rods is only a small part of what can be explained.

It might be objected that, in the case of geometry, surplus meaning is absent at least from the component represented by the g_{ik} tensor, and that the specification of the g_{ik} tensor can be *translated* into a statement of the coincidence properties of solid rods and the properties of clocks. However, this would be a mistake. The g_{ik} tensor is only equivalent to a set of statements about the behavior of rigid rods and clocks which have been *properly corrected for perturbational forces*; and the correction requires laws which presuppose the g_{ik} tensor in question.†

Thirdly, I am disturbed by the attempt in the last few pages of Grünbaum's paper (pp. 520–1) to describe a *discovery procedure* for the correct g_{ik} tensor, one whose dubious workability Grünbaum himself admits in the case of a geometry of variable curvature. The point needing emphasis here is that the empirical character of geometry only requires that given two inductively inequivalent systems of geometry plus physics, we should be able to justify choosing one over the other. A remark by Ernest Nagel on this question seems to me to be to the point:

It is not a simple matter to separate out those components of these numerical values representing features to be counted as 'truly' geometrical properties from components representing the effects of some deforming physical influence. On the other hand, this difficulty is in principle no different than the problem of deciding on the basis of experimental evidence whether light is a vibratory or a corpuscular process. (*The Structure of Science*, pp. 258–9.)

Here Nagel is not talking about the existence or non-existence of 'universal forces' but about the problem of eliminating differential forces. And the comparison he draws is to the choice between vibratory and corpuscular conceptions of light – that is to say, to the choice between two high-level theories both of which have been fully developed as alternative explanations of the phenomena and which, in conjunction with the accepted auxiliary hypotheses, lead to different predictions. I agree with Nagel that the determination of the g_{ik} tensor requires theory construction in this 'high-level' sense, and no cookbook full of recipes for the discovery of these high-level theories can be or ought to be given.‡

† In order to meet this difficulty, Grünbaum (1962) tries to show that it is possible to determine both the g_{ik} tensor and the coincidence behavior of rods which have been compensated for perturbational forces simultaneously by a quasi-operational procedure of successive approximations. This procedure rests, however, on the error of Reichenbach already commented upon; the claim on pp. 520–1 that there is only one geometry of constant curvature with certain properties is incorrect. (See Appendix.)

‡ Even the statement that geometry is 'empirical' can be accepted only if it is understood that an empirical theory in the sense just specified can sometimes enjoy privileged status relative to a body of knowledge. I have argued elsewhere that *prior to the development of non-Euclidean geometry* the laws of Euclidean geometry were in a sense 'necessary'. They were indeed synthetic (and in fact, unknown to those who maintained them, *false*); but their abandonment was nevertheless not a real conceptual possibility for the scientists of that time. They could not have been overthrown by a finite set of experiments *plus* inductions of a simple kind (say, using Reichenbach's 'straight rule'); nor even by a 'crucial' experiment to decide between Euclidean geometry and any alternative theory of a *kind* known to the science of the time.

Fourthly, I think that the extent to which and the way in which the choice of a metric is a matter of 'convention' is vastly exaggerated by Grünbaum.

No one of these disagreements is of great importance by itself, perhaps; but it will be seen that the four of them together constitute a contrast between fundamentally different views of geometry. By failing to see the extent to which anything that could be properly called a *definition of congruence* underdetermines the extension of the relation; by suggesting that geometry just 'articulates' the idealized coincidence behavior of solid bodies; and by exaggerating the extent to which geometrical relations can be ascertained by relatively low-level empirical procedures, Grünbaum has in my opinion failed to give a true picture of one of the greatest scientific advances of all time. The eventual outcome of that advance is anticipated by Wheeler in the following words: 'The vision of Clifford and Einstein can be summarized in a single phrase, "a geometro-dynamical universe": a world whose properties are described by geometry, and a geometry whose curvature changes with time – a *dynamical* geometry' (Wheeler, p. 361).

I have already indicated in broad outline what I take the correct interpretation of dynamical geometry to be: Dynamical geometry or General Relativity Theory is to be understood as a scientific theory introducing a theoretical concept – the metrical field (which is identical with the gravitational field, and which may turn out to be identical with the electromagnetic field as well). However, this 'construct', or 'postulated entity' (depending upon one's favorite philosophical attitude towards scientific theories) is not wholly unprecedented in the history of scientific thought. In a Newtonian world there is no physically significant *space–time* metric; that is to say, the notion of total space-time distance between two point events has no physical importance. What is of more use is to describe the relationship between two events by giving two numbers rather than one: the temporal separation of the two events and their spatial separation. Both numbers are invariant under the transformations which obtain between 'privileged' coordinate systems in a Newtonian world. If we consider space at any one time, however, then in a Newtonian world there is also an objective metrical field, albeit one which is considerably less interesting than the metrical field of dynamical geometry. It is less interesting precisely because it is not dynamical. It is constant (flat Euclidian space), and it does not change with time. Thus it cannot enter either as an effect or as a cause in any physical process. It constitutes rather, in Wheeler's words, 'only an arena within which fields and particles move about as "physical" and "foreign" entities' (Wheeler, p. 361). Because spatial points and relations do not appear as causes or as effect in Newtonian physics, geometry had a rather special status as a scientific theory during the Newtonian epoch. In a Newtonian world, space and time do nonetheless have physical significance, however, inasmuch as physical processes depend upon true temporal and spatial separation in a simple way. Moreover, this is not a matter of mere descriptive simplicity. If we transform to a different metric, or to a different chronometry, so that time, for example, becomes what we would ordinarily call (time)^3, then it will still be true that the behavior of clocks in a Newtonian world depends in a simple way upon the time elapsed, only this invariant fact will now be explained by saying that clocks measure (time)^3. Whether we express this fact by the words 'clocks measure time' or by the words 'clocks measure (time)^3', the fact is the same. And it is the same magnitude that enjoys the same significance.

The argument I shall use to defend the objectivity and non-conventionality of spatio-temporal separation may thus be easily carried over to defend the objectivity and non-conventionality of spatial separation and temporal separation in a *Newtonian* world. Thus this paper is in part a defense of Newton against those who have sought to criticize him, as Riemann and Grünbaum have done, on philosophical grounds. I am in wholehearted agreement with Newton's statement that 'those defile the purity of mathematical and philosophical truth, who confound real qualities with their relations and sensible measures' (Newton, 1947, pp. 7–8). Newton's three-dimensional 'ether' has been replaced by a four-dimensional 'field', but Newton has not thereby lost standing as a philosopher but has only been advanced upon as a physicist.

Having now surveyed the totality of my disagreements with Grünbaum, I shall devote the rest of this paper to exploring one of these disagreements in detail, the disagreement which concerns the allegedly conventional character of the choice of a g_{ik} tensor from the class of Riemannian metrics.

Conventionalism

1.

Grünbaum explains his own position (which he calls *geochronometric conventionalism* or GC, for short) by contrasting it with two theses which he regards as contrary to each other and to GC. Unfortunately, these two other theses are not, in fact, incompatible at all. The first of these alternative theses in the choice of a metric is a matter of convention *only in the trivial sense* that an 'uncommitted noise' may be assigned any meaning we wish. Thus, if we do not require that the customary meaning of the term 'congruent' should be preserved, then we are free to press upon the *noise* 'congruent' any new meaning we wish. If we take 'X and Y are congruent' to mean that X and Y have the same length according to M, where M is an unconventional metric that we have invented, then we will simply have changed the meaning of the English word 'congruent'. Only in the trivial sense that speakers of English have explicitly or implicitly, to decide what meaning to give to the noise 'congruent' if they are to employ it as a meaningful word is the choice of a metric a matter of convention at all.

The other alternative to GC that Grünbaum considers is the thesis that space (or space–time in the relativistic case) possesses an intrinsic metric.

One thing is clear at the outset. No one who wished to maintain that space possessed an intrinsic metric would wish to *deny* the thesis of 'trivial semantic conventionalism' (TSC). An example should make this point clear. Grünbaum regards phenomenal qualities as intrinsic. Suppose that I ascribe a headache to you. Then I may say that I regard it as an objective fact, and in no sense a matter of convention, that you have a headache, and at the same time recognize that that objective fact can be expressed in the way it is expressed, namely by using the English word 'headache', only because certain semantic conventions have been laid down.

This rather obvious remark permits us to dispose of one argument used by Grünbaum. If, Grünbaum argues, the thesis that the choice of a metric is a matter of convention is nothing but a tautology, then Newton was denying a tautology; but Newton clearly did not mean to deny a tautology! It is easy to see what has gone wrong. Newton meant to affirm, not deny, TSC, and to affirm further that, given the received meaning of 'congruent', questions of congruence are matters of objective physical fact. GC similarly affirms that, given the received *definition* of congruence, questions of congruence are matters of objective physical fact; but that there are many alternative admissible *definitions*, leading to different metrics, and that to adopt a noncustomary one is *not* to change the *meaning* of 'congruent' (thus denying *both* TSC *and* Newton). In particular, Grünbaum and Newton differ over the question whether metrical notions acquire their meaning via a 'coordinating definition' that mentions solid rods.

It is necessary to be cautious here. If we take Newton to have been denying that correspondence rules (cf. fourth footnote, p. 93) are necessary in geometry, or even in a physical theory as a whole, then Newton was making an epistemological mistake and not a mistake which is particularly about geometry. He would have been maintaining that theoretical terms do not require to be linked to observables at all. Such an interpretation of Newton would be as unwarranted† and as ridiculous as the interpretation that Newton was denying even semantic conventionality (which is the interpretation that Grünbaum ascribes to me). When Newton wrote that in philosophical disquisitions 'we ought to abstract from our senses, and consider things themselves, distinct from what are only sensible measures of them' (Newton, pp. 7–8), he neither meant to deny that there are sensible measures for theoretical quantities nor that we obtain our notions of theoretical quantities by *abstracting from* those sensible measures.

It is no accident that Newton used the language of the then contemporary empiricism nor that he repeatedly acknowledged his intellectual debt to empiricism. Whatever Newton's error may have been, if it was an error, it was neither denying tautologies nor denying that our notions of theoretical quantities and relations arise by abstraction from sensible measures.

What Newton did maintain is that it is an objective property of a body to be at the same (absolute) place at two different times. The reasons he gave for this were, as is well known, physical and not metaphysical ones: within the framework of Newtonian physics we obtain different predictions in two universes all of whose *relative* positions and velocities are the same at one instant if one of the two universes consists of bodies at rest relative to the ether frame and the other is in motion relative to the ether frame. In short, the ether frame, or absolute space, represents a theoretical notion which Newton introduces in order to explain observable phenomena. Grünbaum's position is that Newton was *a priori* mistaken. No continuous space‡ can have an intrinsic metric. Of course, Grünbaum is not saying that the predictions of Newtonian physics are *a priori* wrong: the world might have been Newtonian. But in that case Grünbaum would still not agree with Newton that bodies possess a nonrelational property which is their sameness of (absolute) location at different times. Possibly what he would say is that this property ought to be analyzed as the property of being at rest relative to a solid body which is such that if that solid body is taken as the reference system, then the laws of nature assume a simplest form. If no such solid

† Newton emphasizes that we *can* determine theoretical ('true') motions from the 'sensible measures'. How? We deduce them 'by means of the astronomical equation' (Newton, 1947, pp. 7–8).
‡ Here Grünbaum means a dense *nondenumerable* space.

body in fact exists (because no solid body is actually at rest in absolute space), then counterfactuals would have to be used.

Note that this way of 'analyzing away' the notion of absolute space has been attempted in the case of every theoretical notion. The statement that macroscopic objects consist of atoms, for example, has been 'analyzed' as 'meaning' that macroscopic objects behave in such a way with respect to all their macroscopically observable properties that we obtain the simplest laws of nature by introducing atoms. One can replace all references to objective magnitudes by references to objectively simplest laws. (Here *both* descriptive *and* inductive simplicity are at issue. No one denies that one could obtain equivalent laws even in a Newtonian world by choosing a system other than the ether frame as the rest system.) It is not clear to me that Newton would have troubled to either affirm or deny this particular 'analysis' of absolute space. That there does exist a unique possible reference frame relative to which the laws of nature assume the Newtonian form would have been enough for him, I think. In the second place, it seems to me that the proposal to analyze *being at absolute rest* as *being at rest relative to a possible system relative to which the laws of nature assume a simplest form* is part of a general proposal to translate all statements about objective magnitudes into statements about the form of natural laws. This is a special form of reductionism. I myself do not believe that such a reduction program can be carried out in detail, i.e. that the notions of *simplest laws, possible objects*, and so forth, which are essential for this program, can ever be made precise in a satisfactory way. But even if they can be made precise in a satisfactory way, the important thing is that the possibility, if it is a possibility, of translating statements about absolute rest into statements about simplicity of laws completely fails to show that the property of being at absolute rest (in a Newtonian world) is not objective in the sense in which other physical properties and relations are objective. What has been said about sameness of absolute position applies equally well to the relation of congruence. Newton would have maintained that congruence is an objective relation among quadruples† of spatial points. Of course, he would not thereby be committed to denying that correspondence rules are necessary in physical geometry, or at any rate in connection with physical theory as a whole. And it would be a gross error to conclude from the fact that these correspondence rules mention, say, solid bodies, that physical geometry is *about* solid bodies. It is misleading to say that physical theories are about the objects mentioned in the correspondence postulates rather than about the theoretical

---

† For the statement that the interval $P_1 P_2$ is congruent to $P_3 P_4$ involves *four* points.

entities and magnitudes mentioned in the theoretical postulates. As we explained above, in connection with the discussion of explanation *versus* articulation, physical theories are said to explain the behavior of the ordinary middle-sized objects mentioned in the correspondence rules by reference to the theoretical entities and magnitudes that they postulate.

We may now explain Grünbaum's position a little more exactly. Grünbaum's position is that any continuum, i.e. any dense nondenumerable space, is *intrinsically metrically amorphous*. Exactly what he means by this is far from clear. However, his discussion indicates the following. Grünbaum supposes that there is an objective topology of physical space determined by the consideration that causal chains shall not be described as discontinuous. Grünbaum moreover seems to assume that we are not free to assume such differential forms as

$$ds^4 = \sum_{u,o,v,d} g_{uovd} \, du \, do \, dv \, dd$$

We assign a Riemannian metric, i.e. one of the form

$$ds^2 = \sum_{ik} g_{ik} \, di \, dk$$

The choice of any such metric is a matter of pure convention; a continuum cannot intrinsically have one metric as opposed to another; but the choice of a non-Riemannian metric would represent a change in the meaning of 'congruent', and would not be the sort of thing that Grünbaum has in mind. It is to be emphasized that the choice of what Grünbaum regards as the true or intrinsic topology of space is justified by the fact that it leads to simpler laws, e.g. causal chains are never said to be discontinuous. It is also to be emphasized that Grünbaum is not claiming that all of the admissible metrics lead to equally simple physical laws. On the contrary, Grünbaum admits that if we choose the customary metric in the general theory of relativity given by the general–relativistic g_{ik} tensor, then we obtain much simpler laws than if we choose any other g_{ik} tensor. But even the choice of a g_{ik} tensor in which the description of physical facts becomes incredibly complex represents a permissible convention for Grünbaum. In short, space has an intrinsic topology; but it does not have an intrinsic metric *even though the consequences of changing the metric would be essentially the same as the consequences of changing the topology*, namely, that physical laws would become incredibly complex. Moreover, although space is said to be intrinsically metrically amorphous, it is admissible only to choose a Riemannian metric. This position seems to me to verge on downright contradiction.

One mistake that Grünbaum frequently makes† must be guarded against from the beginning. This is the mistake of supposing that because it is a tautology that we are free to use an uncommitted word to mean whatever we like, all of the *consequences* or *entailments* of any particular decision must likewise be tautological. That such and such will result as a matter of physical or mathematical fact if we make one choice rather than another may be far from tautological. In particular, then, given the topology of space–time it is a fact that certain definitions of 'congruent' will lead to a Riemannian metric.

Grünbaum argues that this fact, and indeed the fact that any continuous space can be metricized in a great number of different ways, would all be tautologies if TSC were true. This is just a mistake. TSC does not assert, and it is not a tautology, that there exists even one way of using the word 'congruent' that leads to a Riemannian metric in the technical sense; that is a matter of mathematical fact. That any space which possess a Riemannian metric possesses a great many of these is also a matter of mathematical fact. That of these infinitely numerous possible Riemannian metrics for space–time there appears to be exactly one (apart from the choice of a unit, which is not at issue) which leads to simple and manageable physical laws is a matter of objective physical fact. Grünbaum is tremendously impressed by the fact that there is nothing in the nature of space–time that requires us to use the term 'congruent' to mean congruent according to one g_{ik} tensor rather than another. This is true. Similarly, it is true that there is nothing in the nature of space–time or in the nature of the bodies occupying space–time that requires us to use the word 'mass' to mean mass rather than charge. But there is an objective property of bodies which is their mass whether *we call it mass* or call it $(mass)^3$, and there is an objective relation between two spatial locations which we ordinarily call the distance between those spatial locations and which enters into many physical laws (e.g. consider the dependence of *force* on *length* in Hooke's law), whether we refer to that magnitude as *distance* or as *length* or as

† Grünbaum *explicitly* commits the mistake of supposing that according to TSC our *entailed* semantical decisions must be an instance of trivial semantic conventionality on p. 491 of his article, when he writes: 'Sophus Lie then showed that, in the context of this group–theoretical characterization of metric geometry, the conventionality of congruence issues in the following results: (i) the set of all the continuous groups in space having the property of displacements in a bounded region fall into three types which respectively characterize the geometrics of Euclid, Lobachevski-Bolyai, and Riemann, and (ii) for *each* of these metrical geometries, there is *not* one but an *infinitude* of difference congruence classes ... On the Eddington–Putnam thesis, Lie's profound and justly celebrated results no less than the relativity of simultaneity and the conventionality of temporal congruence must be consigned absurdly to the limbo of trivial semantical conventionality ... '

(*distance*)³ or as (*length*)³ or as any other function of distance. If we decide by 'distance' to mean distance according to some other metric, then in stating Hooke's law we shall have to say that force depends not on length but on some quite complicated function of length; but that quite complicated function of length would be just what we ordinarily mean by 'length'.

What Grünbaum means when he says that the metric is not intrinsic is that whereas there *really is* such a thing as color, whether we call it 'color' or not, there is not really any such thing as distance except a complicated relational property among solid bodies. To put it positively, he believes that distance really is a relation to a solid body which has been properly corrected for differential forces. Unfortunately, this has no clear sense apart from the contention, which Grünbaum rejects, that statements about distance are translatable into statements about the behavior of solid bodies. Grünbaum rejects this for the quite correct reason that statements about distance are translatable only into statements about the behavior of properly corrected solid bodies, and since the laws used in correcting for the action of differential forces must themselves employ the notion of 'distance', to offer such a 'translation' as an *analysis* of the notion of 'distance' would be circular.

Grünbaum attempts to break the circularity by suggesting, in effect, that we choose a metric arbitrarily, then discover the set of true physical laws (assuming we can do this), expressing them in terms of that metric, and then finally determine that remetricization which agrees with what he takes to be the congruence definition we actually employ, namely, that the length of a solid body should remain constant when transported after we have properly corrected for the action of differential forces. As I have already indicated, this is not successful. Even if we were *given* the 'true' laws of nature stated in terms of some arbitrary metric, there would still be no mechanical procedure for remetricizing in such a way as to obtain the g_{ik} tensor customarily employed in the general theory of relativity. To obtain the g_{ik} tensor that is customarily considered to be correct, it is not enough to remetricize in such a way as to satisfy any finite set of theoretical conditions, e.g. the condition that according to the combined physics plus geometry it should be said that a solid body retains a constant length when not acted upon by differential forces, and it is not enough to satisfy any number of operational conditions – e.g. that it should be said that the measurements we get from our rigid bodies are ordinarily at least approximately correct; but over and above all this it is necessary to seek the descriptively simplest possible total system of laws. But even if we could do this, even if we could discover the metric, as Grünbaum says, 'empirically', i.e. by following a prescribed set of *rules of discovery*, what would follow? Certainly not that physical geometry is about the behavior of solid bodies 'apart from substance-specific distortions'; for even on Grünbaum's account we arrive at the physical geometry of our world not *merely* by considering the behavior of solid bodies but by formulating and reformulating hypotheses and by testing and retesting hypotheses until we have succeeded in formulating a set of 'true' physical laws, and this means a set of laws formulated, not necessarily in terms of the customary notion of 'distance' but in terms of *some notion which is interdefinable with it.* In short, Grünbaum is telling us to *take the notions of space–time point and space–time distance as primitive*, to formulate laws in terms of them, and then to transform those laws in an equivalent way, if necessary, until we eventually obtain an equivalent system of laws satisfying certain conditions, e.g. that causal chains should be continuous, that solid bodies should not be said to change their lengths when not acted upon by differential forces, etc. Even if we could discover the geometry of our world by faithfully following some such prescription, this would not at all tend to show that distance is *really* a relation to a solid standard 'apart from substance-specific distortions' or even to bestow a sense upon that proposition.

2.

Discussing an example which is due to Eddington (the possibility of changing the meaning of the word 'pressure'), Grünbaum writes,

Of what *structural* features in the domain of pressure phenomena does the possibility of Eddington's above linguistic transcription render testimony? The answer clearly is *of none.* Unlike GC, the thesis of the 'conventionality of pressure,' if put forward on the basis of Eddington's example...is thus merely a special case of TSC. (Grünbaum, 1962, pp. 490–1.)

Here the notion of structural features of the domain in question gives the entire show away. What Grünbaum means by saying that the conventionality of geometry reveals important structural features of space–time is that that conventionality, *when restricted in the way that Grünbaum arbitrarily restricts it*, reveals the fact that space–time is a nondenumerable, dense continuum. It reveals that fact because in the case of a denumerable continuum Grünbaum does not allow us to choose any metric except the one based on a measure which assigns zero to point sets.† According to Grünbaum a denumerable dense space

† More precisely he insists on a *countably additive measure.* I wish to call attention here to a major error in Grünbaum's writings. Grünbaum has asserted that Zeno's paradox *depends for its solution* on countable additivity, and hence on the nondenumerability of space–time. I submit that this is something that the mathematical

has an intrinsic metric, namely, the zero metric, whereas a non-denumerable dense space has no intrinsic metric. To say of a space that it has no intrinsic metric is in fact exactly to say that it is neither finite nor denumerable but nondenumerable and dense. *On Grünbaum's usage of the terms*, the statement 'space S is intrinsically metrically amorphous' is not an instance of TSC in that he has *forced* upon these words the meaning 'space S is nondenumerable and dense'. To say that GC then reveals important structural properties of space is just to say that Grünbaum finds it an *interesting* fact that a nondenumerable and dense continuum possesses an infinitude of Riemannian metrics.

Suppose now that I stipulate that in remetricizing pressure (P) one may only mean by the word 'pressure' $f(P)$, where f is an arbitrarily chosen bicontinuous mapping. There will then be infinitely many permissible 'pressure metrics'. That there are infinitely many is a mathematical theorem, not an instance of trivial semantical conventionality. The theorem indeed is a rather trivial one, namely, the theorem that there exist infinitely many one-one bicontinuous functions. But this is nonetheless a theorem. Deliberately mocking Grünbaum, I may now go on to say that the important thesis of the conventionality of pressure is not an instance of trivial semantic conventionality but reveals an important structural property of the pressure domain, namely, its intrinsic metrical amorphousness.

3. A-conventionality and B-conventionality

Grünbaum's most important attempt to distinguish GC from the trivial semantical conventionality allowed by TSC occurs on p. 489 (Grünbaum).

community simply knows to be false. Even if we assume there are only space–time points with rational coordinates 'in reality', world lines, and hence *motion*, are perfectly possible. Moreover, Achilles can catch the tortoise in the conventional way: by being at $x = \frac{1}{2}$ at $t = \frac{1}{2}$, at $x = \frac{3}{4}$ at $t = \frac{3}{4}$, at $x = \frac{7}{8}$ at $t = \frac{7}{8}$, $\cdots$ (note that only rational space–time points are involved). Grünbaum's mistake appears to be a simple quantifier confusion: from the fact that the countable sum $\frac{1}{2} + \frac{1}{4} + \frac{1}{8} + \cdots = 1$ it follows that the measure is *sometimes* countably additive, not that it must *always* be countably additive.

Let us assume that only points with rational coordinates exist (using some specified coordinate system), and that the physically significant notion of distance can be taken to be given by the square root of the sum of the squares of the coordinate differences (Pythagorean theorem). Then this measure is only *finitely* additive in this denumerable space; but no serious reason exists for supposing that we would not use this measure for that reason. The 'intrinsic measure' of every distance, according to Grünbaum, is *zero* in any denumerable space, but to accept this would be to abandon the notion of distance altogether.

Grünbaum also suggests that to use finitely additive measures would require going over to mathematical intuitionism. This is an incorrect remark.

To state my objections to the Eddington–Putnam thesis, I call attention to the following two sentences:

(A) Person X does not have a gall bladder.

(B) The platinum-iridium bar in the custody of the Bureau of Weights and Measures in Paris (Sèvres) is 1 meter long everywhere rather than some other number of meters (after allowing for 'differential forces.')

I maintain that there is a *fundamental difference* between the senses in which each of these statements can possibly be held to be conventional, and I shall refer to these respective senses as 'A-conventional' and 'B-conventional.'

He goes on to point out that (A) is a factual statement and conventional only in the trivial sense that the truth-value of every sentence depends on the meanings of the words it contains, whereas (B) is conventional in (roughly) the sense of being an analytic statement. I don't happen to believe that (B) is true by convention; but that is not relevant here. What is relevant here is that the distinction between A-conventionality and B-conventionality cannot possibly help Grünbaum.

Grünbaum does maintain that (B) is conventional, and this would be interesting if true. But this is not the thesis of GC, which is what is at issue. Indeed, asserting without qualification that the 'customary congruence definition' is *analytic* runs directly counter to GC. For if the 'customary congruence definition' is analytic, and if in addition, as Grünbaum mistakenly believes, it determines the g_{ik} tensor, then there is no freedom to choose an alternative metric and there is no such thing as the 'choice' of a g_{ik} tensor! The thesis of GC is not that the 'customary congruence definition' is *uniquely* analytic in the way in which the usual definition of 'bachelor' is analytic; but the thesis of GC is rather that the use of geochronometrical language pre-supposes that we render either the sentence (B) *or some other sentence tying geometrical statements to statements about the behaviour of rods and clocks* true by stipulation. It further asserts that any convention consistent with the alleged intrinsic topology of space–time is admissible if it leads to a Riemannian metric. And it finally asserts that the choice of an admissible convention according to which the rod is said to change its length when transported is *not* a change of *meaning*. On this one page the reader is surprised to find that Grünbaum suddenly restricts his claim to the claim that the customary congruence definition is a stipulation. But this claim, although it happens to be false, is not at all incompatible, as we have already seen, with the Newtonian view that the metric of space is something perfectly objective or with the analogous view of Wheeler in connection with the metric of space–time. If (B) were an analytic statement, it would be an intra-theoretical analytic statement and would not imply any trans-latability of statements about distance into statements about the actual behavior of solid rods. In fact, as we already noted, it is compatible with the adoption of any metric whatsoever, although one has to change the physical laws appropriately (in a manner explained in the Appendix) if one changes the metric.

If GC only asserted the analyticity, or conventionality in some sense, of the *customary* congruence definition, then it would not be an interesting conventionality thesis. What makes GC an interesting conventionality thesis is the fact that it *emphasizes the admissibility* of other congruence definitions and asserts that this existence shows the 'intrinsic metrical amorphousness' of space notwithstanding the admitted fact that the *adoption* of one of these other definitions would infinitely complicate the statement of physical laws.

At this point, I should like to digress and say why I do not regard the definition (B) as analytic, or true by convention (or stipulation) in *any* sense. Suppose that the standard meter stick in Paris doubled in length although not subject to any unusual differential forces at all. It would then be twice its previous length relative to everything else in the universe. I am not supposing that this (a stick doubling its length in the absence of forces) could in fact happen (I wish to ignore quantum mechanics, in which it in fact could happen); our question is whether I have described a *self-contradictory state of affairs*. If now (B) were true by convention, we would be required to do one of two things:

(1) *Invent* a differential force to explain the expansion of the meter stick in Paris notwithstanding the non-existence of any physical source for that force, or any law obeyed by that force, or any reason to suppose that it would not have affected any other body that might have been there in place of the meter stick in exactly the same way. (In this case (B) becomes completely empty, of course.)

(2) *Say* that the rod did *not* change length, and that everything else in the universe underwent a shrinking. It seems to me that neither of these descriptions of the hypothetical case is required by our language, or our stipulations, or by anything else. And this shows that (B) is a *framework principle of physics*, and not a stipulation.

On Grünbaum's view, refusing to give either of the two descriptions mentioned would be *changing the definition* of 'congruent'. But it seems to me that this is a *reductio ad absurdum* of the view in question.

Suppose that a solid object does change its shape or size. Let us ask why we are required to say that the body was acted upon by any forces at all. The answer is that such changes presuppose accelerations and that, according to Newton's second law, accelerations require forces to produce them. Reichenbach and Grünbaum are tacitly assuming that

Newton's second law is analytic, or at least that one of its consequences the Law of Inertia, is built into the customary definition of 'congruent'. On this view, we are changing the definition of 'congruent' (or possibly of 'force' and hence *indirectly* of 'congruent') if we ever say for any reason whatsoever that an acceleration occurred although no appropriate force was present. Thus, the Newtonian physicist would have been held to be revising the customary 'congruence definition' if he gave up either the second law or the principle that causal chains are continuous (which is held to define the topology of physical space); but the physicist is said not to have changed the 'definitions' of geometrical terms when he abandons the principles of Euclidean geometry. In Newtonian physics the second law was 'conventional' whereas Euclidean geometry was synthetic, on this account.

I have criticized this view in detail in 'The Analytic and the Synthetic', chapter 2, volume 2, of these papers and I will not go into the details again here. But I think it is clear to common sense that this account is a distortion. Newton's second law and the principles of Euclidean geometry functioned on a par in Newtonian physics, and no physicist in his right mind ever considered giving up either one. Indeed, a Newtonian physicist would not have been able to make anything of the question: 'Which would you rather give up, the second law or geometry?' (If he had answered at all, he would probably have answered, 'the second law'.) Grünbaum severely berates Nagel for suggesting that the postulating of universal forces (or of deformations not accounted for by differential forces) would be anything other than a change of definition.†
It is only because he holds the erroneous view that the principles of Euclidean geometry enjoyed a different status from the second law, that the second law was 'conventional' and that the second law together with the decision to exclude universal forces uniquely determines the metric, that Grünbaum is able to maintain that the shattering discovery that Euclidean geometry was false was merely an 'empirical' discovery. In fact, as I have emphasized elsewhere (chapter 15 of this volume), it was *not* 'empirical' in the narrow sense in which Grünbaum uses the term (although it was empirical in the sense of being about the world), but represented rather an unprecedented revision of *framework principles* in science. This point is not affected by the point that we might

---

† Nagel suggested that this would be an *ad hoc* assumption, which is in agreement with the standpoint taken here. Even if Grünbaum were right in supposing that what is *really* happening, in some sense, is that the proponent of universal forces is tacitly changing the meaning of distance, I do not see why he so vigorously disagrees with Nagel here. For surely *ad hoc* hypotheses, in the strong sense in which Grünbaum himself has employed the term (implying *logical* immunity to revision) can always be *regarded* as tacit meaning-changes.

now decide *ex post facto* to stipulate that Newton's second law should be regarded as analytic, and that the principles of Euclidean geometry should not be. The important point is that there was no rule in effect in the Newtonian time which would determine that giving up one would be to change the 'definition' of 'congruent' whereas giving up the other would be to construct an admissible empirical hypothesis.†

Let us return for a moment to our hypothetical case of the doubling stick. The reader may feel inclined to say that this much of GC is correct: that there are *two* perfectly equivalent admissible descriptions, one according to which the stick doubled its length, and one according to which everything else in the universe underwent a shrinking. Grünbaum may be wrong in maintaining that adopting the former description is *changing* the *definition* of 'congruent'; but it would be equally wrong (one might suppose) to say that adopting the latter description is 'changing the meaning'.

It is necessary to be cautious here, or one can easily fall into downright factual errors. The two descriptions are not nearly as symmetrical as one at first supposes. If one adopts the description of the hypothetical case according to which the stick doubled in length, then it is necessary to change the laws of physics only to the extent of admitting this one 'fluke'; and if the laws were statistical to begin with, it may not be necessary to make any change at all. If one adopts the description according to which the stick retained its old length while everything else in the universe (including the 'star shell') underwent a shrinking, then the laws of physics must also be radically revised to obtain an 'equivalent description'. For, if the laws of physics did not change except for the admission of one 'fluke' that *everything* except the stick underwent a shrinking, then very strange gravitational and electromagnetic fields should have been set up by the shrinking process. But these did not occur. Also, physical constants – g, Planck's constant h, etc. – must have changed their values during the shrinking, otherwise false predictions will result. In short, this description is equivalent to the intuitive one only if the laws of physics are vastly complicated.

We can now see the status of (B) more clearly: within the context of the rest of physics, (B) is falsifiable! If we accept the view of Professor Feigl, that the meaning of theoretical terms in science is partly determined by the whole 'network' of laws, then we can say: the present

† I wish to concede here Grover Maxwell's point (cf. Maxwell, 1962), that there are *some* rational reconstructions in which a sharp analytic–synthetic distinction can be drawn in this *ex post facto* way. I also wish to point out that we would have to make a *lot* analytic (not just 'there are no universal forces') to even *begin* to restrict the 'choice' of a metric (cf. Appendix).

meaning (not 'definition') of 'congruent' is such that under certain circumstances (B) might turn out to be false. It is adopting the 'crazy' description (*everything* shrinking) and infinitely complicating the laws to do it, that *is* 'changing the meaning'.

4. English, spenglish, and color words

Grünbaum cites an earlier argument of mine,

> ...H. Putnam maintains that instead of using phenomenalist (naive realist) color words as we do customarily in English, we could adopt a new usage for such words – to be called the 'Spenglish' usage – as follows: we take a white piece of chalk, for example, which is moved about in a room, and we lay down the rule that depending (in some specified way) upon the part of the visual field which its appearance occupies, its color will be called 'green', 'blue', 'yellow', etc., rather than 'white' under constant conditions of illumination. (Grünbaum, 1962, p. 487.)

The purpose of the cited example was to show that one could argue (paralleling Grünbaum) that the color properties of things are not 'intrinsic' either, and hence that *no* property is, or can be, 'intrinsic' in Grünbaum's sense. Grünbaum goes on to reply,

> But it is *not* conventional whether the various chalk appearances do have the same phenomenal color property...and thus are *'color congruent'* to one another or not! Only the *color words* are conventional, *not* the *obtaining* of specified color properties and of color congruence. And the *obtaining* of color congruence is *non*conventional quite independently of whether the various occurrences of a particular shade of color are denoted by the same color word or not. (Grünbaum, 1962, pp. 489–90.)

And Grünbaum goes on to remark,

> And the alternative color descriptions *do not render any structural facts of the color domain* and are therefore purely trivial. Though failing in this decisive way, Putnam's chalk color case is falsely given the *semblance* of being a bona fide analogue to the spatial congruence case by the device of laying down a rule which makes the use of color names *space dependent:* the rule is that *different noises* (color names) will be used to refer to the same *de facto* color property occurring in different portions of visual space. But this stratagem cannot overcome the fact that while the assertion of the possibility of assigning a space-dependent length to a transported rod reflects linguistically the objective non-existence of an intrinsic metric [*sic*!], the space-dependent use of color names does *not* reflect a corresponding property of the domain of phenomenal colors in visual space. (Grünbaum, 1962, p. 490.)

This is an interesting example of the way in which Grünbaum

deploys his concepts. The reference to 'structural properties of the domain' has already been dealt with in the case of *pressure*. But note the retreat to the phenomenal level here! The general point at issue is that the criteria for the application of an 'uncommitted' word at one place are *logically* independent of the criteria for the application of that word at any other place (or time). Grünbaum points to the case of phenomenal color properties in phenomenal space (which I wasn't discussing). Let us examine this. I shall now introduce the word 'tred', which is an 'uncommitted noise', and which I can, therefore, feel free to use as I like. 'Tred' will be used in such a way that an object is called 'tred' if it is at one place in my visual field and red, or at a certain other place in my visual field and green, or if it is at a certain third place in my visual field and blue, etc. In addition to introducing the term 'tred', I shall introduce quite a number of similar terms, e.g. 'grue' (with apologies to Goodman), 'breen', etc. Since I am not free to call tred, grue, breen, etc. *colors*, I will call them *grullers*. In addition, two objects in my visual field which possess the same *quale* will be said to possess different '*shmalia*'. The definition of a *shmale* will be left to my readers' imagination; but, roughly, two objects irrespective of their location in my visual field which are both *tred*, or both *grue*, or both *breen*, will be said to possess the same *gruller shmale*, they don't, of course, possess the same *color quale*. Conversely, two objects in my visual field which are both red, or both blue, or both green, will be said to possess the same *color quale*, although they don't possess the same *gruller shmale*.

Now, let us for the moment treat 'red' as an 'uncommitted noise'. Suppose we decide on the criteria to be employed for assigning a truth-value to the assertion that an object at a certain particular place in the visual field is *red*. Then regardless of its color and gruller, it is still open whether any object at any other place in the visual field will or will not be called 'red'. Note that whether by the words 'same color' I decide to mean what we ordinarily mean by 'same color' or rather to mean 'same gruller', the formal properties of an identity relation – transitivity, symmetry, and congruence – will all be preserved. Moreover, the gruller of a color patch is just as much an objective property of that color patch as its color is. Is it an intrinsic property? Anyone familiar with psychology would certainly venture the conjecture that if our experience were such that patches *moving across our visual field* normally *retained the same gruller* rather than the same color, then the *gruller* of a patch would *quickly seem real and intrinsic* to us. We would begin to *see* grullers. Thus the notion of an intrinsic property seems to be relative to a particular conceptual frame. If we normally describe a patch as e.g. 'that red patch moving across my visual field from left to right', then in that conceptual frame *red* is an 'intrinsic' property of the patch in question. If, on the other hand, we *normally* employ such descriptions as 'that tred patch moving across my visual field from left to right', then *tred* is an 'intrinsic' property of the patch in question.

We have seen that insofar as the existence of infinitely many Riemannian metrics is used by Grünbaum as an argument for the 'intrinsic metrical amorphousness' of space–time (and insofar as that argument bestows a sense on the assertion in question) we could just as well argue that 'phenomenalist colors' are *intrinsically color amorphous*, for there are a large number (though not an infinite number) of possible 'color metrics' of the kind discussed. Moreover, this fact *does* reflect 'structural properties of the color domain': both psychophysical properties (the existence and number of locations in visual space) and mathematical theorems (the *number* of color metrics is a certain combinatorial function of the number of discriminable locations in the visual field).

One difference between the 'phenomenalist color' case and the space–time case must be admitted, however: there are 'correspondence rules' in connection with physical geometry, and none in connection with colors. But this difference separates *all* theoretical entities and magnitudes from all observables. One might as well conclude that electrical charge is 'impressed from without' from the fact that there are correspondence rules in connection with *charge*, and hence that electrons are 'intrinsically electrically amorphous' as that space–time is 'intrinsically metrically amorphous'. Would not a better conclusion be that this distinction between attributes which are 'intrinsic' and attributes which are 'impressed from without' is untenable?

5. *Special relativity*

I believe that the disagreement between Grünbaum and myself can be revealingly brought out by contrasting our interpretations of the Special Theory of Relativity. Permit me to repeat to you a rather lengthy quotation from Grünbaum:

In discussing the definition of simultaneity, Einstein italicized the words '*by definition*' in saying that the *equality* of the to and fro velocities of light between two points A and B is a matter of definition. Thus, he is asserting that metrical simultaneity is a matter of definition or convention. Do the detractors really expect anyone to believe that Einstein put these words in italics to convey to the public that the *noise* 'simultaneous' can be used as we please? Presumably they would recoil from this conclusion. But how else could they solve the problem of making Einstein's avowedly conventionalist

conception of metrical simultaneity compatible with their semantical triviali-
zation of GC? H. Putnam, one of the advocates of the view that the conventionality of congruence is a subthesis of TSC, has sought to meet this
difficulty along the following lines: in the case of the congruence of intervals,
one would never run into trouble using the *customary* definition; but in the
case of simultaneity, actual contradictions would be encountered upon using
the customary classical definition of metrical simultaneity, which is based on
the transport of clocks and is vitiated by the dependence of the clock rates
(readings) on the transport velocity. But Putnam's retort will not do. For the
appeal to Einstein's recognition of the inconsistency of the classical definition
of metrical simultaneity accounts only for his abandonment of the latter but
does *not* illuminate – as does the thesis of the conventionality of simultaneity –
the logical status of the *particular set* of definitions which Einstein put in its
place. Thus, the Putnamian retort does *not* recognize that the logical status of
Einstein's synchronization rules is not at all adequately rendered by saying
that whereas the classical definition of metrical simultaneity was inconsistent,
Einstein's rules have the virtue of consistency. For what needs to be elucidated
is the *nature of the logical step* leading to Einstein's particular synchronization
scheme within the wider framework of the *alternative consistent sets of rules for*
metrical simultaneity, any one of which is allowed by the nonuniqueness of
topological simultaneity. Precisely this elucidation is given, as we have seen,
by the thesis of the conventionality of metrical simultaneity. (Grünbaum,
1962, pp. 423–4.)

Let us consider precisely what was 'the nature of the logical step' (a) on
Grünbaum's view, and (b) on my view.

Grünbaum and I agree that the customary definition of simultaneity
led to actual contradictions in view of the dependence of the clock rates
on the transport velocity – only I would say 'customary rule of correspondence' instead of 'customary *definition*'. And we agree *up to a point*
on the nature of the ensuing logical step. Namely, Einstein realized
that simultaneity of distant events has to be *defined* and not *discovered*.
We agree, moreover, that certain qualifications have to be made. Einstein
did not mean that we are free to define simultaneity in such a way that
Caesar's crossing the Rubicon is simultaneous with my lecturing to you
right now. Grünbaum meets this difficulty by reading into Einstein the
implicit stipulation that a particular law of nature shall be preserved,
namely, that causes shall not be said to be simultaneous with their effects.
As we shall see, however, this is only a small part of the restriction that
Einstein actually imposed upon our freedom to define simultaneity.
Also, it goes without saying that Einstein never considered definitions of
simultaneity which would disturb the customary usage when only small
('non-relativistic') relative velocities and distances are involved. In short,
when Einstein asserted that we have to define simultaneity and not

discover it, what he meant was that we have to repair an inconsistent notion, not throw it away altogether. But there can be no talk of empirically ascertaining the simultaneity of distant events until the notion of simultaneity has *first* been repaired.

What Grünbaum and I differ on, however, is the amount of latitude that we interpret Einstein as giving us when we redefine simultaneity. According to Grünbaum, we are free to redefine simultaneity in any way whatsoever as long as we do not conflict with the principle that a cause is never said to be simultaneous with or later than its effect. In particular, then, Einstein is held to have been asserting even in the special theory of relativity that we are free to define the to and fro velocities of light to be equal even when the reference system is not an inertial system. This is contrary to the way in which *everyone*, including Einstein himself, has always interpreted the special theory of relativity. As is well known, the demand of the special theory of relativity is that our reference system *must* be an inertial system. We are free to *define* the to and fro velocities of light to be equal relative to any admissible reference system. We are also free to define the to and fro velocities to be unequal but in a constant proportion to one another: this is equivalent to changing the reference system. We are not free, and Einstein has never suggested that we are free, in the context of the *special* theory of relativity, to stipulate that the to and fro velocities of light are in a variable proportion to one another, or that the to and fro velocities of light are equal relative to a reference system which is not inertial. Only in 1915 did Einstein finally arrive at the Theory of General Relativity according to which the principle of equivalence holds and the distinction between inertial systems and physical systems in accelerated motion loses all meaning. I shall argue that the nature of the logical step from the special theory of relativity to the general theory of relativity is precisely what is obscured and not illuminated if one accepts Grünbaum's account.

First, however, I wish to clarify Grünbaum's account to some extent. In order to do this I shall introduce a special kind of fictitious universe, to be called a *quasi-Newtonian* universe. The quasi-Newtonian world obeys all the laws of Newtonian physics except the principle of universal gravitation. In place of that principle one substitutes the principle that gravity is a central force obeying an inverse square law but propagating with the speed of light *relative to absolute space*. In this world Maxwell's equations also hold relative to absolute space. In particular, the velocity of light is constant only relative to absolute space. Thus the Michelson–Morley experiment would have had its expected outcome, and not the famous outcome it actually did have, if performed in such a world.

In his book *The Philosophy of Space and Time*, Reichenbach correctly points out that, as we have already discovered, absolute space is a legitimate theoretical construction, not an illegitimate one, and that the question whether it exists or not is an empirical question (pp. 214–16). He mistakenly asserts, however, that although empirical findings might compel us to accept absolute space, we would not then be able to determine its state of motion relative to the bodies occupying it. His argument is that the centrifugal forces appearing in a rotating body could be accounted for either by asserting that the body is rotating relative to absolute space, or by saying that the body is at rest in absolute space and that the forces appear because the star shell is rotating in absolute space and producing the effects somehow gravitationally. This interpretation could be falsified, however, by verifying the counter-factual that the centrifugal forces would continue even if the star shell were removed. This counterfactual could, of course, only be confirmed indirectly – either by verifying a theory, Newton's theory, from which the counterfactual follows as a consequence, or by performing certain kinds of experiments, e.g. removing the star shell to great distances or speeding it up so that it rotates along with the rotating earth. If we find that changing the state of motion of the star shell relative to the rotating earth, and, presumably, relative to absolute space as well, has no effect on the centrifugal forces, and if in addition it follows from a well-confirmed theory that these centrifugal forces are causally independent of the relative state of the earth and the star shell, then there is no reason not to accept the Newtonian explanation of these centrifugal forces, namely, that they appear because the earth is rotating relative to absolute space.

The dynamical phenomena emphasized by Newton enable us, however, only to determine absolute *accelerations*, and not to distinguish a system at absolute rest from a system with a *constant* velocity through absolute space. This can be done, however, by performing the Michelson–Morley experiment once on a body at rest relative to absolute space and once on a body moving with a constant velocity through absolute space. The to and fro velocities will be found to be equal only on the first body.

I will now take the further step of supposing that the quasi-Newtonian world contains only a finite amount of matter. The total energy of this universe is then well-defined, and we shall suppose that this total energy is too small to accelerate even the smallest particle to the speed of light. The quasi-Newtonian world would thus share with the actual world the feature that no causal signal can be transmitted with a velocity greater than that of light.

Now then, what is the situation with respect to simultaneity in the quasi-Newtonian world? In the quasi-Newtonian world, the customary correspondence rule for simultaneity involving the transported clocks does not lead to inconsistencies that cannot be explained as due to the actions of differential forces upon the clocks. There is then, in my view, no reason to regard simultaneity as a notion needing a definition except in the trivial sense in which every notion requires a definition (TSC). It is, in my view, an empirical fact in such a world that the to and fro velocities of light are equal relative to absolute space; and they are in fact unequal relative to anything else. We *could*, of course, define the to and fro velocities of light to be equal relative to some system in motion and put up with the incredible complication that would result in the statement of all physical laws. This is, however, just an instance of the possibility of altering the nomenclature in connection with any physical magnitude and as such is an instance of TSC. This point is explicitly granted by Reichenbach (Reichenbach, 1958, p. 37), who uses the analogy of arbitrarily altering the temperature scale – possibly even in a noncontinuous but one–one fashion. And this case is completely analogous with Eddington's pressure case.

Grünbaum's interpretation of the 'logical step' can now be succinctly stated. According to Grünbaum, what Einstein realized under the circumstances described above was that (a) in any universe in which there exists a limit to the velocity with which causal signals can travel (in particular, even in a quasi-Newtonian universe) we are free to define simultaneity as we wish subject to one important restriction, (b) the only important restriction is that a cause may not be said to be simultaneous with or later than its effect, and (c) it is *logically* irrelevant that some 'definitions' would lead to vastly more complicated physical laws, although it is required that the system of physical laws should not become more complicated in the *one* respect prohibited by (b), i.e. that causes should not be said to be simultaneous with or later than their effects.

I shall now state my interpretation of the special theory of relativity (or perhaps I should say 'rational reconstruction', since the question is not a question as to what went on in Einstein's mind, but a logical question as to how the special theory of relativity may reasonably be understood) – and leave it to my readers to decide which of the two interpretations is correct.

I take it that Einstein was not pointing out the possibility of using metrics and coordinate systems which lead to unnecessarily complicated physical laws. He realized not only that simultaneity has to be defined before we can talk of empirically ascertaining simultaneity at a distance; but he realized furthermore that *as a matter of empirical fact* there exists a class of privileged space–time coordinate systems, i.e. coordinate systems relative to which the laws of nature assume not just a simplest form but the same form for all; and further that the class of transformations from one of these coordinate systems to another is not the Galilean group of transformations but the group of Lorentz transformations. *As a matter of empirical fact* again, these 'privileged' coordinate systems are exactly the ones we get if we choose any inertial system as our rest body and *define* the to and fro velocities of light to be equal, or in any constant proportion to one another.

In this interpretation, the realization that we are free to define the to and fro velocities of light to be equal (assuming that we are in an inertial system) was simultaneously a logical and a physical insight – not *merely* an insight into the logical possibility of using space–time metrics which complicate the form of physical laws. In other words, we are free to define the to and fro velocities of light to be equal relative to *any* inertial system in the very strong sense that we will not change the form of physical laws by doing this.

If we interpret Einstein as emphasizing merely the logical possibility of using nonstandard definitions of simultaneity, then we miss the whole significance of the relation between temporal order and causal order. In Reichenbach's and Grünbaum's writings this relation is made quite mysterious. It is asserted that in some sense spatio-temporal order (topological properties) *really is* causal order, and this view has led to a series of attempts to *translate* all statements about spatio-temporal order into statements about causal order. In fact, the relation between the two is quite simple. According to the special theory of relativity, we are free to use any *admissible* reference system we like; however, the admissible reference systems are singled out by the two conditions that (*a*) the laws of nature must assume a simplest form relative to them, and (*b*) they must assume in fact the same form relative to all of them. It is, of course, assumed that our customary coordinate system is to be one of the admissible ones. Now, one of the *laws of nature* in our customary reference system is that *a cause precedes its effects*. (Einstein was assuming that gravitation would turn out to have a finite velocity of propagation.) Since this law holds in one admissible reference system, it must hold in all. Hence the admissible reference systems are restricted to be such that a cause always does, indeed, precede its effect (according to the time order defined by that reference system).

The above interpretation is amply supported by Einstein's own writings. Thus Einstein writes,

One had to understand clearly what the spatial coordinates and the temporal duration of events meant in physics. The physical interpretation of the spatial coordinates presupposed a fixed body of reference, which, moreover, *had to be in a more or less definite state of motion (inertial system)*.

And again,

...the insight which is fundamental for the Special Theory of Relativity is this: the assumptions (1) and (2) [of (1) the constancy of the light velocity, and (2) *the independence of the laws of physics of the choice of the inertial system*] are compatible if relations of a new type (Lorentz transformations) are postulated for the conversion of coordinates and times of events.

The step from special relativity to general relativity involved the formulation of the principle of equivalence, in consequence of which the concept of the *inertial system* becomes completely empty. Here we have a purely physical, not a logical insight, namely, the insight that the laws of nature can be written in a simple covariant form, which is then the same no matter what system we choose as the 'fixed body of reference'. In other words, even if bodies A and B are in accelerated motion relative to one another, the laws of nature are the same whether we use a coordinate system in which A is always at rest and the velocity of light is constant, or one in which B is always at rest and the velocity of light is constant. Here it is necessary to be completely clear that a further step is involved. It is involved precisely because Einstein did *not* immediately (i.e. with the advent of the special theory of relativity) permit the selection of a body in accelerated motion with respect to the fixed stars as the 'body of reference'. If, as Grünbaum claims, Einstein's 'insight' *from the beginning* was that we are 'free' to define the to and fro velocities of light to be equal, even relative to a system in accelerated motion relative to the fixed stars, then an important part of general relativity was already included in the special theory of relativity. In fact, however, Einstein allowed the use of systems in accelerated motion relative to the fixed stars as 'fixed bodies of reference' only *after* he became convinced that *as a matter of physical fact* no complication in the form of the laws of nature results thereby.

Given this conceptual setting it is completely clear that, if the world had been discovered to be quasi-Newtonian, Einstein would certainly have insisted on the requirement of invariance of the laws of nature under the Galilean group of transformations as the appropriate requirement, and would not have stressed this 'freedom' to adopt non-customary definitions of simultaneity which so impresses Grünbaum. We also see that the requirement imposed by Grünbaum, that all the admissible metrics should agree on the topology (the topology is 'intrinsic') makes sense only within the framework of a desire to single out a class of permissible reference systems which will lead to the same laws of nature. If we are willing to change our definitions in a way which involves a change of the laws of nature, then no justification whatsoever can be given for insisting that the principle must be retained that causal signals should not be discontinuous.

I conclude that the alleged insight that we are 'free' to employ a non-customary 'definition' of simultaneity even in a quasi-Newtonian universe is correct only in the sense allowed by TSC, and that Grünbaum's interpretation of Einstein therefore *trivializes* Einstein's profound logical–physical insight.

Finally, I wish to emphasize that just as we are *not* free in special relativity to employ a body in accelerated motion relative to the fixed stars as the 'body of reference' so we are *not* free in general relativity to employ any metric tensor other than g_{ik}. It is possible to choose the reference system arbitrarily and still arrive at the same covariant laws of nature in general relativity *only* because we are *not* allowed to 'choose' a space–time metric. Grünbaum obscures this situation by asserting that in general relativity we sometimes *do* employ nonstandard 'definitions of congruence'; he further asserts that our freedom to do this is a consequence of the alleged 'insight' that we discussed above. This seems puzzling in view of the fact that all observers employ the same g_{ik} tensor in general relativity, until one realizes how Grünbaum employs his terms. What he is pointing out is the fact that the customary rule of correspondence (employing a clock at rest in the reference system to determine local time) is not valid under exceptional circumstances (on a rotating disk). But correspondence rules always break down under some exceptional circumstances. Thus no special insight is needed to allow for the fact that the local time of the theory is to be preserved over the readings of a clock situated on a rotating disk! In sum: our alleged 'freedom' to *choose a different g_{ik} tensor* (a different space–time metric) *at the cost of complicating the laws of nature is in fact never employed in the general theory of relativity*. All observers are required to 'choose' the *same* space–time metric.

Reichenbach on space and time

Since Reichenbach's views on space and time differ in certain respects from Grünbaum's, a discussion of those views may be of some interest.

In the first place, Reichenbach does not insist, as Grünbaum does, that the dependence of geometrical statements upon 'arbitrary definitions' (in the sense of 'arbitrary' relevant to TSC) applies only to geometry. On the contrary, he denies this. I have already mentioned the example of *temperature* which occurs in Reichenbach's book, and on the last page of that book there occurs the sentence, 'it may suffice at this place to remark that the problem concerning space and time is not different from the description of any other physical state as expressed in physical laws'. Why then is Reichenbach so concerned to emphasize the conventional character of our definitions? The answer is twofold. TSC only asserts that the meaning that we give to words is arbitrary. It does not assert that words are given meaning by means of *definitions*. Thus what Reichenbach is asserting is a quite special epistemological thesis. That empirical laws cannot be confirmed or disconfirmed, and, indeed, cannot be empirical laws at all unless the words occurring in them have a meaning is trivial. That the words must *first* be given a meaning by the laying down of *definitions* is not trivial, and indeed, in the opinion of most philosophers of science today, is not true. Yet it is just this that Reichenbach is concerned to assert and in no uncertain terms. He asserts *both* that *before* we can discuss the truth or falsity of any physical law all the relevant theoretical terms must have been *defined* by means of 'coordinating definitions' *and* that the definitions must be *unique*, i.e. must uniquely determine the extensions of the theoretical terms. Such views were quite common when Reichenbach wrote (in 1928). What distinguished Reichenbach's work was its greater sophistication than the work of, say, Bridgeman, who was writing at the same time, with respect to the role of theory. Reichenbach did recognize that what he called the 'interposition of theories' must play an essential role in the actual use of scientific concepts. He was thus caught between two somewhat conflicting desires: the desire to do justice to this 'interposition of theories', and the desire to retain his epistemological conviction that all theroetical terms must possess coordinating definitions which uniquely define them.

These 'coordinating definitions' do not have to be given for the primitive terms of a theory. Reichenbach emphasizes that these primitive terms may be implicitly defined by their connection with complex (molecular) terms which themselves receive coordinating definitions. But even the primitive terms must have a unique interpretation, even if it be indirectly specified. In this respect Reichenbach is quite at variance with the currently fashionable idea that theoretical terms are only 'partially interpreted'.

Grünbaum's and Reichenbach's *motives* are thus entirely different. Grünbaum is inclined to emphasize the 'intrinsic metrical amorphousness' of space–time, while Reichenbach is concerned to emphasize the dependence of all knowledge on logically arbitrary coordinating definitions.

Did Reichenbach recognize the fact that our statements can remain 'objective' notwithstanding the fact that they contain words whose 'definitions' are 'arbitrary'? There is some evidence of confusion on this point. That our knowledge *as a whole* remains 'objective' notwithstanding the fact that it contains an element of convention Reichenbach clearly sees and emphasizes again and again. With respect to individual statements he is less sure. There occurs on p. 21 the astounding assertion that the statements that the floors and ceilings of our rooms are plane, and that the corners of our rooms are rectangular, are not synthetic statements but definitions, and 'have nothing to do with cognition as one might at first believe'. This is clearly just a mistake. Our 'coordinating definitions' may be arbitrary, but *given* our 'coordinating definitions' it is an objective fact that we live in houses with rectangular corners and not in igloos. On p. 37 there occurs a desperate attempt to straighten out this confusion concerning the 'objectivity' of our statements. Reichenbach writes,

as long as it was not noticed at what points of the metrical system arbitrary definitions occur, all measuring results were undetermined; only by discovering the points of arbitrariness, by identifying them as such, and by classifying them as definitions can we obtain objective measuring results in physics. *The objective character of the physical statement is thus shifted to a statement about relations* [Italics Reichenbach's]. A statement about the boiling point of water is no longer regarded as an absolute statement, but as a statement about the relation between the boiling point of water and the length of the column of mercury. There is a similar objective statement about the geometry of real space: *it is a statement about the relation between the universe and rigid rods.*

Clearly, this will not do. Either the statement that the water boils at 100°C is an 'absolute statement', notwithstanding the dependence of the notion of temperature upon conventions, or the statement of the *relation* of the boiling water and the length of the column of mercury is not an 'absolute statement' either; for the meaning of a two-place predicate is 'conventional' in exactly the sense in which the meaning of a one-place predicate is conventional. Reichenbach may have meant, in part, that 'water boils at 100°C' is analytic (by definition of the centigrade scale), but that there is a synthetic statement to be made about the boiling, point of water; but he surely expressed this in a most unfortunate way and he saw it as an instance of a general fact about 'objective statements' (that they concern *relations*) which is not so.

This is easy to straighten out, however. The points at which I take Reichenbach to have been *importantly* mistaken are these: it is incorrect to refer to correspondence rules as 'arbitrary definitions' for at least two reasons:

(a) correspondence rules are always used in conjunction with theories, and normally in conjunction with theories of existential import. If the theory of the electromagnetic field had turned out to be radically wrong, then even if its correspondence rules were consistent by themselves, we would not say that *there exists such a magnitude as voltage and it is measured by a voltmeter because that's how we define it*; we might very well have said that *there is no such thing as voltage* (compare 'phlogiston'). *Theoretical entities and magnitudes do not 'exist by definition'*.

(b) Correspondence rules by themselves underdetermine the extensions of theoretical terms. The extensions of theoretical terms are in practice determined only by the correspondence rules together with the theoretical postulates together with the requirement that further postulates and singular statements be accepted only to the extent that they are compatible with the requirements of inductive *and* descriptive simplicity.

Reichenbach saw this difficulty in 1928, and tried to meet it by including a portion of physical theory in what he called the 'coordinating definitions'. In the case of geometry, the portion he attempted to include was (tacitly) Newton's second law, and (explicitly) the principle that 'universal forces do not exist'. He hoped that one could determine a unique metric from this combination of theoretical assumptions and correspondence rules, with the aid of a limiting induction which would determine how a body would behave if all external and internal forces were set equal to zero (Reichenbach, 1958, p. 23). In fact, in order to carry out the limiting induction it is necessary not just to make interior forces weaker and weaker (which means reducing the masses, charges, etc. making up the body indefinitely), but to make them more and more *constant* – otherwise one may get larger deformations from smaller interior forces. However, this idea fails because there is no empirical way of determining when the interior forces are becoming constant (assuming external forces have been eliminated) apart from the absence of deformations in the body, which themselves depend upon the metric. Moreover, the proposed limiting induction would run into trouble from both general relativity and atomic theory.†

These criticisms are not intended to be a denial of the great merits of Reichenbach's work. Reichenbach's account would not have been influential if it were not in many ways correct. Unlike Grünbaum, Reichenbach does not derive from the possibility of remetricizing

---

† It would run into trouble from general relativity because one cannot set external forces equal to zero, i.e. remove all other masses from the universe, without changing the metric. It would run into trouble from atomic theory because the masses and charges of the elementary particles are fixed and cannot be made to approach zero.

space–time any epistemological moral which does not apply to every physical state. And although he shares with most other analytic philosophers writing at that time the belief that theoretical terms must be completely defined, he recognizes that it is necessary to widen the notion of 'definition' if this is to be accomplished, and that the 'interposition of theories' must play an essential role. One can scan the other philosophers writing in the same period – Wittgenstein in the *Tractatus*, Carnap in the *Logische Aufbau*, Russell in *Our Knowledge of the External World* or in the *Analysis of Matter* – in vain for a similar awareness. Moreover, Reichenbach was right in maintaining that the standard of congruence in physical theory is the corrected solid body.† But this is an *intratheoretic* criterion of congruence. Once we abandon talk of a 'congruence definition' almost all of the other problems in the philosophy of geometry fall into place of themselves. In particular, we are then prepared to understand the senses in which the falsity of Euclidean geometry was and was not 'empirical', and to appreciate just how important it is for the whole theory of knowledge that even conceptions as 'necessary' as those of Euclidean geometry are subject to revision.

Appendix

Let g_{ik} be the customary metric tensor and let P be the customary physics. The combination of the two I shall refer to as a world system. Suppose we introduce an alternative metric tensor g'_{ik} leading to a quite different geometry and modify the physics from P to P', so that the new world system $g'_{ik} + P'$ is 'equivalent' to the world system $g_{ik} + P$, i.e. the two world systems are thoroughly intertranslatable. Reichenbach points out that if the standard of congruence in the system $g_{ik} + P$ is the solid rod corrected for differential forces, then going over to $g'_{ik} + P'$ it will be necessary to postulate that the rod undergoes certain additional deformations. These 'additional' deformations will be the same for all solid rods independently of their chemical composition *relative to the original description* $g_{ik} + P$. Overlooking the fact that it is only *relative to the original description* that these additional forces affect all bodies in the same way, Reichenbach supposes that the new physics P' must be obtained by simply taking over the physics P and postulating an additional force U which affects all bodies the same way. Thus if the total

† In a geometry of constant curvature only, however. In the Appendix to 'The refutation of conventionalism' (chapter 9, volume 2, of these papers) we explain why even this criterion fails in spaces of variable curvature.

force acting on a body according to the world system $g_{ik} + P$ is $F = E + G + I$, where E is the total electromagnetic force acting on the body, G is the total gravitational force acting on the body, and I is the sum of the interactional forces acting on the body, i.e. the forces due to the interaction of the electromagnetic and the gravitational field, if such exist, then he supposes that according to the physics P', the total force F' acting on the body will be given by $F' = E + G + I + U$, where U is the 'universal force'. There are a number of things wrong with this account. For example, 'universal forces' *must depend on the chemical composition of the body they act on*, since no force which is independent of the chemical composition of the body will have the effect of producing the same *deformations* in all solid bodies *independently of their resistance to deformation*. For example, gravitation is not a 'universal' force in this sense, since the surface gravity of Jupiter will produce quite *different* deformations in (a) a man and (b) an iron bar.

However, the main point overlooked by Reichenbach is this. When we construct P', we cannot in general simply take over the physics, P, and introduce an additional force, U. This is not, in general the simplest way of obtaining 'equivalent' world systems based on the metric g'_{ik}. What will in general happen is that P' *changes the laws obeyed by the differential forces*. In other words, if the total electromagnetic force acting on a body according to P is E, then the total electromagnetic force acting on the same body according to P' will not be E but $E' = E + \Delta E$. Similarly, the gravitational force acting on a body according to P' will not be G but $G' = G + \Delta G$, and the interactional force will not be I but $I' = I + \Delta I$. It is perfectly possible that the additional force $\Delta E + \Delta G + \Delta I$ introduced by the change in the laws for the differential forces will be perfectly sufficient to account for the additional deformations required to take us from the world system $g_{ik} + P$ to the world system $g'_{ik} + P'$. According to $g_{ik} + P$, the total force acting on a body at any given time is $F = E + G + I$; according to $g'_{ik} + P'$, the total force acting on a body at any time is $F' = E' + G' + I'$. According to both systems, there are no universal forces in Reichenbach's sense and bodies change their lengths as they are moved about only slightly and only due to the differential forces including differential forces accounted for by their own atomic constitutions.

If we change from the world system $g_{ik} + P$ to the world system $g'_{ik} + P'$, then we have to postulate universal deformations in the sizes and shapes of measuring rods as they are moved about which are the same for all measuring rods independently of their chemical composition relative to our original description, i.e., the description provided by $g_{ik} + P$; but symmetrically, if $g'_{ik} + P'$ was our original system and we change to $g_{ik} + P$, then we'll have to postulate additional deformations in the sizes and shapes of measuring rods which will be the same for all measuring rods independently of their chemical composition relative to what we described their size and shape as being in $g'_{ik} + P$. Thus the so-called *congruence definition* – '*the standard of congruence is the solid body corrected for differential forces* – may be compatible with every admissible metric tensor. Moreover, even if we specify a metric tensor g_{ik} and a world system $g_{ik} + P$ to which all admissible world systems are to be 'equivalent', the physics P that goes with the metric tensor g_{ik} is not uniquely determined. We can use, by hypothesis, the physics P to get a correct world system, $g_{ik} + P$; but there will be many other systems of physics, including systems P^* which postulate universal forces, such that $g_{ik} + P^*$ is 'equivalent' to $g_{ik} + P$. In other words, even if g_{ik} is the normal metric, there will be systems of physical theory P^* according to which universal forces are present and according to which we have to obtain the corrected length of the measuring rod by correcting not only for differential forces but also for some universal force. Thus, once again we see that in fact it may be the case that for every metric g_{ik} there is both a correct physics P such that universal forces are absent and an 'equivalent' physics P^* (i.e. a physics P^* such that $g_{ik} + P$ and $g_{ik} + P^*$ are 'equivalent descriptions') such that universal forces are present according to $g_{ik} + P^*$.

It is difficult to show the full nature and extent of Reichenbach's error without going into details on the question of transformations between metrics. But the essence of the mistake has already been indicated. If Reichenbach's universal force, U, can be expressed as a sum $U = \Delta E + \Delta G + \Delta I$ of increments in the differential forces introduced by the new physics P', then the new physics P' will not have to postulate any such universal force U at all. Indeed, the simplest way of going over from the physics P to the new physics P' required by the change in the metric tensor from g_{ik} to g'_{ik} is in general *not* to introduce a universal force, since it is very difficult to construct a force which will produce the same deformations in bodies independently of their chemical constitution, but rather to suitably modify the equations for the differential forces. Conversely, if, according to the normal physics P, $F = E + G + I$, then it is always possible to introduce new quantities E', G', I' such that $E = E' + \Delta E'$, $G = G' + \Delta G'$, $I = I' + \Delta I'$ and such that the sum $U = \Delta E' + \Delta G' + \Delta I'$ is a universal force in Reichenbach's sense, and thus to obtain a physics P^* according to which the forces acting on any body are not the forces E, G, and I but rather the differential forces E', G', and I' plus the universal force U. This will lead to the same total resultant force $U + E' + G' + I' = E + G + I$. In sum, the forces acting on a body can always be broken up so that the total resultant force includes a *component* universal force; and a component universal force acting on a body can always be broken up into components which can be combined with the differential forces. That this is indeed the case is the content of the following.

THEOREM. Let P be a system of physics (based on a suitable system of coordinates) and E be a system of geometry. Then the world described by E *plus* P can be redescribed in terms of an arbitrarily chosen metric g_{ik} (compatible with the given topology) *without postulating 'universal forces'*, i.e. forces permanently associated with a spatial region and producing the same deformations (over and above the deformations produced by the usual forces) independently of the composition of the body acted upon. In fact, according to the new description g_{ik} *plus* P' (which is an 'equivalent description' in the sense explained):

(1) All deformations are ascribed to three sources: the electromagnetic forces, the gravitational forces, and gravitational–electromagnetic interactions.

(2) All three types of forces are dependent upon the composition of the body acted upon.

(3) If there are small deformations constantly taking place in solid bodies according to E *plus* P (as there are, owing to the atomic constitution of matter), then no matter what geometry may be selected, the new g_{ik} can be so chosen that the deformations according to g_{ik} *plus* P' will be of the same order of magnitude. Moreover, it will be impossible to transform them away by going back to E *plus* P.

(4) If it is possible to construct rods held together by only gravitational forces or only electromagnetic forces, then (in the absence of the other type of field) the interactional forces of the third type (postulated by P') will vanish.

(5) If there are already 'third type forces' according to E *plus* P, then the situation will be thoroughly symmetrical, in the sense that (i) going from the old metric to g_{ik} involves postulating additional deformations (relative to the description given in E *plus* P) which are the same for all bodies, and (ii) going from g_{ik} back to E *plus* P involves postulating additional deformations which are also the same for all bodies, *relative to the description given in g_{ik} plus P'*; and the same number and kind of fundamental forces are postulated by both P and P'.

Proof: It suffices to retain the original coordinate system, and replace the original notion of distance, wherever it occurs in physical laws, by the appropriate function of the coordinates. The second law of motion is now destroyed, since it now reads

$$F = m\ddot{\textbf{x}}$$

and $\ddot{\mathbf{x}}$ (the second derivative of the position vector) is no longer 'acceleration', owing to the 'arbitrary' character of the coordinate system as viewed from the new g_{lk} tensor. (We assume the new g_{lk} is compatible with the original topology.) But we can restore the second law by construing 'force' in the old laws as not force at all, but some other quantity – say 'phorce' (P). Then the above law is rewritten as

$$P = m\ddot{\mathbf{x}}$$

and the law $F = ma$ is reintroduced as a definition of 'force'. (Here a must be defined in terms of the new g_{lk} tensor.) The difficulty is that so far we have only defined *total resultant force*. To obtain a resolution into component forces, we proceed as follows: obtain the total resultant force F on the body B by determining its mass and acceleration (the latter can be found from the law $P = m\ddot{\mathbf{x}}$, once we know the total 'phorce' acting on B and the g_{lk} tensor). Now set the gravitational field equal to zero, determine the total 'phorce' that would now be acting on B, and determine from this the total force that would be acting on B. Call this E (electromagnetic force). Similarly, set the electromagnetic field equal to zero, and obtain the total force that would be acting on B. Call this G (gravitational force). Finally define I (interactional force) from the equation $F = E + G + I$.

One can now verify all five parts of the theorem. (1) is immediate. (4) is likewise immediate, since from the definition of E and G we have that $F = E$ whenever $G = 0$ and that $F = G$ whenever $E = 0$. (2) is clear, since E and G depend upon the composition of the body acted upon, and so does F. (That the dependence of I on the composition of B is genuine follows from (4) which shows that I can be altered by altering either E or G, even without altering F). The first part of (3) only asserts that the 'metrical deformations' (*relative* to the description of length given by E *plus* P) induced by going over to g_{lk} can be made small in comparison with, say, the electronic 'radius' (10^{-13} centimeters) by using a g_{lk} tensor that is Euclidean in the small, provided 'in the small' is large enough. If there were *no deformations at all* according to E *plus* P, then E *plus* P could be singled out as *the* system in which all deformations can be transformed away;† but it is easily shown (by considering inconsistencies that would arise if we defined all solid bodies to remain *exactly* unchanged under transport) that in the presence of even the smallest fluctuations this cannot be done. If there are fluctuations in P there will also be fluctuations in every P'; and since there are

† In that case, however, congruence would be defined operationally (in terms of the transport of these rods) and the whole problem of correcting for 'differential forces' would not arise.

no universal forces in P' these are entirely ascribed to differential forces, no matter what g_{ik} we are using. Only the exact amount of the fluctuations is differently described in different P's. The only difference between P and P' is (one might suppose) that there are no interactional forces in P (forces arising from the interaction of the electromagnetic and gravitational fields). But this need not be true; indeed, in general relativity theory, it *isn't* true. And if there are already such forces, according to P, then we have the symmetry situation asserted in (5).

Comment. If E *plus* P is the 'normal' system, one may attempt to rescue the claim that 'every other system g_{ik} *plus* P'' contains universal forces' in various ways. For example, Reichenbach considers letting the external *and interior* forces approach zero 'in the limit'. But then our forces I go to zero as well.† Alternatively, one might eliminate the external forces, and then let the interior forces become *constant* (instead of zero). Universal forces are present if deformations *still* take place. But the difficulty is that whether the interior forces are constant or not depends upon whether P or P' is used. (An experimental criterion is ruled out, beyond a certain point, by the atomic constitution of matter.) One might say that the coordinating definition should be changed to: 'The differential forces obey laws which do not depend upon the coordinates but only upon coordinate differences'. But this requirement can be formally complied with as well. For example, if the curvature of space varies from place to place (as in general relativity), then any reference to a particular place can be made coordinate independent by *describing* the place in questions as e.g. 'the place where the metrical field has such-and-such values'. (There is a tendency in Grünbaum to emphasize the constant curvature case, which is not physically realistic, since general relativity is well confirmed.) Finally, we might try to stipulate that 'Phorce' = Force, i.e. $ma = m\ddot{x}$, so that $a = \ddot{x} = \ddot{x}i +$ $\ddot{y}j + \ddot{z}k$. But this is possible *only* in Euclidean space, so this would just be to stipulate that Euclidean geometry shall be used.‡

† This depends upon the identity of gravitational and inertial mass. The point is that in order to make the interior *gravitational* forces approach zero, we must make the mass of B approach zero and hence $\mathbf{F} = m\mathbf{a} \to 0$.

‡ In the Appendix to 'The refutation of conventionalism' (chapter 9, volume 2, of these papers) a better proof of this theorem is given and we generalize to remetricization of *space–time*.

7

A philosopher looks at quantum mechanics*

Those defile the purity of mathematical and philosophical truth, who confound real quantities with their relations and sensible measures.

—Isaac Newton, *Principia*

Before we say anything about quantum mechanics, let us take a quick look at the Newtonian (or 'classical') view of the physical universe. According to that view, nature consists of an enormous number of particles. When Newtonian physics is combined with the theory of the electromagnetic field, it becomes convenient to think of these particles as dimensionless (even if there is a kind of conceptual strain involved in trying to think of something as having a *mass* but not any *size*), and as possessing electrical properties – negative charge, or positive charge, or neutrality. This leads to the well-known 'solar system' view of the atom – with the electrons whirling around the nucleons (neutrons and protons) just as the planets whirl around the sun. Out of atoms are built molecules; out of molecules, macroscopic objects, scaling from dust motes to whole planets and stars. These latter also fall into larger groupings – solar systems and galaxies – but these larger structures differ from the ones previously mentioned in being held together exclusively by gravitational forces. At every level, however, one has trajectories (ultimately that means the possibility of continuously tracing the movements of the elementary particles) and one has causality (ultimately that means the possibility of extrapolating from the history of the universe up to a given time to its whole sequence of future states).

When we describe the world using the techniques of Newtonian physics, it goes without saying that we employ *laws* – and these laws are stated in terms of certain *magnitudes*, e.g. distance, charge, mass. According to one philosophy of physics – the so-called *operationalist* view so popular in the 1930s – statements about these magnitudes are mere shorthand for statements about the results of measuring operations.

* First published in Robert G. Colodny (ed.), *Beyond the Edge of Certainty: Essays in Contemporary Science and Philosophy* (1965). Reprinted by permission of Prentice-Hall, Inc., Englewood Cliffs, New Jersey.

Statements about distance, for example, are mere shorthand for statements about the results of manipulating foot rulers. I shall here assume that this philosophy of physics is *false*. Since this is not a paper about operationalism, I shall not defend or discuss my 'assumption' (although I *do* refer the interested reader to the investigations of Carnap, Braithwaite, Toulmin, and Hanson for a detailed discussion of this issue). I shall simply state what I take the correct view to be.

According to me, the correct view is that when the physicist talks about electrical charge, he is talking quite simply about a certain magnitude that we can distinguish from others partly by its 'formal' properties (e.g. it has both positive and negative values, whereas mass has only positive values), partly by the structure of the system of laws this magnitude obeys (as far as we can presently tell), and partly by its *effects*. All attempts to *literally* 'translate' statements about, say, electrical charge into statements about so-called observables (meter readings) have been dismal failures, and from Berkeley on, all *a priori* arguments designed to show that all statements about unobservables must ultimately reduce to statements about observables have contained gaping holes and outrageously false assumptions. It is quite true that we 'verify' statements about unobservable things by making suitable *observations*, but I maintain that without imposing a wholly untenable theory of meaning, one cannot even *begin* to go from this fact to the wildly erroneous conclusion that talk about unobservable things and theoretical magnitudes *means the same* as talk about observations and observables.

Now then, it often happens in science that we make inferences from measurements to certain conclusions couched in the language of a physical theory. What is the nature of these inferences? The operationalist answer is that these inferences are *analytic* – that is, since, say, 'electrical charge' *means by definition* what we get when we measure electrical charge, the step from the meter readings to the theoretical statement ('the electrical charge is such-and-such') is a purely conventional matter. According to the nonoperationalist view, this is a radical distortion. We know that this object (the meter) measures electrical charge, *not* because we have adopted a 'convention', or a 'definition of electrical charge in terms of meter readings', but because we have accepted a body of theory that includes a *description of the meter itself in the language of the scientific theory*. And *it follows from the theory*, including this description, that the meter measures electrical charge (approximately, and under suitable circumstances). The operationalist view disagrees with the actual procedure of science by replacing a probabilistic inference within a theory by a nonprobabilistic inference based on an unexplained linguistic stipulation. (Incidentally, this anti-operationist view has sometimes been termed a 'realist' view, but it is espoused by some positivists – in particular by Carnap in his well-known article 'The Interpretation of Physical Calculi'. It is well, for this reason, to avoid these '-ist' words entirely, and to confine attention to the actual methodological theses at issue in discussing the philosophy of the physical sciences.)

If the nonoperationist view is generally right (that is to say, correct for physical theory in general – not just for Newtonian mechanics), then the term 'measurement' plays *no fundamental role in physical theory as such*. Measurements are a subclass of physical interactions – no more or less than that. They are an important subclass, to be sure, and it is important to study them, to prove theorems about them, etc.; but 'measurement' can never be an *undefined* term in a satisfactory physical theory, and measurements can never obey any 'ultimate' laws other than the laws 'ultimately' obeyed by *all* physical interactions. It is at this point that it is well for us to shift our attention to quantum mechanics.

The main fact about quantum mechanics – or, at any rate, the main fact as far as this paper is concerned – is that the state of a physical system can be represented by a set of waves (more precisely, by a 'ψ-function'; however, nontechnical language will be employed throughout this paper, so I shall stick to the somewhat inaccurate, but more popularly intelligible, phrase, 'set of waves'). Of course any set of waves (superimposed on one another) may also be regarded as a single wave; thus if we have a system of, say, three particles, we may represent the 'state' of the entire system by a single wave. A 'wave' is simply a magnitude, with an intensity at every point of space, whose intensity has certain periodicity properties (normally expressed by differential equations of a certain kind). For example, the intensity may (moving in space in a certain direction) rise to a peak, then fall, then rise to a peak again, etc. Normally the places at which these peaks are located are more or less evenly spaced – at least in the case of simple harmonics; all regularity may be lost when a great many different waves are superimposed in a jumbled way – and these places change location with time in the case of a moving 'wave front'. All of this (which is, of course, purely qualitative) should be familiar to the reader from a consideration of sound waves in air or pressure waves in water.

The quantum mechanical story becomes complicated immediately, however. In the first place, the waves treated in quantum mechanics – the waves used to represent the state of a physical system – are not waves in ordinary space, but waves in an abstract mathematical space. If we are dealing with a system of three particles, for example (with position coordinates x_1, y_1, z_1 for the first particle, x_2, y_2, z_2 for the second particle, and x_3, y_3, z_3 for the third particle), then we employ a space with nine 'dimensions' to represent the system (one dimension for each of the coordinates $x_1, \ldots, z_3$). In the second place, the amplitude of the waves employed in quantum mechanics can have such recherché values as the square root of minus one. (In technical language, the ψ-function is a *complex* valued function of the dimensions of the space, whereas 'ordinary' waves are real valued.) Finally, the dimensions of the space need not be thought of as corresponding to the position coordinates of the various particles – one can instead think of the nine dimensions (in the case of a three-particle system) as corresponding to, say, the nine momentum coordinates (or to certain other sets of physical magnitudes, or 'observables' in the jargon of quantum mechanics); this will make a difference, in the sense that one will employ a different *wave* to represent the state. (This is called using 'momentum representation' instead of 'position representation'. The two representations are equivalent, in the sense that there exist mathematical rules for going from one to the other.)

At any rate, ignoring these complications for one moment, let us repeat the 'simple-minded' statement we made at the outset of this part of the paper: the state of a physical system can be represented by a set of waves. What are we to make of this fact? Why is this technique of representation successful? What is it about physical systems that makes them lend themselves to representation by systems of waves? In short, what is the significance of the 'waves'? Answers to this question are usually known as 'interpretations' of quantum mechanics; and we shall now proceed to consider the most famous ones.

Note at the outset that there is one answer that we can dismiss, namely, that the success of this technique is just an accident. No physicist, in fact, no person in his right mind, has ever proposed *this* answer – and for good reason. Not only does the quantum mechanical formalism yield correct answers to too many decimal places, but it also yields too many predictions of whole classes of effects that would not have been anticipated on the basis of older theory, and these predictions are correct.

The historically first answer to our question might be called the 'De Broglie interpretation of quantum mechanics'. It is simply this: physical systems *are* sets of waves. The waves spoken of in quantum mechanics do not merely 'represent' the state of the system; they *are* the system.

Unfortunately, this interpretation runs at once into insuperable difficulties (at least, almost all physicists consider them insuperable; a small minority, including De Broglie, continue to defend it). We have, in effect, stated these difficulties in pointing out the differences between quantum mechanical waves and ordinary waves: that the amplitudes are complex and not real; that the space involved is a mathematical abstraction, and not ordinary space: that the waves depend not just upon the system, but also upon the 'representation' used – i.e. upon the set of 'observables' we are interested in measuring (position, or momentum, or whatever). One further difficulty is important: the so-called 'reduction of the wave packet'.

The 'state' of a physical system changes, of course, through time. This change is represented in two ways (not just one – motion along a trajectory, or collection of trajectories – as in classical physics). When one wishes to represent the state of a system as it is changing during an interval in which the system remains isolated, one allows the waves to 'expand' in a continuous fashion. This process is called 'motion' (this 'expansion' or 'spreading out' of the waves is governed by the famous Schrödinger equation). When one wishes to represent the change in the state of a system induced by a *measurement*, one simply 'puts in' the new state (as determined by the measurement). In von Neumann's axiomatization of quantum mechanics, this putting in of a state is allowed for in the simplest (one is tempted to say 'crudest') way possible: an axiom is introduced which says, in effect, that a measurement throws a system *discontinuously* into a new state. Axiomatization aside, the fact remains that in the quantum mechanical formalism one sometimes speaks of the waves as undergoing certain kinds of continuous changes, and sometimes as *abruptly* or *discontinuously* jumping into a new configuration.

An example may make this clear: uncertainty concerning the position of a particle is represented by the volume occupied by the wave corresponding to the particle. The particle may be located (if a position measurement comes to be made) anywhere in the region corresponding to the volume occupied by the wave. Thus a particle whose position is known with high accuracy corresponds to a wave that is concentrated in a very small volume – a wave packet – whereas a particle whose position is known with low accuracy corresponds to a wave that is spread out over a large volume. Now suppose a position measurement is made on a particle of the latter kind – one whose position was known with low accuracy before the measurement. After the measurement, the position will be known with high accuracy. So the change of state produced by the measurement will be represented as follows: a very spread out wave suddenly jumped into the form of a wave packet. In other words, the wave suddenly vanished almost everywhere, but at one place the intensity suddenly increased (to compensate for the vanishing elsewhere). Needless to say, this reduction of the wave packet constitutes very strange behavior if this is really to be thought of as a physical wave.

The second 'interpretation of quantum mechanics' we shall consider will be called the 'Born interpretation' (perhaps I should call it the *original* Born interpretation, to distinguish it from the 'Copenhagen interpretation', which also depends in part on Born's ideas, and which is discussed below). This interpretation is as follows: the elementary particles *are* particles in the classical sense – point masses, having at each instant both a definite position and a definite velocity – though not obeying classical laws. The wave corresponding to a system of particles does *not* represent the state of the system (simultaneous position and velocity of each particle), but rather our *knowledge* of the state, which is always incomplete. That our knowledge of the state must always be incomplete, that one cannot, for example, simultaneously measure position and momentum with arbitrarily high accuracy (the famous 'uncertainty principle'), is not an independent assumption. It is a mathematical consequence of the basic assumption that any possible state of a system – or, rather, any physically obtainable *knowledge* of the state of a system – can be represented by a set of waves according to the rules of quantum mechanics, for a particle whose position and momentum were both known with virtually perfect accuracy would have to correspond to a wave that had the property of being 'packet-like' (occupying a very small volume) in *both* 'position representation' and 'momentum representation', and as a fact of pure mathematics, there are no such waves.

This interpretation is able to deal very easily with all the difficulties mentioned in connection with the De Broglie interpretation. For example, it is no difficulty that the spaces used in quantum mechanics†† are multi-dimensional, for they are not *meant* (according to this interpretation) to represent real physical space; the so-called 'dimensions' are nothing but sets of 'observables' (position coordinates, momentum coordinates, or whatever). The complex amplitudes are handled by a simple device: *squaring* (or, rather, taking the square of the absolute value). In other words, the probability that the particle is inside a given volume is measured not by the intensity of the wave inside that volume, but by the squared absolute value of the intensity, which is always a non-negative real number not greater than one (and so can be interpreted as a probability, as the square root of minus one obviously cannot be).

† Spaces of yet another kind – 'Hilbert spaces' – are also used in quantum mechanics, but these will be avoided here.

This way of calculating probabilities – squaring the intensity of the wave – leads to experimentally correct expectation values, and has become fundamental to quantum mechanics (as we shall see, it is taken over in the Copenhagen interpretation). Since the wave is not supposed to be a physical wave, but only a device for representing knowledge of probabilities, it is not serious that we have to perform this operation of squaring to get the probabilities, or that the intensity (amplitude) is a complex number before we square.

Finally, the reduction of the wave packet is no puzzle, for it represents *not* an instantaneous change in the state of a whole, spread out physical wave (which is supposed to take place the moment a measurement is over), but an instantaneous change in our *knowledge* of the state of a physical system, which takes place the instant we learn the result of the measurement. If I know nothing concerning the position of a particle except that it is somewhere in a huge volume of space, and then I make a position measurement that locates the particle *here*, the wave that represents my knowledge of the position of the particle, and that occupies the entire huge volume until the result of the measurement is learned, has to be replaced by a wave packet concentrated in the appropriate submicroscopic place. But a physical 'something' does not thereby contract from macroscopic to submicroscopic dimensions; all that 'contracts' is human ignorance.

In view of all the attractive features of the (original) Born interpretation, it is sad that it, too, encounters insuperable difficulties. These difficulties have to do with two closely related phenomena (mathematically, they are virtually one phenomenon) – 'interference' and the 'superposition of states'. Interference may be illustrated by the famous 'two-slit experiment'. Here a particle is allowed to strike a surface – say, a sensitive emulsion – after having first passed through one or the other of two suitably spaced narrow slits (we do not know which one it passes through in such an experiment, but it *has* to pass through one or the other, on the particle interpretation, to reach the emulsion, because the emitter is placed on the other side of the barrier from the emulsion). What the experiment reveals is that the mathematical interference of the quantum mechanical waves shows up *physically* in this case as an interference pattern. The various particles – say, photons – are not able to interact (this can be insured by reducing the intensity of the radiation until only one or two photons are being emitted per second), so this is a case in which the wave that represents the whole system can be obtained by simply superimposing the waves of the individual particles. Also, each particle is represented by a wave in three-dimensional space (since there are three position coordinates), so the whole system of particles (in this special case) can be represented by a wave in three-dimensional space. This facilitates comparing the wave with a physical wave – in fact, if we identify the space used in this representation with ordinary physical space, we are led to predict exactly the correct interference pattern. Thus we have the following difficult situation: the reduction of the wave packet makes no sense unless we say that the waves are not physical waves, but only 'probability waves' (which is why they collapse when we obtain more information); but the interference pattern makes no sense unless we say the waves (in this very special three-dimensional case) *are* physical waves.

If we analyze the interference mathematically (which we are able to do, since it is correctly predicted by the quantum mechanical formalism), we find that what the *mathematics* reveals is even stranger than what the experiment reveals. Mathematical analysis rules out even more conclusively than the experiment the intuitive explanation of the phenomenon – that particles going through one slit somehow interfere with *different* particles going through the other slit. Rather, different particles correspond to 'incoherent' or unrelated waves, and these produce no detectable interference. Each particle corresponds to a wave that is split into two halves by passing through the two slits; the two halves of the wave belonging to a *single* particle are 'coherent', or have intimately related wave properties, and all the detectable interference is produced by the interference between the two coherent halves of the waves corresponding to single particles. In other words, we get an interference pattern because *each* photon *interferes with itself* and *not* because different photons somehow interfere.

Superposition of states may be described as follows: let S be a system that has various possible states A, B, C,... according to classical physics. Then in *addition* to these states, there will also exist in quantum physics certain 'linear combinations' of an arbitrary number of these states, and a system in one of these may, in certain cases, behave in a way that satisfies no classical model whatever.

I have stressed from the beginning the *unity* of the quantum mechanical formalism. Mathematically, almost all there is is a systematic way of representing physical systems and situations by means of a *wave* in a suitable *space*. The uncertainty principle, we noted above, is not an independent assumption, but follows directly from the formalism. Similarly, superposition of states is not something independent, but corresponds to the fact that any two permissible waves (representing possible states of a physical system) can literally be superimposed to obtain a new wave, which will also (according to the theory) represent a possible state of the system. More precisely, if ψ_1 and ψ_2 are permissible wave functions, then so is $c_1\psi_1 + c_2\psi_2$, where c_1 and c_2 are arbitrary complex constants.

To illustrate the rather astonishing physical effects that can be obtained from the superposition of states, let us construct an idealized situation. Let S be a system consisting of a large number of atoms. Let R and T be properties of these atoms which are incompatible. Let A and B be states in which the following statements are true according to both classical and quantum mechanics:

(1) When S is in state A, 100 per cent of the atoms have property R.

(2) When S is in state B, 100 per cent of the atoms have property T – and we shall suppose that suitable experiments have been performed, and (1) and (2) found to be correct experimentally. Let us suppose there is a state C that is a 'linear combination' of A and B, and that can somehow be prepared. Then classical physics will not predict anything about C (since C will, in general, not correspond to any state that is recognized by classical physics), but quantum mechanics can be used to tell us what to expect of this system. And what quantum mechanics will tell us may be very strange. For instance we might get:

(3) When S is in state C, 60 per cent of the atoms have property R, and also get:

(4) When S is in state C, 60 per cent of the atoms have property T – and these predictions might be borne out by experiment. But how can this be?

The answer is that, just as it turns out to be impossible to measure *both* the position and the momentum of the same particle at the same time, so it turns out to be impossible to test *both* statement (3) *and* statement (4) experimentally in the case of the same system S. Given a system S that has been prepared in the state C, we can perform an experiment that checks (3). But then it is physically impossible to check (4). And similarly, we can check statement (4), but then we must disturb the system in such a way that there is then no way to check statement (3).

We can now see just where the Born interpretation fails. It is based (tacitly) on the acceptance of the following principle:

THE PRINCIPLE OF NO DISTURBANCE (ND)

The measurement does not disturb the observable measured – i.e. the observable has almost the same value an instant before the measurement as it does at the moment the measurement is taken.

But this assumption is incompatible with quantum mechanics.

Applied to statements (3) and (4) above, the incompatibility is obvious, and Heisenberg and Bohr have given quite general arguments (Heisenberg's of a more precise and mathematic nature, Bohr's of a more general and philosophical nature) to show this incompatibility. Even in the case of the two-slit experiment, the falsity of ND is indirectly involved. This enters in the following way: if the Born interpretation were correct and ND true, it would make no difference if we modified the experiment by placing in each slit a gadget that determined through *which* slit the particle passed – for the interference pattern would be the same whether we used two slits or two emitters (assuming the history of the particle *before* reaching whichever slit it went through to be irrelevant), and the gadget could be so designed that the distribution of 'hits' on the emulsion would be the same in the case of a *single* slit whether the gadget was used or not (which shows that the history of the particle *after* leaving the slit is not being affected statistically). But, in fact, such a gadget always destroys the interference pattern. This argument is not rigorous, because even if the measurement determines the slit through which the particle was going to pass 'in any case' ('even if the measurement had not been made'), the gadget might disturb its *subsequent* behavior in a way too subtle to show up in the case of an experiment with a *single* slit. However, the impossibility of (3) and (4)'s *both* being true *before* any R-measurement or T-measurement is made is a matter of simple arithmetic (plus the incompatibility of the properties R and T), and if the Born interpretation were correct, the very meaning of the wave (considered in two different 'representations' corresponding to R-measurement and T-measurement) would be that, in the case of a system S in the state C, (3) and (4) are *both* true. Thus the principle ND and the (original) Born interpretation must both be abandoned.

In the literature of quantum mechanics, interpretations according to which the elementary particles have *both* position and momentum at every instant (although one can know only the position or the momentum, but never both at the same instant) are called 'hidden variable' interpretations. The falsity of ND has serious consequences for these hidden variable theories. They are required to postulate strange laws whereby each measurement somehow disturbs the very thing it is measuring – e.g. letting a particle collide with a plate produces a speck on the plate, but at a place where the particle *would not have been* but for the presence of the plate. Actually, such a disturbance of the thing observed by the measurement need not be postulated in the case of *every* measurement, but it does have to be introduced in a great many cases. For example, in the two-slit experiment, *one* of the two measurements – the measurement that takes place when a speck appears on the emulsion, showing that a particle has hit, or the measurement that takes place when a gadget is introduced at the slits to determine which particles go through which slit – must disturb the particle. Similarly, in the case of (3) and (4), at least one of the two measurements – R-measurement or T-measurement – must disturb the system. In the best-known hidden variable theory – that due to David Bohm – an unknown physical force (the 'quantum potential') obeying strange laws is introduced to account for the disturbance by the measurement.

The most famous interpretation of quantum mechanics, and the one that 'works' (in the opinion of most contemporary physicists), is usually referred to as the 'Copenhagen interpretation' hereafter abbreviated 'CI'), after Bohr and Heisenberg (who worked in Copenhagen for many years), although it is in many ways a modification of the Born interpretation, and Born's principle (that the squared amplitude of the wave is to be interpreted as a probability) is fundamental to it. What the CI says, in a nutshell, is that 'observables' such as position, *exist* only when a suitable measurement is actually being made. Classically, a particle is thought of as having a position even when no position measurement is taking place. In quantum mechanics (with the CI) a particle is something that has, at most times, no such property as a definite position (that could be represented by a set of three position coordinates), but only a *propensity* to have a position if a suitable experimental arrangement is introduced.

The first effect of the CI is obvious: principle ND has to be abandoned. Principle ND says that an observable has the same value (approximately) just *before* the measurement as is obtained by the measurement; the CI denies that an observable has *any* value before the measurement. Born's principle can be retained, but with a modification: the squared amplitude of the wave measures not the probability that the particle *is* in a certain place (whether we look or not), but the probability that it *will be* found in that place if a position measurement is made at the appropriate time. (If no position measurement is made at the relevant time, then it *does not* make sense to ascribe a position to the particle.) It has often been said that this view is intermediate between views according to which the particle is a (classical) particle and views according to which the particle is really a physical wave, since sometimes the particle (say, an electron) has a definite, sharply localized position (when a precise position measurement is made), whereas at other times it is spread out (can only be assigned a large region of space as its position).

The effect of the CI on statements (3) and (4) is also straightforward. Under the CI, these statements get replaced by:

(3′) When S is in state C *and an* R-*measurement is made*, 60 per cent of the atoms have property R, and

(4′) When S is in state C *and a* T-*measurement is made*, 60 per cent of the atoms have property T.

Thus the incompatibility vanishes. Of course incompatibility is avoided only because the following statement is also true:

(5) An R-measurement and a T-measurement cannot be performed at the same time (i.e. the experimental arrangements required are mutually incompatible).

This replacement of classically incompatible statements such as (3) and (4) by corresponding compatible statements such as (3′) and (4′), as a result of the CI, is often referred to by the name *complementarity*:†
the most famous case is one in which the one of the two incompatible properties is a 'wave' property – such as a momentum‡ – while the other is a 'particle' property – especially (sharply localized) *position*.
In all cases of complementarity, the idea is the same: the principle ND would require us to assign incompatible properties to one and the same system at the same time, whereas the incompatibility disappears if we say that a system has one or the other property only when a measurement of that property is actually taking place.

Even if the CI does meet all the difficulties so far discussed, it seems extremely repugnant to common sense to say that such observables as position and momentum exist only when we are measuring them. A hidden variable theory would undoubtedly fit in better with the pre-conceptions of the man in the street. It seems worth while for this reason to describe in a little more detail the difficulties that have led almost all physicists to abandon the search for a successful hidden variable theory and to embrace the extremely counterintuitive CI.

First of all, consider the phenomena for which a hidden variable theory must account. On the one hand, there are diffraction experiments, interference experiments, etc., which suggest the presence of physical waves (although it should be emphasized that such waves are never directly detected – the interference patterns, for example, in the case of a two-slit experiment, can be shown to consist of a myriad of tiny specks built up by individual particle collisions). On the other hand, when we emit some particles, and then interpose a 'piece of flypaper' – i.e. a plate with a suitably sensitized surface – we do, indeed, get sharply localized

† The formulation of complementarity is due to Paul Oppenheim and Hugo Bedau.
‡ Recalling that $p = h/\lambda$ (the De Broglie relation)– i.e. the momentum p of a particle is just the reciprocal of the wave length of the corresponding wave, in units of h.

collisions with the flypaper, which seems to confirm the view that what we emitted *were* 'particles', with sharply localized positions at each instant and with continuous trajectories. (If the position were a discontinuous function of the time, the particle would not have to hit the flypaper at all; it could first exist on one side of it and then on the other side without leaving a mark.) In view of these phenomena, it is not surprising that existing hidden variable theories all make the same assumption: that there are *both* waves *and* particles. Some success has been encountered in elaborating this idea, particularly by De Broglie and his students at the Institut Henri Poincaré. Before explaining what goes wrong (or seems to go wrong), let me first describe a case in which the 'pilot wave' idea – the idea that there are both particles (which are what we ultimately detect) and (indetectable) waves 'guiding' the particles – has been successfully worked out. This case is the two-slit experiment that we have already described.

Briefly, the idea is that the particle is nothing but a singularity in the wave – that is, a point in the wave at which a certain kind of mathematical discontinuity exists. The energy of the wave-particle system is almost wholly concentrated in the particle, and the relation between the wave and the particle is such that the *probability* that the particle is at any given place is proportional to the squared amplitude of the wave at that place. (Thus on this theory, the original Born interpretation is correct for 'position representation'. One feature of quantum mechanics that is sacrificed by this theory – and, indeed, by any hidden variable theory – is the nice symmetry between position representation and any other representation; only the wave used in position representation is a physical wave according to this interpretation, and then only in the three-dimensional case.) Singularities having the right properties – the ratio between the energy of the singularity and the energy of the wave is of the order c^2, where c is the speed of light – have been constructed by De Broglie.

Let us now consider how this interpretation affects the two-slit experiment. The fact that we can detect the wave only indirectly – through its effect on the particle that it guides – is explained on this theory by the fantastically low energy of the wave. The interference we detect is *real* interference – the wave corresponding to a single particle is split on passing through the slits, and the two halves of the wave (which still cohere, as explained above) then interfere. But what about the reduction of the wave packet?

De Broglie's answer is that the wave packet is *not* reduced. A wave that has lost the particle it guides does not collapse – it just goes on. But although it goes on, we have no way of detecting it, which is why in the usual theory we treat it as no longer existing. We cannot detect it directly because of its low energy. And we cannot detect it indirectly (through interference effects) because it does not cohere with the wave of any other particle, and there is no experimental procedure to detect the random interference caused by the interference of two mutually incoherent waves.

My opinion is that De Broglie, Bohm, and others who are working along these lines *have* indeed constructed a classical model for the two-slit experiment. But the model runs into difficulties – insuperable difficulties, I believe – in connection with *other* experiments.

In the first place, observe that, according to the theory just outlined, the principle ND is correct for the special case of position measurement. The principle ND cannot be true for an arbitrary measurement, as we showed before. But if we are really dealing with a particle whose position varies continuously with time, then, of course, we can measure the position of that particle by allowing it to smack a suitable kind of flypaper and then looking to see where the mark is on the flypaper. Wherever the mark may appear, we can say, 'an instant before, the particle cannot have been very far away from here because of the assumed continuity of the particle's motion'. However, I believe that the principle ND cannot be correct even for the case of position measurement. In order to explain why I believe this, I have to describe another phenomenon – 'passage through a potential barrier'.

Imagine a population P that is simply a huge collection of hydrogen atoms, all at the same energy level e. Let D be the relative distance between the proton and the electron, and let E be the observable, 'energy'. Then we are assuming that E has the same value, namely e, in the case of every atom belonging to P, whereas D may have different values, d_1, d_2,... in the case of different atoms A_1, A_2,...

The atom is, of course, a system consisting of two parts – the electron and the proton – and the proton exerts a central force on the electron. As an analogy, one may think of the proton as the earth and of the electron as a satellite in orbit around the earth. The satellite has a potential energy that depends upon its height above the earth, and that can be recovered as usable energy if the satellite is made to fall. It is clear from the analogy that this potential energy P, associated with the electron (the satellite), can become large if the distance D is allowed to become sufficiently great. However, P cannot be greater than E (the total energy). So if E is known, as in the present case, we can compute a number d such that D cannot exceed d, because if it did, P would exceed e (and hence P would be greater than E, which is absurd). Let us imagine a sphere with radius d and whose center is the proton. Then if all that we know about the particular hydrogen atom is that its energy E has the value e, we can still say that wherever the electron may be, it cannot be outside the sphere. The boundary of the sphere is a 'potential barrier' that the electron is unable to pass.

All this is correct in classical physics. In quantum physics, if we use the (original) Born interpretation, then in analogy to (3) and (4) we get (the figure 10 per cent has been inserted at random in the example):

(6) Every atom in the population P has the energy level e.
(7) 10 per cent of the atoms in the population P have values of D which *exceed d*.

These statements are, of course, in logical contradiction, since, as we have just seen, they imply that the potential energy can be greater than the *total* energy. If we use the CI, then, just as (3) and (4) went over into (3′) and (4′), so (6) and (7) go over into:

(6′) If an energy measurement is made on any atom in P, then the value e is obtained, and
(7′) If a D-measurement is made on any atom in P, then in 10 per cent of the cases a value greater than d will be obtained.

These statements are consistent in view of

(8) An E-measurement and a D-measurement cannot be performed at the same time (i.e. the experimental arrangements are mutually incompatible).

Moreover, we do not have to accept (6′) and (7′) simply on the authority of quantum mechanics. These statements can easily be checked, not, indeed, by performing both a D-measurement and an E-measurement on each atom (that is impossible, in view of (8)), but by performing a D-measurement on every atom in one large, fair sample selected from P to check (7′) and an E-measurement on every atom in a different large, fair sample from P to check (6′). So (6′) and (7′) are both known to be true.

In view of (6′), it is natural to say of the atoms in P that they all 'have the energy level e'. But what (7′) indicates is that, paradoxically, some of the electrons will be found on the wrong side of the potential barrier. They have, so to speak, 'passed through' the potential barrier. In fact, quantum mechanics predicts that the distance D will assume arbitrarily large values even for a fixed energy e.

Since the measurement of D is a distance measurement, on the hidden variable theory, the assumption ND holds for it (in general, distance measurement reduces to position measurement, and we have already seen that position measurement satisfies ND on this theory). Thus we can infer from the fact that D is very much greater than d in a given case, that D must also have been greater than d before the measurement – and hence before the atom was disturbed. So the energy could not have been e. In order to square this with (6'), the hidden variable theory assumes that the E we measure is not really the *total* energy. There is an additional energy, the so-called 'quantum potential', which is postulated just to account for the fact that the electron gets beyond the potential barrier. Needless to say, there is not the slightest direct evidence for the existence of quantum potential, and the properties this new force must have to account for the phenomena are very strange. It must be able to act over arbitrarily great distances, since D can be arbitrarily large, and it must not weaken over such great distances – in fact it must get stronger.

On the other hand, the CI takes care of this case very nicely. In order to get a population P with the property (6'), it is necessary to make a certain 'preparation'. This preparation can be regarded as a 'preparatory measurement' of the energy E; so it is in accord with the CI, in this case, to replace (6') by the stronger statement (6). However, a D-measurement disturbs E (according to both classical and quantum physics). So when a D-measurement is made, and a value greater than d is obtained, we can say that D exceeds d, but we can make no statement about the energy E. So there is no potential barrier, and a fortiori no passing of a potential barrier. If we had the assumption ND (for distance measurement), we could conclude from the fact that D exceeded d by some large amount that D must also have exceeded d *before* the disturbance of E by the D-measurement, and then we would be back in trouble. But the CI rejects the idea that there is such a thing as the 'value' of D an instant before the D-measurement. Thus no quantum potential is necessary (on the CI). If the energy of the atom exceeds E (when D is found to be greater than d), it is not because some mysterious quantum force is at work, but simply because energy has been *added by the D-measurement*.

In view of this rather disappointing experience with the hidden variable theories, I believe that it would be natural to impose the following three *conditions of adequacy* upon proposed interpretations of quantum mechanics:

A. The principle ND should not be assumed even for position measurement.
B. The symmetry of quantum mechanics, represented by the fact that one 'representation' has no more and no less physical significance than any other, should not be broken. In particular, we should not

treat the waves employed in one representation (position representation in the case of the hidden variable theorists) as descriptions of physically real waves in ordinary space.

C. The phenomena of superposition of states, described at the beginning of this paper, must be explained in a unitary way.

The hidden variable theories violate all three of these principles. In particular, both the two-slit experiment and the passage of a potential barrier represent superpositions of classical states, as do all the so-called anomalies encountered in quantum mechanics. Yet the theories just described handle the one (the two-slit experiment) by means of a classical model with waves guiding particles, and the other (passage of a potential barrier) by means of quantum potential, in flagrant violation of (C). (The classical states superimposed in the two-slit experiment correspond to the two situations: (a) just the left-hand slit open, and (b) just the right-hand slit open. These are 'classical' in the sense that the same predictions can be derived from classical and from quantum physics in these states. The nonclassical state that is represented by superimposing the wave representations of (a) and (b) – i.e. represented in the formalism of quantum mechanics as a linear combination of the states (a) and (b) – is the state (c), which is prepared by having both slits open. In the potential barrier experiment we have an *infinite* superposition of states – for the state corresponding to the energy level e is a linear combination of all possible states of the distance D, in the formalism.)

What about the CI? As already noted, the CI rejects the principle ND in all cases. Thus (A) is satisfied. Also, the waves are interpreted 'statistically', much as in the Born interpretation, and one representation *is* treated just like any other. Thus (B) is satisfied. (C) is the doubtful case; superposition of states is not *explained* at all, but simply assumed as primitive. However, this is certainly a unitary treatment, and the CI theorist would say that our demand for some further explanation is just an unsatisfiable hankering after classical models.

We see now, perhaps, why the CI has been so widely accepted in spite of its initial counterintuitiveness. In view of the need for rejecting ND, it seems pointless to talk of 'values' of observables at moments when no measurement of these observables is taking place. For if such 'values' exist, they cannot be assumed to be related to the values we find upon measurement even by the very weak assumption of *continuity*. So why assume that these values (hidden variables) exist? Even if such an argument may appear convincing to most physicists, I shall now try to show that the CI runs into difficulties of its own which are just as serious as the difficulties that beset previous interpretations.

To begin with, it is necessary to grant at once that the CI has made an important and permanent contribution to quantum mechanics. That the principle ND must be rejected, and that such statements as (3) and (4) must be reformulated in the way illustrated by (3') and (4'), is a part of quantum mechanics itself. To put it another way, it is a part of quantum mechanics itself as it stands today that the proper interpretation of the wave is statistical in this sense: the square amplitude of the wave is the probability that the particle will be found in the appropriate place *if a measurement is made* (and analogously for representations other than position representation). We might call this much the *minimal statistical interpretation* of quantum mechanics, and what I am saying is that the minimal statistical interpretation is a contribution of the great founders of the CI – Bohr and Heisenberg, building, in the way we have seen, on the earlier idea of Born – and a part of quantum mechanical theory itself. However, the minimal statistical interpretation is much less daring than the full CI. It leaves it completely open whether there are any observables for which the principle ND is correct, and whether or not hidden variables exist. The full CI, to put it another way, is the minimal statistical interpretation *plus* the statement that hidden variables do not exist and that the wave representation gives a *complete* description of the physical system.

Before actually discussing the CI, let us consider for a moment what a formalization of quantum mechanics might look like. The famous von Neumann axiomatization takes the term 'measurement' as primitive, and postulates that each measurement results in a 'reduction of the wave packet'. That the term 'measurement' is taken as primitive is no accident. Any formalization of quantum mechanics must either leave the question of interpretation open – in which case no testable predictions whatsoever will be derivable within the formalization, and we will have formalized only the *mathematics* of quantum mechanics, and not the physical theory – or we must include in the formalization at least the minimal statistical interpretation, in which case the term 'measurement' automatically enters.

As we remarked at the outset, however, it is not something that we can accept in the long run that the term 'measurement' should simply remain primitive in physical theory. Measurements are only a certain subclass of physical interactions; we have to demand that this subclass should be specified with sufficient precision, and without reference to anything subjective. To define a measurement as the apprehension of a fact by a human consciousness, for example, would be to interpret quantum mechanics as asserting a dependence of what exists on what human beings happen to be conscious of, which is absurd.

Bohr has proposed to meet this difficulty by saying, in effect, that an interaction (between a micro-system and a macro-system) is a measurement if the interaction is a measurement according to *classical physics*. Besides pushing the problem back to exactly the same problem in classical physics (where it was never solved because it never had to be solved – it never became necessary to have a definition of 'measurement' in *general*), the suggestion is completely unacceptable in the context of axiomatization. We can hardly refer to one theory (classical physics) in the axiomatization of another if the first theory is supposed to be incorrect and the second is designed to supersede it.

A more acceptable alternative might be as follows: let us define a measurement as an interaction in which a system A (the 'measured' system), which was previously isolated, interacts with a system B (the 'measuring' system) in such a way as to cause a change in B that affects some 'macro-observable' – some macroscopically observable property of B. This definition is, of course, incomplete. We also need to be told how to *interpret* measurements – i.e. how to determine what the change in a particular macroscopically observable property of B signifies. It is possible to do this, however, and in a way that does not go outside the formalism of quantum mechanics itself.† Thus we might arrive at a formalization of quantum mechanics (including the minimal statistical interpretation) that does not contain the term 'measurement' as primitive. But the term 'macro-observable' *will* now appear as primitive. Perhaps, however, this term will prove easier to define. It seems plausible that the macro-observables can be defined as certain averages of large numbers of micro-observables in very big systems.

Two more points before turning to the CI itself:

i. Quantum mechanics must, if correct, apply to systems of arbitrary size, since it applies both to the individual electrons and nuclei and also to the interactions between them. In particular, it must apply to macro-systems.‡

ii. Since we are considering only elementary, nonrelativistic quantum mechanics, we have no reason to consider the nature of the universe 'in the large'. We can perfectly well idealize by neglecting all force outside our solar system and by considering that we are dealing with

---

† Cf. chapter 8 of this volume.

‡ This has sometimes been denied on the ground that we cannot know the exact state of a macro-system. However, our *incomplete* knowledge of the state of a macro-system can be represented in quantum mechanics exactly as it was in classical physics – by giving a 'statistical mixture' of the various possible states of the system weighted by their probabilities (in the light of our information).

experiments that in the last analysis take place in *one* isolated system – our solar system.

Let us now consider the question: how might the CI be incorporated in the formalization of quantum mechanics itself? The simplest method is to modify the formation rules so that an observable does not necessarily exist (have any numerical value) at a time, and then to add a postulate specifying that an observable exists in a system A at a time if and only if a measurement of that observable takes place at that time. However, if A is the solar system, then no observable exists in A since we are assuming that our solar system is all there is. In fact, since not even macro-observables exist, if this is taken literally, then the solar system itself does not exist.

Let us consider the difficulty a little more closely. The difficulty is that measurement, as we proposed defining it, requires interaction with an *outside* system. If we introduce the assumption that all measurements ultimately take place in some 'big', isolated system (the 'universe', or the solar system, or whatever), then we immediately arrive at a contradiction, for macro-observables in the 'big' system are supposed to exist (and thus be measured), but there is no *outside* system to measure them.

We might try to avoid the difficulty by rejecting the idea of a 'biggest system', although given the present formalism of quantum mechanics, this also leads to mathematical contradictions.† But this is not much happier. What of macro-observables that are isolated for a long time, say, a system consisting of a rocket ship together with its contents out in interstellar space? We cannot seriously suppose that the rocket ship begins to exist only when it becomes once again observable from the earth or some other outside system.

In view of these difficulties, we very quickly see that there is only one reasonable way of formalizing quantum mechanics that does justice to the ideas of the founders of the CI. We cannot really suppose that the founders of the CI meant that *no* observable has a sharp value unless measured by an outside system. Indeed the repeated insistence of Bohr and Heisenberg on macrophysical realism, combined with their equally strong insistence that realism is untenable as a philosophical point of view with respect to micro-observables, leads us to the following, more careful formulation of the CI. Instead of saying henceforth that, according to the CI, observables do not exist unless measured, we shall have to say that, according to the CI, *micro*-observables do not exist unless measured. We shall take it as an assumption of quantum mechanics that *macro*-observables retain sharp values (by macroscopic

† Cf. chapter 8 of this volume.

standards of sharpness) at all times. The formulation of the CI that I am now suggesting, then, comes down to just this: that macro-observables retain sharp values at all times in the sense just explained, while micro-observables have sharp values only when measured, where measurement is to be defined as a certain kind of interaction between a micro-observable and a macro-observable.

This formulation would in all probability be unacceptable to Bohr and Heisenberg themselves. They would accept the statement that micro-observables do not have sharp values unless a measurement takes place. And they would agree that, as we have just seen, this principle has to be restricted to micro-observables. In this way they differ sharply from London and Bauer and, possibly, von Neumann, who would hold that all observables can have unsharp values unless measured by a 'consciousness'. Bohr and Heisenberg are macrophysical realists who hold that the *nature of the microcosm* is such that we cannot succeed in thinking about it in realistic terms. We are dealing with what has been called a 'nonclassical relation between the system and the observer'. But this has to do, in their view, with the special nature of the microcosm, not with any special contribution of human consciousness to the determination of physical reality. What Bohr and Heisenberg do not see, in my opinion, is that their interpretation has to be restricted to *micro*-observables from the very outset, and that the relative sharpness of macro-observables is then an underived assumption of their theory.

I shall say something later about the rather ambiguous position of the Copenhagen theorists (especially Heisenberg) with respect to the reduction of the wave packet. The central question in connection with the CI is whether the special character of macro-observables can be *derived* from some plausible definition of macro-observable together with a suitable formulation of the laws of quantum mechanics. In classical physics macro-observables were simply certain averages defined in very big systems. This definition will not do in quantum mechanics because the quantities of which macro-observables are supposed to be averages are not required always to 'exist', i.e. have definite numerical values, in quantum mechanics. A macro-observable in quantum mechanics has to be defined as an observable whose corresponding operator is in a certain sense an average of a number of operators. The question we face is whether from such a quantum-mechanical characterization of a macro-observable together with the laws of quantum mechanics it is possible to deduce that macro-observables always retain sharp values whether a measurement interaction involving them is going on or not. If we can do this, then the appearance of paradox and the *ad hoc* character of the CI will disappear.

In spite of a number of very ingenious attempts, it does not appear that this can be done. Briefly, what is needed is some demonstration that superpositions of states in which macro-observables have radically different values cannot be physically prepared. Unfortunately, it does not seem that there is anything in quantum mechanics to give this conclusion.

One fallacious attempt to obtain this conclusion has been via an appeal to what are known as the 'classical limit theorems' of quantum mechanics. These say that a system obeying the laws of classical physics and not subject to relevant microscopic uncertainties will behave in the same way in quantum mechanics as in classical physics. In other words, if we take a wholly classical system, say, a machine consisting of gears and levers, and translate the description of that system with great pains into the mathematical formalism of quantum mechanics, then the wave representation will yield the information that the various macro-observables defined in that system have sharp values. Moreover, as the state of the system is allowed to change with time, although the wave will spread out in the relevant space, this spreading out will happen in such a way that the macro-observables in question will retain sharp values for long periods of time. In short, if all macro-observables had sharp values to begin with, then they will retain sharp values during the period of time under consideration without our having to assume that any measurement took place.

This result shows that *in some cases* it follows from quantum mechanics that macro-observables will retain sharp values whether interacted with or not. Unfortunately, there are other cases in which a diametrically opposed conclusion follows, and these are of great importance. The most famous of these cases is the so-called 'Schrödinger's cat' case. Schrödinger imagined an isolated system consisting of an apparatus that contains a cat together with a device for electrocuting the cat. At a certain preset time, an emitter emits a single photon directed toward a half-silvered mirror. If the photon is deflected, then the cat will live; if the photon passes through the half-silvered mirror, then the switch will be set off and the cat will be electrocuted. The event of the photon passing through the half-silvered mirror is one whose quantum mechanical probability is exactly $\frac{1}{2}$. If this entire system is represented by a wave function, then prior to the time at which the emitter emits the photon – say, 12.00 – the wave that represents the state of the system will evolve in accordance with the classical limit theorems, that is to say, all macro-observables, including the ones that describe the behavior of the cat, will retain sharp values 'by themselves'. When the photon is emitted, however, the effect of the half-silvered mirror will be to split the wave corresponding to the photon into two parts. Half of the wave will go through the half-silvered mirror and half will be reflected. From this point on, the state of the system will be represented by a linear combination of two waves, one wave representing the system as it would be if the photon passed through the half-silvered mirror and the cat were electrocuted, and the other wave representing the state of the system as it would be if the photon were reflected and the cat were to live. Thus if the system is not interfered with prior to 1.00 p.m., then we will predict that at 1.00 p.m. the system will be in a state that is a superposition of 'live cat' and 'dead cat'. We then have to say that the cat is *neither alive nor dead* at 1.00 p.m. unless someone comes and looks, and that it is the act of looking that *throws the cat into a definite state*. This result would, of course, be contrary to the macro-physical realism espoused by the CI.

It should be observed that for all its fanciful nature, Schrödinger's cat involves a physical situation that arises very often, in fact one that arises in all quantum mechanical experiments. Almost every experiment in which quantum mechanics is employed is one in which some micro-cosmic uncertainty is amplified so as to affect something detectable by us human beings – hence detectable at the macro-level. Consider, for example, the operation of a Geiger counter. A Geiger counter affects a macro-observable, the audible 'clicks' that we hear, in accordance with phenomena (individual elementary particle 'hits') that are subject to quantum mechanical uncertainties. It is easily seen that if we describe any situation in which a Geiger counter is employed by a wave, then the result is analogous to the Schrödinger's cat case. Since the Geiger counter will click if a particle hits it and will not click if a particle does not hit it, then its state at the relevant instant will be represented by a wave that is a superposition of the waves corresponding to 'click' and 'no-click'.

It must be admitted that most physicists are not bothered by the Schrödinger's cat case. They take the standpoint that the cat's being or not being electrocuted should itself be regarded as a measurement. Thus in their view, the reduction of the wave packet takes place at 12.00, when the cat either feels or does not feel the jolt of electric current hitting its body. More precisely, the reduction of the wave packet takes place precisely when if it had not taken place a superposition of different states of some macro-observable would have been predicted. What this shows is that working physicists accept the principle that macro-observables always retain sharp values (by macroscopic standards of sharpness) and deduce when measurement *must* take place *from* this principle. But the intellectual relevance of the Schrödinger's cat case is not thereby impaired. What the case shows is that the principle that macro-observables retain sharp values at all times is not *deduced* from the foundations of quantum mechanics, but is rather dragged in as an additional assumption.

A number of physicists, most notably Ludwig, have reasoned as follows. 'All right', they say, 'let us suppose that it is possible to prepare a superposition of states in which macro-observables have radically different values. We know, as already remarked in connection with the two-slit experiment, that detectable interference effects require mutually coherent waves. Since the waves corresponding to different states of a macro-observable will be incoherent, no *detectable* interference effects will result, and hence (sic) macro-observables do always retain sharp values'.

To make this argument a little more respectable, we may construe Ludwig as unconsciously advocating that certain further assumptions be added to quantum mechanics. The CI involves what has been called 'the completeness assumption', that is to say, the assumption that the wave representation gives a *complete* description of the state of the physical system. If the completeness assumption is accepted without any restriction, then a state that is a superposition of two states assigning different values to some macro-observable could not and should not be confused with a state in which the system is really in one of the two substates but we do not know in which. *Physical* states should not be confused with states of ignorance, and there is all the difference in the world between a state that is a superposition of two states A and B (the superposition being construed as a third state C in which the macro-observable in question has no sharp value) and a *state of knowledge*, call it D, in which *we* are when we know that the system in question is either in state A or in state B, but we do not know which. Ludwig in effect argues: 'If a system is in state C, then since the interference effects are not detectable, we will get the same predictions for the outcomes of all practically possible observations as if we suppose that the system is either in state A or in state B, but we do not know which. So in that case, being pragmatic, let us say the system either is in state A or is in state B but we don't know which.' This argument is incoherent since it tacitly assumes the usual statistical interpretation of quantum mechanics, which in turn assumes that looking at the system throws it into some definite state.

If we say, for example, that the cat might have been in a neither-alive-nor-dead state until 1.00 p.m., but that due to the mutual incoherence of the waves corresponding to 'dead cat' and 'live cat' this leads to exactly the same predictions as the theory that the cat was thrown into a definite state by the event that took place at 12.00, then we are still left with the question with which we started: how does our looking affect the state of the cat? What is it about us and about our looking that makes this interaction so special? If the answer is that it is just that we are macrophysical systems, and that macro-observables always retain sharp values, then it is necessary to object that it is just this assumption which is abandoned when we assume that it may be possible to prepare superpositions of states that assign different values to some macro-observable. In other words, the assumption is rejected when it is assumed that prior to 1.00 p.m. the cat really is in a superposition of states – 'live cat' and 'dead cat' – but then this very assumption is used when it is assumed that the observer, upon looking, will find the cat either alive or dead. Why should not the result of this new interaction – my looking – rather be to throw me into a superposition of two states: 'Hilary Putnam seeing a live cat' and 'Hilary Putnam seeing a dead cat'?

The 'coherence' or 'destruction of interference' theory might perhaps be repaired in the following way. Suppose that we add to quantum mechanics the following assumption: that some wave descriptions are *complete* descriptions of the state of a physical system while others are *not*. In particular, *if a wave is a superposition of mutually incoherent waves* corresponding to two or more possible states of a system, *then that wave description gives incomplete information concerning the state of the system in question.* What we must say is that the system in question is really in one of the more precisely specified states, only we do not know which. Applied to the Schrödinger's cat case, this would mean saying that the interaction that took place at 12.00 sent the system into a superposition of two states, 'live cat' and 'dead cat'. This is a superposition of two states A and B whose corresponding waves are incoherent. Since the wave function that represents the state of the system after 12.00 is a superposition of this kind, we are required to say, by the new theory, that this wave function is an *incomplete* description of the state of the system, and that the cat is really either alive or dead. I think this is the most respectable interpretation of what the coherence theorists, Ludwig *et al.*, are really trying to do.

But there are grave difficulties with this suggestion. In particular, to say that a wave gives an incomplete description of the state of a system is to say that systems whose state can be described by that wave function may really be in several quite different physical conditions. Completeness versus incompleteness is not a matter of *degree*. On the other hand, coherence and incoherence of waves *are* a matter of degree. It seems that in principle a postulate of the kind suggested is going to be unworkable. Exactly what degree of incoherence of the superimposed waves is required before we are to say that a wave function gives an incomplete description of the state of a system? Secondly, it would have to be shown that the proposed added principle is consistent. Perhaps every wave function can be written as the superposition of two mutually incoherent waves. What is worse is that we might well be required to say that the descriptions were incomplete in *mutually incompatible ways*.

At this point, I would like briefly to discuss Heisenberg's remarks on the subject of the reduction of the wave packet. These remarks are puzzling, to say the least. First of all, Heisenberg emphatically insists that at least when macro-observables are involved the reduction of the wave packet is not a physical change, but simply an acquisition of information. What Heisenberg says is that any system which includes a macroscopic measuring apparatus *is never known to be in a particular pure state but only known to be in a mixture*. He does not elaborate the exact significance of this remark, and therefore it becomes incumbent upon us to ask just what it might mean and what its relevance is supposed to be in the present context.

Usually the statement that a system is in a mixture of states *A*, *B*, *C* is interpreted as meaning that it is really in one of the states *A*, *B*, or *C*, but we do not know which. I shall henceforth refer to this as the 'ignorance interpretation' of mixtures. If the ignorance interpretation of mixture is correct, then Heisenberg's remark is simply irrelevant. What Heisenberg is saying is that Schrödinger's cat starts out in some pure state and that we can give a large set of pure states *A*, *B*, *C*,... so that the state it starts out in must be one of these, but we cannot say which one. This is irrelevant because Schrödinger's argument did not depend on our *knowing* the state in which the cat started out. Suppose, however, the ignorance interpretation of mixtures is *incorrect*. In that case quantum mechanics involves two fundamentally different kinds of states, *pure states* and *mixtures*. We may also say that states may be combined in essentially different ways: on the one hand, two pure states *A* and *B* may be superimposed to produce a new state that will be a linear combination of *A* and *B*; and, on the other hand, the two states *A* and *B* may be 'mixed' to produce a state that will depend on *A* and *B*. To say that a system is in some particular mixture of the two states *A* and *B* will not be to say that it is really in one of the cases of the mixture, but to say that it is in a new kind of state not exactly representable by any wave but only by this new mathematical object – a mixture of wave functions. It is easily seen that giving up the ignorance interpretation of mixtures will not help us, for now we have the difficulty that when the observer looks at 1.00 p.m. he does not find the cat in a mixture of the states 'live cat' and 'dead cat' but in *one state or the other*.

If to say that the cat is in a mixture does not *mean* that the cat is already either alive or dead, then we have again the result that the observer's looking throws the cat from one kind of state – a mixture of superpositions of 'live cat' and 'dead cat'† – into another state, say 'live cat'. Put in another way, *selection* from a mixture is not usually regarded as analogous to reduction of the wave packet, because it is generally accepted that if a system is in a mixture then it must already be in one of the pure cases of the mixture. On this interpretation, if a system is known to be in a mixture and we look to see which of the cases of the mixture it is in, our looking does not in any way disturb the system. As soon as we give up the ignorance interpretation of mixtures, however, this process of selection becomes a discontinuous change in the state of the system, and all the anomalies connected with the reduction of the wave packet now arise in connection with the idea of selection from a mixture.

Let us now conclude by seeing what has happened. The failure of the original Born interpretation, we recall, led us to the conclusion that the principle ND is incorrect. The falsity of the principle ND was in turn the crucial difficulty for the hidden variable interpretations. In order to account for the falsity of the principle ND, these interpretations had to introduce special forces, e.g. quantum potential, for which no real evidence exists and with quite unbelievable properties. Since we cannot assume that the principle is true and that micro-observables, if they exist, are related to measured values even by the very weak assumption of continuity, we decided not to assume that micro-observables exist – i.e.

† This point needs emphasis. Suppose that the mixture representing the possible states of the cat at 12.00 p.m. is $M_1 = [s_1, s_2, \ldots]$. (Each s_i should be weighted by a probability, of course.) Then the mixture representing the state of the cat at 1.00 p.m. will be $M_2 = [U(s_1), U(s_2), \ldots]$, where U is a certain transformation function. Each $U(s_i)$ has the form '$\frac{1}{2}$ (Live cat)$+\frac{1}{2}$ (Dead cat); putting it roughly. Thus M_2 does *not* contain the information that 'the cat is either alive or dead, but we don't know which until we look'. *That* information would rather be given by a third mixture $M_3 = [t_1, t_2, \ldots]$, where each t_i describes a pure case in which the cat is alive or a pure case in which the cat is dead.
It can sometimes happen that two mixtures $[A_1, A_2, \ldots]$ and $[B_1, B_2, \ldots]$ have the same expectation values for *all* observables (same 'density matrix'). Such mixtures might be regarded as physical equivalents (although this would involve giving up the ignorance interpretation, and hence involve finding a new interpretation of *selection* from a mixture). However, the mixtures $M_2 = [U(s_1), U(s_2), \ldots]$ and $M_3 = [t_1, t_2, \ldots]$ do not have the same density matrix, no matter what probabilities may be assigned to the t_i (I plan to show this in a future paper). Thus this is not relevant here.
Lastly, it might be argued that M_2 and M_3 lead to *approximately* the same expectation values for all *macro*-observables, and hence [*sic*] describe the same physical states of affairs. It is easily seen that this is unsatisfactory (since the special status of *macro*-observables is presupposed, and not explained), and that otherwise this reduces to Ludwig's proposal, discussed above.

have sharp numerical values – at all times and to modify such statements as (3) and (4) in the way illustrated in (3′) and (4′) – that is to say, to adopt at least the minimal statistical interpretation of quantum mechanics. At this point we found ourselves in real difficulty with macro-observables. The result we wish is that although micro-observables do not necessarily have definite numerical values at all times, macro-observables do. And we want this result to come out of quantum mechanics in a natural way. We do not want simply to add it to quantum mechanics as an *ad hoc* principle. So far, however, attempts to derive this result have been entirely unsuccessful.

What, then, is to be done? We might try giving up the idea that micro-observables can have such things as unsharp values. However, this would not be of much help, for if we accept the minimal statistical interpretation and the falsity of principle ND, then we are not much better off than with the Copenhagen interpretation. It is true that we would then know that micro-observables always exist, and hence that certain macro-observables will always exist, namely, such macro-observables as are simply averages of large numbers of micro-observables; but not all macro-observables are so simple. The surface tension of a liquid, for example, is a macro-observable that ceases to exist if the swarm of particles in question ceases to be a liquid. If we cannot say anything about the values of micro-observables when we are not making measurements except *merely* that they exist, then we will not be able to say that the swarm of particles constituting, let us say, a glass of water, will continue to be a liquid when no measurement is made. Or in the Schrödinger's cat case, we will be able to say only that the swarm of particles making up the cat and the apparatus exists and that it will certainly take the form 'live cat' or the form 'dead cat' if a measurement is made, i.e. if somebody looks. We will not be able to say that the cat is either alive or dead, or for that matter that the cat is even a cat, as long as no one is looking. If we go back to the idea that micro-observables have numerical values at all times, then if we are to be able to handle the Schrödinger's cat case in a satisfactory way, we will have to say *more* about these numerical values of micro-observables at times when no measurement is made than merely that they exist. In short, we will be back in the search for a hidden variable theory.

In conclusion, then, *no* satisfactory interpretation of quantum mechanics exists today. The questions posed by the confrontation between the Copenhagen interpretation and the hidden variable theorists go to the very foundations of microphysics, but the answers given by hidden variable theorists and Copenhagenists are alike unsatisfactory. Human curiosity will not rest until those questions are answered, but whether they will be answered by conceptual innovations within the framework of the present theory or only within the framework of an as yet unforeseen theory is unknown. The first step toward answering them has been attempted here. It is the modest but essential step of becoming clear on the nature and magnitude of the difficulties.

8

Discussion: comments on comments on comments: a reply to Margenau and Wigner*

The Margenau and Wigner 'Comments' (1962) on my 'Comments on the Paper of David Sharp' (Putnam, 1961; Sharp, 1961) is a strange document. First the authors say, in effect, 'had anything been wrong (with the fundamentals of quantum mechanics) we should certainly have heard'. Then they issue various *obiter dicta* (e.g. the 'cut between observer and object' is unavoidable in quantum mechanics; the – highly subjectivistic – London–Bauer treatment of quantum mechanics is described, along with von Neumann's book, as 'the most compact and explicit formulation of the conceptual structure of quantum mechanics'). My assumption 2 (that the whole universe is a system) is described as 'not supportable', because 'the measurement is an interaction between the object and the observer'. The 'object' (the closed system) cannot *include* the observer.

The issues involved in this discussion are fundamental ones. I believe that the conceptual structure of quantum mechanics today is as unhealthy as the conceptual structure of the calculus was at the time Berkeley's famous criticism was issued. For this reason – as much to emphasize the seriousness of the present situation in the foundations of quantum mechanics as to remove confusions that may be left in the mind of the general reader upon reading the Margenau and Wigner 'Comments' – I intend to restate the main points of my previous 'Comments', and to show in detail why the Margenau and Wigner remarks fail completely to meet them.

1. The main point

Let S be a system which is 'isolated' (as well as possible) during an interval $t_0 < t < t_1$, and whose state at t_0 is known, let M be a measuring system which interacts with S so as to measure an observable O at t_1, and let T be the 'rest of the universe'. In quantum mechanics, a physical situation is described by giving two things: a Hamiltonian and a state function. The usual way of obtaining an approximate description of the

* First published in the *Philosophy of Science*, 31:1 (January 1964).

situation of the system S is simply to set the interaction of $M + T$ with S equal to zero for the interval $t_0 < t < t_1$. This, of course, is only an approximation – *rigorously*, the interaction between S and $M + T$ never completely vanishes, as Sharp and I both pointed out in our papers. What then is the *rigorous* description of the system S?

The answer, surprisingly, is that usual quantum mechanics provides *no* rigorous, contradiction-free account at all! (The parallel with the eighteenth century situation in the foundations of the calculus is surprisingly close: setting $dx = O$ *after* one has divided by dx 'works'. But *mathematically* this procedure is wholly unjustified, and it took the work of Weierstrauss and the development of the concept of a *limit* to provide a rigorous, thoroughly justifiable procedure.) In fact, if we take account of the fact that S is not *strictly* isolated (i.e. Hamiltonian (interaction between $M + T$ and S) $\neq O$), then, by an elementary calculation,† S cannot be assigned *any* state function. Also, since $M + T$ generates a field (however weak) which would have to be exactly known to describe the situation of S by means of a Hamilton an, and by quantum mechanics itself, one *cannot* exactly know this field, since one cannot know the simultaneous positions *and* momenta of its sources, S cannot be assigned a Hamiltonian either. So the 'approximation' made in quantum mechanics – setting Hamiltonian (interaction $M + T$ and S) $= O$ – is like the 'approximation' setting $dx = O$, and not like the legitimate approximations in classical mechanics, which can always in principle be dispensed with. It is an algorithm which 'works', but which has not, to date, been grounded in a consistent and, in principle, mathematically rigorous theory.

2. The Margenau–Wigner reply

Margenau and Wigner reply: 'Overall consistency of all parts of quantum mechanics, especially when that theory is forced to make reference to "the entire universe" has never been proven or claimed'. This is the only reference to the main point of my 'Comments', and it gives the erroneous impression that the point we have just reviewed depends on treating 'the entire universe' $(S + M + T)$ as a system with a ψ-function of its own.

3. Cosmological problems not relevant

Margenau and Wigner's phraseology – 'especially when that theory is *forced* to make reference to "the entire universe"' (italics mine) –

† Cf. Sharp 1961, p. 227, equation (4), and p. 230ff.

suggests that by 'the entire universe' I must have meant the cosmological universe and that I sought to embroil quantum mechanics in the problems of cosmology. Nothing could be wider of the mark. Footnote 1 of my paper made it clear that the question is whether quantum mechanics can consistently treat measurement as an interaction taking place within a single closed system (containing the observer). There is no objection to 'idealizing' by setting Hamiltonian $(T, M+S) = O$. After all, it is purely contingent that T is not just empty space. But it is *not* purely contingent that M is not just empty space: empty space cannot make measurements.

If we do attempt to treat all measurements – that is to say, all the measurements we are interested in – as taking place within one closed system (as we would in classical physics), then we *must* imagine that the 'rest of the universe', T, is just empty space, or at least that no measurements are carried out by observers in T upon $M+S$. Otherwise (1) the, main point (see above) is not taken care of at all, and (2) we are not imagining that *all* measurements relevant in the context take place in *one* closed system (which is the question at issue).

Margenau and Wigner write, 'In fact, if one wants to ascertain the result of the measurement, one has to observe the measuring apparatus, i.e. carry out a measurement on it'. As an argument against the 'one closed system' view this is worthless, since it *presupposes* that the observer is not *part* of M.

4. It is not true that 'the object cannot be the whole universe'

Margenau and Wigner also state that *von Neumann's axioms* for quantum mechanics are incompatible with the assumption that a closed system which contains the observer (the 'entire universe') is a system in the sense of quantum mechanics. It is true that if we make the assumption that 'measurement' involves the interaction of the system under consideration with an *outside* system, *then* we cannot also assume that 'the entire universe' is a system. Must we make this assumption? In my 'Comments', I suggested that it might be possible to give it up, but I did not give details. Since this is the point (the 'cut' between observer and object) that Margenau and Wigner say is central to all of quantum mechanics, I will now be more explicit on this point.

Let M and S be as before, and let T be empty (so that the 'entire universe' consists of $M+S$ for present purposes). Von Neumann postulates that when M measures an observable O in S, then S is thrown into a new state, an eigenstate of the observable O. *Which* eigenstate of O S is in is determined by M. According to Bohr, this is done in a wholly classical manner – that is, the process by which some macro-observable in M (say a pointer reading) comes to register the value corresponding to the O-state S is in can be explained by classical physics. In particular, M can be treated using *classical physics alone* – only S has to be described quantum mechanically. Of course, the 'cut' can be shifted – that is, a proper part of M (always including the observer) can be taken as the measuring system M', while the rest of M can be adjoined to S to make the new observed system S'. But, *however* we make the 'cut', the measured system S is thought of as obligingly 'jumping' into an eigenstate of O so that a classical system M can measure O in a purely classical way. This is not only implausible on the face of it, but inconsistent since S *cannot*, strictly speaking, have states of its own, as has already been pointed out. What *is* consistent and what also seems to avoid the whole difficulty, is to say that the interaction between M and S causes the *entire system $M + S$* to go into an eigenstate of O. In other words, assume: *O-measurement causes the entire universe to go into an eigenstate of O.*

This assumption is consistent with the mathematical formalism of quantum mechanics – in fact, more consistent than the assumption that *S alone* jumps into an eigenstate of O, as we have seen – and expresses the view that the measuring system is a *part* of the total system under consideration, and not an 'outside' system.

5. Quantum mechanics and classical physics

In the preceding section, I referred to a well-known peculiarity of the received interpretation of quantum mechanics (the so-called 'Copenhagen interpretation') – namely, S is ascribed a ψ-function, and treated according to the laws of quantum mechanics, while the measuring system or 'observer', M, is treated as a *classical object*. Thus quantum mechanics 'treats the world as consisting of two kinds of objects – classical objects and quantum mechanical objects – with the former measuring the latter', as I wrote in my previous paper.

Of course, any classical system can also be taken as the *object*, and *then* the laws of classical physics are forthcoming as special cases of the laws of quantum mechanics, in an appropriate limiting sense. However, according to the usual account, some *other* classical system has then to play the role of 'observer', and this other system has then to be treated classically. In their paragraph on this point, Margenau and Wigner refer to the classical limit theorems and the Bohr correspondence principle. But these imply only that any classical system can be treated as the object, which was never at issue. Indeed, this point has been made sharply and again and again by Bohr himself. E.g. 'The account of the experimental arrangement and the results of the observations must be expressed... with suitable application of the terminology of classical physics' (Bohr, 1951, p. 209). 'The quantum mechanics formalism represents a purely symbolic scheme permitting only predictions, on the lines of the correspondence principle, as to results obtainable *under conditions specified by means of classical concepts*' (p. 211, italics mine).

Specifically, the point that we neglect the 'atomic' (quantum mechanical) structure of the 'observer' is made by Bohr: 'The neglect of the atomic constitution of the measuring instruments themselves, in the account of actual experience, is equally characteristic of the applications of relativity and of quantum theory' (Bohr, 1951, p. 238). The Russian physicist Landau has recently gone so far as to argue that *it is not strictly true* (as is usually maintained – e.g. by Margenau and Wigner) that classical physics is reducible to quantum mechanics, for just this reason – that, although classical physics is *deduced* on the 'object' side of the 'cut', it is *assumed* on the 'observer' side. As we saw before (cf. 'the main point'), neglect of the 'atomic constitution' of $M+T$ is fundamental in even setting up a Hamiltonian for S.

How can we overcome this unsatisfactory state of affairs? London and Bauer would like to reduce the 'observer' to a disembodied 'consciousness', but Margenau and Wigner admit this is not yet successful. 'Present-day physics' [*sic!*] is not applicable to the 'consciousness'. The alternative suggested in the preceding section is much more direct and unmetaphysical. Namely, we should treat $M+S$ as a single closed system obeying the laws of quantum mechanics. If O is the observable being measured, and O' is the correlated *macro*-observable (e.g. the position of the pointer), then at the end of the interaction O and O' (considered now as observables in $M+S$, even though O depends only on S and O' only on M) will have *the same spectrum of eigen-functions*. These eigen-functions will have (approximately) the form $\psi_i \chi_i$, where ψ_i is an eigen-function of O in S and χ_i is the corresponding eigen-function of O' in M.† This is a purely *quantum mechanical* characterization of measurement – no use is made at all of classical physics or of the classical description of M. To complete the account, we need only postulate that the entire system $M+S$ goes into the state $\psi_i \chi_i$ with the corresponding probability $|c_i|^2$ – but no reference to 'classical concepts' is thereby introduced.

---

† Note that whether ψ_i or $\psi_i \chi_i$ is called an eigen-function of O depends on which Hilbert space one is using – the Hilbert space of S or of $S+M$. The statement that O and O' have the same spectrum is true only in the Hilbert space of $S+M$.

6. Remark on 'quantum jumps'

The standard interpretations of quantum mechanics accept the so-called 'Projection Postulate' – that measurement 'throws' a system into an eigenstate of the observable measured. In my paper, I included a brief argument for the necessity of this principle. Margenau and Wigner of course accept the conclusion – that one must postulate a process of measurement, distinct from and not reducible to 'motion' (continuous change of state, governed by the Schrödinger equation) in a single closed system – indeed, this is just their 'cut between the observer and the object'. However, they misunderstood my argument, which was, indeed, too briefly stated in my 'Comments'. My argument was just this: Let E be an electron which is *free*† during the interval $t_0 < t < t_1$. Suppose E has a sharp position at t_0 and at t_1, and that 'the whole business' – the position measurement at t_0, the free movement of the electron in the interval $t_0 < t < t_1$, and the position measurement at t_1 – is treated as a case of 'motion' in one closed system. Then the state function of the whole system must be an eigen-function of the position of E at t_0 and at t_1. On the other hand, the electron E is not interacting with the rest of the system during $t_0 < t < t_1$, so the state function of the whole system must have the form $\psi\phi$, during the interval $t_0 < t < t_1$, where ψ is the state function of a free electron (subject to the constraint that $\psi(q_1, q_2, q_3, t_0)$ is a δ-function), and ϕ is the state function of the rest of the system. But $\psi(q_1, q_2, q_3, t)$ is spread out over all space at every $t > t_0$, (in nonrelativistic quantum mechanics) so that the state function of the whole system cannot be an eigen-function of the position of E at t_1 ('reduction of the wave packet') except by possessing a discontinuity at t_1. In a nutshell, the Schrödinger equation is first order in the time, and thus we are not free to impose boundary conditions at two or more different times. This is a perfectly correct argument to a conclusion Margenau and Wigner accept (the need for the Projection Postulate) from premises they accept.

However, the argument loses part of its force if we renounce my assumption 1 (which is just the 'cut' assumption) and revise the Projection Postulate (as suggested above) to say that measurement sends $M+S$, and not just S, into an eigenstate of the observable measured. In this case it is still true that we cannot say the Schrödinger equation is obeyed when $t_0 < t < t_1$, except at the price of introducing by 'fiat'

---

† In my 'Comments' (Putnam, 1961) I neglected to say that E should not interact with the rest of the system during the interval $t_0 < t < t_1$, and my use of the phrase 'any physically realizable Hamiltonian' unfortunately gave rise to the contrary impression. The point was that the above argument is quite independent of the nature of the interaction between M and E at t_0 and at t_1.

(the revised Projection Postulate) a 'reduction of the wave packet' at t_1; however, we *can* say that 'the whole business' – *including* the applications of the Projection Postulate – takes place in a single closed system which contains the observer.

A defect of my interpretation is that it does not *explain* just why and how measurement (construed as a physical process, in my interpretation) causes a 'reduction of the wave packet'. However, the London–Bauer interpretation is subject to even worse defects. On their interpretation the measuring system is always outside the system S and includes a 'consciousness'. However London and Bauer do not go so far as to make it *just* a 'consciousness' – it must also have a 'body', so to speak. Thus the *main point* applies in full force to this interpretation. Ignoring the interaction between M and S prior to the measurement is not just a useful 'approximation', but is indispensable in this theory. Secondly, the 'reduction of the wave packet' depends on 'measurement' which is *ultimately just the 'direct awareness' of a fact by a 'consciousness'*, in this interpretation. *Subjective* events (the perceptions of an 'observer') *cause* abrupt changes of *physical* state ('reduction of the wave packet'). *Questions:* What evidence is there that a 'consciousness' is *capable* of changing the state of a physical system except by interacting with it physically (in which case an automatic mechanism would do just as well)? By what *laws* does a consciousness cause 'reductions of the wave packet' to take place? By virtue of what *properties*† that it possesses is 'consciousness' able to affect Nature in this peculiar way? No answer is forthcoming to any of these questions.

Ideally, perhaps, we would prefer a theory which was free of the need for postulating 'quantum jumps'. However, if we *are* going to accept the Projection Postulate, the theory suggested here will do – there is neither reason for nor plausibility in making quantum mechanics dependent upon an inconsistency producing 'cut between observer and object' or upon 'consciousness'.

† I am indebted to Abner Shimony for raising this question.

9
Three-valued logic*

Let us make up a logic in which there are three truth-values, T, F, and M, instead of the two truth-values T and F. And, instead of the usual rules, let us adopt the following:

(a) If either component in a disjunction is true (T), the disjunction is true; if both components are false, the disjunction is false (F); and in all other cases (both components middle, or one component middle and one false) the disjunction is middle (M).

(b) If either component in a conjunction is false (F), the conjunction is false; if both components are true, the conjunction is true (T); and in all other cases (both components middle, or one component middle and one true) the conjunction is middle (M).

(c) A conditional with true antecedent has the same truth-value as its consequent; one with false consequent has the same truth-value as the denial of its antecedent; one with true consequent or false antecedent is true; and one with both components middle (M) is true.

(d) The denial of a true statement is false; of a false one true; of a middle one middle.

These rules are consistent with all the usual rules with respect to the values T and F. But someone who accepts *three* truth values, and who accepts a notion of tautology based on a system of truth-rules like that just outlined, will end up with a different stock of tautologies than someone who reckons with just two truth values.

Many philosophers will, however, want to ask: *what could the interpretation of a third truth-value possibly be?* The aim of this paper will be to investigate this question. It will be argued that the words 'true' and 'false' have a certain 'core' meaning which is *independent* of *tertium non datur*, and which is capable of precise delineation.

(1)

To begin with, let us suppose that the word 'true' retains at least this much of its usual force: if one ever says of a (tenseless) statement that it is true, then one is committed to saying that it was always true and will

* First published in *Philosophical Studies*, 8 (Oct. 1957), 73–80.

always be true in the future. E.g. if I say that the statement 'Columbus crosses† the ocean blue in fourteen hundred and ninety-two' is true, then I am committed to the view that it *was* true, e.g. in 1300, and will be true in A.D. 5000. Thus 'true' cannot be identified with *verified*, for a statement may be verified at one time and not at another. But if a statement is *ever* accepted as verified, then at that time it must be said to have been true also at times when it was not verified.

Similarly with 'false' and 'middle'; we will suppose that if a statement is ever called 'false', then it is also said never to have been true or middle; and if a statement is ever said to be middle, it will be asserted that it was middle even at times when it may have been incorrectly called 'true' or 'false'. In other words, we suppose that 'true' and 'false' have, as they ordinarily do have, a *tenseless* character; and that 'middle' shares this characteristic with the usual truth-values.

This still does not tell one the 'cash value' of calling a statement 'middle'. But it does determine a portion of the syntax of 'middle', as well as telling one that the words 'true' and 'false' retain a certain specified part of *their* usual syntax. To give these words more content, we may suppose also, that, as is usually the case, statements that are accepted‡ as verified are called 'true', and statements that are rejected, that is whose denials are accepted are called 'false'. This does not determine that any particular statements must be called 'middle'; and, indeed, someone could maintain that there are some statements which have the truth-value middle, or some statements which could have the truth-value middle, without ever specifying that any particular statement has this truth-value. But certain limitations have now been imposed on the use of the word 'middle'.

In particular, statements I call 'middle' must be ones I do not accept or reject at the present time. However, it is not the case that 'middle' *means* 'neither verified nor falsified at the present time'. As we have seen, 'verified' and 'falsified' are *epistemic* predicates, – that is to say, they are relative to the *evidence* at a particular time – whereas 'middle', like 'true' and 'false' is not relative to the evidence. It makes sense to say that 'Columbus crosses the ocean blue in fourteen hundred and ninety-two' was verified in 1600 and not verified in 1300, but not that it was true in 1600 and false in 1300.

Thus 'middle' cannot be *defined* in terms of 'verified', 'falsified', etc. What difference does it make, then, if we say that some statements – in particular some statements not now known to be true or known to be

---

† 'crosses' is used here 'tenselessly' – i.e. in the sense of 'crossed, is crossing, or will cross'.

‡ More precisely: S is accepted if and only if 'S is true' is accepted.

false – may *not* be either true or false because they are, in fact, middle? The effect is simply this: that one will, as remarked above, end up with a different stock of tautologies than the usual.

Someone who accepts the 3-valued logic we have just described will accept a disjunction when he accepts either component, and he will reject it when he rejects both components. Similarly, he will accept a conjunction when he accepts both components, and he will reject it when he rejects either component. This is to say that the behavior of the man who uses the particular 3-valued logic we have outlined is not distinguishable from the behavior of the man who uses the classical 2-valued logic in cases wherein they know the truth or falsity of all the components of the particular molecular sentences they are considering.

However, they will behave differently when they deal with molecules some of whose components have an unknown truth-value.† If it is known that snow is white, then the sentence 'snow is white *v*. ~ snow is white' will be *accepted* whether one uses classical 2-valued logic or the particular 3-valued logic we have described. But if one does not know whether or not there are mountains on the other side of the moon, then one will *accept* the sentence 'there are mountains on the other side of the moon *v*. ~ there are mountains on the other side of the moon' if one uses the classical 2-valued logic, but one will say 'I don't know whether that's true or not' if one uses 3-valued logic, or certain other non-standard logics, e.g. 'Intuitionist' logic.‡

(2)

At this point the objection may be raised: 'but then does this notion of a "middle" truth-value make sense? If having a middle truth-value does not mean having what is ordinarily called an *unknown* truth value; if, indeed, you can't tell us *what* it does mean; then does it make sense at all?'

Analytic philosophers today normally reject the demand that concepts be translatable into some kind of 'basic' vocabulary in order to be meaningful. Yet philosophers often reject the possibility of a 3-valued logic (except, of course, as a mere formal scheme, devoid of interesting

† The distinction between sentences and statements will be ignored, because we have passed over to consideration of a formalized language in which it is supposed that a given sentence can be used to make only one statement.

‡ Church (1956), p. 141. Intuitionist logic is not a truth-functional logic (with any finite number of truth-values). However, the rules given above hold (except when both components are 'middle' in the case of rules (b) and (c)) provided truth is identified with intuitionist 'truth', falsity with 'absurdity' and middlehood with being neither 'true' nor 'absurd'.

interpretations), just on the ground that no satisfactory *translation* can be offered for the notion of having a 'middle' truth-value. Indeed, if the notion of being a statement with a middle truth-value is defined explicitly in terms of a 2-valued logic or metalogic, then one usually obtains a *trivial* interpretation of 3-valued logic.

Does a middle truth-value, within the context of a system of 3-valued logic of the kind we have described, have a use? The answer is that it does, or rather that it belongs to a *system* of uses. In other words, to use 3-valued logic makes sense in the following way: to use a 3-valued logic means to adopt a different way of using logical words. More exactly, it corresponds to the *ordinary* way in the case of molecular sentences in which the truth-value of all the components is known (i.e. we '2-valued' speakers say it is known); but a man reveals that he is using 3-valued logic and not the ordinary 2-valued logic (or partially reveals this) by the way he handles sentences which contain components whose truth-value is not known.

There is one way of using logical words which constitutes the ordinary 2-valued logic. If we are using 3-valued logic,† we will behave in exactly the same way except that we will employ the 3-valued rules and the 3-valued definition of 'tautology'. Thus 'using 3-valued logic' means adopting a systematic way of using the logical words which agrees in certain respects with the usual way of using them, but which also disagrees in certain cases, in particular the cases in which truth-values are unknown.

(3)

Of course one might say:

'Granted that there is a consistent and complete way of using logical words that might be described as "employing a 3-valued logic". But this alternative way of using logical words – alternative to the usual way – doesn't have any *point*.'

And perhaps this is what is meant when it is said that 3-valued logic does not constitute a real alternative to the standard variety: it exists as a calculus; and perhaps as a non-standard way of using logical words; but there is no *point* to this use. This objection, however, cannot impress anyone who recalls the manner in which non-Euclidean geometries were first regarded as absurd; later as mere mathematical games; and are today accepted as portions of fully interpreted physical hypotheses. In exactly

† In this paper, '3-valued logic' means the system presented at the beginning. Of course, there are other systems, some of which represent a more radical change in our way of speaking.

the same way, 3-valued logic and other non-standard logics had first to be shown to exist as consistent formal structures; secondly, uses have been found for some of them – it is clear that the Intuitionist school in mathematics, for example, *is*, in fact, systematically using logical words in a non-standard way, and it has just been pointed out here that one might use logical words in still another non-standard way, corresponding to 3-valued logic (that is, that this would be a form of linguistic behavior reasonably represented by the formal structure called '3-valued logic'). The only remaining question is whether one can describe a physical situation in which this use of logical words would have a point.

Such a physical situation (in the microcosm) has indeed been described by Reichenbach (Reichenbach, 1944). And we can imagine worlds such that even in *macroscopic* experience it would be physically impossible to either verify or falsify certain empirical statements. E.g. if we have verified (by using a speedometer) that the velocity of a motor car is such-and-such, it might be impossible in such a world to verify or falsify certain statements concerning its position at that moment. If we *know* by reference to a physical law together with certain observational data, that a statement as to the position of a motor car can *never* be falsified or verified, then there may be some point to not regarding the statement as true or false, but regarding it as 'middle'. It is only because, in macroscopic experience, everything that we regard as an empirically meaningful statement seems to be at least potentially verifiable or falsifiable that we prefer the convention according to which we say that every such statement is either true or false, but in many cases we don't know which.

Moreover, as Reichenbach shows, adopting a 3-valued logic permits one to preserve both the laws of quantum mechanics and the principle that no causal signal travels with infinite speed – 'no action at a distance'. On the other hand, the laws of quantum mechanics – both in the form of 'wave' mechanics and in the form of statistical 'particle' mechanics – are logically incompatible with this principle if ordinary 2-valued logic is used (Reichenbach, 1944, pp. 29–34). This inconsistency is not usually noticed, because no causal signal is ever *detected* travelling faster than light, in quantum mechanics. Nevertheless it can be shown – as Einstein and others have also remarked (Einstein, 1935, p. 777) – that the mathematics of quantum mechanics entails that in certain situations a causal signal *must have* travelled faster than light.

A working physicist can dismiss this as 'just an anomaly' – and go on to accept *both* quantum mechanics and the 'no action' principle. But a logician cannot have so cheerful an attitude towards logical inconsistency. And the suggestion advanced by Bohr, that one should classify the trouble-making sentences as 'meaningless' (complementarity) involves its own complications. Thus the suggestion of using a 3-valued logic makes sense in this case, as a move in the direction of simplifying a whole system of laws.

Returning to the macrocosmic case (i.e. the 'speedometer' example), Bohr's proposal amounts to saying that a syntactically well-formed sentence (e.g. 'my car is between thirty and thirty-one miles from New York') is in certain cases *meaningless* (depending on whether or not one looks at the speedometer). Reichenbach's suggestion amounts to saying that it is meaningful, but neither true nor false (hence, 'middle'). There seems little doubt that it would be simpler in practice to adopt Reichenbach's suggestion. And I suspect that beings living in a world of the kind we have been describing would, in fact, regard such statements as *neither true nor false*, even if no consideration of preserving simple physical laws ('no action at a distance') happened to be involved. This 'suspicion' is based on two considerations: (a) the sentences admittedly have a very clear cognitive use; hence it is unnatural to regard them as 'meaningless'; (b) there is no reason why, in such a world, one should even consider adopting the rule that every statement is either true or false.

On the other hand, in our world (or in any world in which Planck's constant h has a small value) it would be very unnatural to adopt 3-valued logic for describing ordinary macrocosmic situations. For suppose we did. Then there would be two possibilities: (i) we maintain that certain sentences are 'middle', but we never say which ones – but this seems disturbingly 'metaphysical'; (ii) we say that some particular sentence S is middle.

This last course is, however, fraught with danger. For, although 'S is middle' does not *mean* 'S will never be either verified or falsified', it *entails* 'S will never be either verified or falsified'. And the prediction that a particular sentence will never be either verified or falsified is a strong empirical prediction (attention is confined to synthetic sentences for the sake of simplicity); and one that is itself always potentially falsifiable in a world where no physical law prohibits the verification of the sentence S, regardless of what measurements may have antecedently been made.

Thus, the reason that it is safe to use 3-valued logic in the Reichenbachian world (the microcosm) but not in the 'actual' world (the macrocosm) is simply that in the Reichenbachian world one can, and in the 'actual' world one cannot, *know in advance* that a particular sentence will never be verified or falsified. It is not that in a 'Reichenbachian' world one *must* call sentences that will never be verified or falsified 'middle' – but, rather, that in *any* world *only* (but not necessarily *all*) such sentences must be classified as 'middle'. This follows from the fact that sentences that are said to be verified are also said to be true; sentences that are said to be falsified are also said to be false; and the truth values are 'tenseless'. Thus it would be a contradiction to say that a sentence is middle, but may someday be verified.

These features of the use of 'true' and 'false' seem indeed to be constitutive of the meaning of these words. *Tertium non datur* might also be said to be 'true from the meaning of the words "true" and "false"' – but it would then have to be added that these words have a certain core meaning that can be preserved even if *tertium non datur* is given up. One can abandon 2-valued logic without changing the meaning of 'true' and 'false' *in a silly way*.

(4)

Analytic philosophers – both in the 'constructivist' camp and in the camp that studies 'the ordinary use of words' – are disturbingly unanimous in regarding 2-valued logic as having a privileged position: privileged, not just in the sense of corresponding to the way we *do* speak, but in the sense of having no serious rival for *logical* reasons. If the foregoing analysis is correct, this is a prejudice of the same kind as the famous prejudice in favor of a privileged status for Euclidean geometry (a prejudice that survives in the tendency to cite 'space has three dimensions' as some kind of 'necessary' truth). One can go over from a 2-valued to a 3-valued logic without *totally* changing the meaning of 'true' and 'false'; and not just in *silly* ways, like the ones usually cited (e.g. equating truth with high probability, falsity with low probability, and middlehood with 'in between' probability).

Indeed, so many strange things have been said about 2- and 3-valued logic by philosophical analysts who are otherwise of the first rank that it would be hopeless to attempt to discuss them all in one short paper. But two of these deserve special mention:

(i) It has often been said that 'even if one uses a 3-valued object language, one must use 2-valued logic in the metalanguage'. In the light of the foregoing, this can hardly be regarded as a *necessary* state of affairs. 3-valued logic corresponds to a certain way of speaking; there is no difficulty in speaking in that way about any particular subject matter. In particular, one may assign truth-values to molecular sentences in the way we have discussed, whether one is talking about rabbits or languages or metalanguages.

(Of course, if one is *explaining* 3-valued logic to someone who only uses 2-valued logic one will employ a 2-valued language as a medium of communication. This is like remarking that one uses French to teach Latin to French schoolboys.)

(ii) It has been argued† that the meaning of 'true' has been made clear by Tarski for the usual 2-valued system, but that no analogous clarification is available for 'true' in 3-valued systems. The obvious reply is that the famous biconditional: *'snow is white' is true if and only if snow is white* – is perfectly acceptable even if one uses 3-valued logic. Tarski's criterion has as a consequence that one must accept *'snow is white' is true* if one accepts *snow is white* and reject *'snow is white' is true* if one rejects *snow is white*. But these (along with the 'tenseless' character of the truth-values) are just the features of the use of 'true' and 'false' that we have preserved in our 3-valued logic. It is, for instance, just because *tertium non datur* is independent of these features that it is possible for intuitionist logicians to abandon it without feeling that they are changing the 'meaning' of 'true' and 'false'.‡

† E.g. Hempel writes in his review of the previously cited work by Reichenbach (*Journal of Symbolic Logic*, x, p. 99): 'But the truth-table provides a (semantical) interpretation only because the concept of *truth* and *falsity*, in terms of which it is formulated, are already *understood*: they *have the customary meaning* which can be stated in complete precision by means of the *semantical definition of truth*.' (Italics mine.)

‡ I no longer believe Reichenbach's proposal is satisfactory (cf. p. 83 of this volume); but the idea of using a non-standard logic to avoid the antinomies is, I believe, correct. The appropriate logic is described in chapter 10.

10

The logic of quantum mechanics*

I want to begin by considering a case in which 'necessary' truths (or rather 'truths') turned out to be falsehoods: the case of Euclidean geometry. I then want to raise the question: could some of the 'necessary truths' of logic ever turn out to be false *for empirical reasons*? I shall argue that the answer to this question is in the affirmative, and that logic is, in a certain sense, a natural science.

I. The overthrow of Euclidean geometry

Consider the following assertion (see Figure 1): two straight lines AB and CD are alleged to come in from 'left infinity' in such a way that, to the left of EF, their distance apart is constant (or, at any rate, 'constant on the average'), while after crossing EF they begin to converge – i.e. their distance apart diminishes – without its being the case that they bend at E or F (i.e. they are really straight, not just 'piecewise straight').

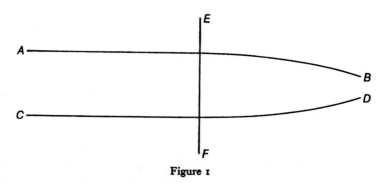

Figure 1

Is it not 'intuitively clear' that this is a contradiction? If one did not know anything about non-Euclidean geometry, relativity, etc., would the intuitive evidence that this is a contradiction, an impossibility, a complete absurdity, etc., be any less than the intuitive evidence that no

* First published as 'Is logic empirical?' in R. Cohen and M. Wartofsky (eds.), *Boston Studies in the Philosophy of Science*, 5 (Dordrecht, Holland, D. Reidel 1968).

surface can be scarlet (all over) and bright green at the same time? Or that no bachelor is married? Is the intuitive 'feeling' of contradiction really different in the following three cases?

(1) Someone claims that *AB* and *CD* are both *straight lines*, notwithstanding their anomalous behavior.

(2) Someone claims that he has a sheet of paper which is scarlet (all over, on both sides) and green (all over, on both sides) at once.

(3) Someone claims that some men are married (legally, at the present time) and nonetheless still *bachelors*.

It seems to me that it is not. Of course, (1) does not involve a 'contradiction' in the technical sense of formal logic (e.g. '$p \cdot -p$'); but then neither does (2), nor, unless we stipulate the definition 'Bachelor = man who has never married', does (3). But then 'contradiction' is often employed in a wide sense, in which not all contradictions are of the form '$p \cdot -p$', or reducible to that form by logic alone.

The important thing, for present purposes, is this: according to General Relativity theory, the claim mentioned in (1) is *possible*, not *impossible*. *AB* and *CD* could both be *straight paths* in the strict sense: that is, for no *P*, *P'* on *AB* (or on *CD*) is there a shorter way to travel from *P* to *P'* than by sticking to the path *AB* (respectively, *CD*). If we are willing to take 'shortest distance between two points' as the defining property of straight lines, then *AB* and *CD* could both be straight lines (in technical language: *geodesics*), notwithstanding their anomalous behavior.

To see this, assuming only the barest smattering of relativity: assume space is Euclidean 'in the large' – i.e. the average curvature of space is zero. (This is consistent with the General Theory of Relativity.) Then two geodesics could well come in from 'left infinity' a constant distance apart on the average ('on the the average' mind you! – and I am speaking about straight lines!). Suppose these two geodesics – they might be the paths of two light rays approaching the sun on opposite sides† – enter the gravitational field of the sun as they cross *EF*. Then, according to GTR, the geodesics – not just the light, but the very geodesics, whether light is actually travelling along them or not – would behave as shown in Figure 1.

Conclusion: what was yesterday's 'evident' impossibility is today's possibility (and even *actuality* – things as 'bad' as this actually happen,

---

† The physics of this example is deliberately oversimplified. In the GTR it is the *four-dimensional* 'path' of the light-ray that is a geodesic. To speak of (local) 'three' dimensional space' presupposes that a local reference system has been chosen. But even the geodesics in three-dimensional space exhibit non-Euclidean behavior of the kind described.

according to the GTR – indeed, if the average curvature of space is *not* zero, then 'worse' things happen!)

If this is right, and I believe it *is* right, then this is surely a matter of some philosophical importance. The whole category of 'necessary truth' is called into question. Perhaps it is for this reason that philosophers have been driven to such peculiar accounts of the foundations of the GTR: that they could not believe that the obvious account, the account according to which 'conceptual revolutions' can overthrow even 'necessary truth', could possibly be correct. But it *is* correct, and it is high time we began to live with this fact.

II. Some unsuccessful attempts to dismiss the foregoing

One way to dismiss the foregoing is to deny that 'straight line' ever meant 'geodesic'. But then one is forced to give up 'a straight line is the shortest path', which was surely a 'necessary truth' too, in its day. Moreover, if the geodesics are not the 'straight' paths in space, then which paths in space are 'straight'? Or will one abandon the principle that there are straight paths?

Again, one might try the claim that 'distance' (and hence 'shortest path') has 'changed its meaning'. But then what is the 'distance' between, say, A and B *in the old sense*? And what path between A and B is *shorter* than AB *even in the old sense*? No matter what path one may select (other than AB) as the 'really' straight one (i.e. the one that was 'straight' before the alleged 'meaning change'), it will turn out that one's path AGB will not *look* straight *when one is on it*, will not *feel* straight, as one travels along it, and will measure *longer*, not shorter, than AB by any conventional method of measurement. Moreover, there will be no nonarbitrary ground for preferring AGB over $AG'B$, $AG''B$ (Figure 2) ... as the 'really straight' path from A to B.

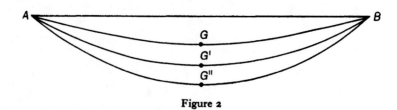

Figure 2

This brings out, indeed, one of the important facts about a conceptual revolution: such a revolution cannot successfully be dismissed as a mere case of *relabeling*. It is easy to see why not. The old scientific use of the term 'straight line' rested on a large number of laws. Some of these laws were laws of pure geometry, namely, Euclid's. Others were principles from physics, e.g. 'light travels in straight lines', 'a stretched thread will lie in a straight line', etc. What happened, in a nutshell, was that this cluster of laws came apart. If there are any paths that obey the pure geometrical laws (call them 'E-paths'), they do not obey the principles from physics, and the paths that do obey the principles from physics – the geodesics – do not obey the old principles of pure geometry. In such a situation, it scarcely makes sense to ask 'what paths are straight in the old sense'? We cannot say that the E-paths are straight, because they are not *unique*; there is no *one* way of picking out paths in space and calling them 'straight' which preserves Euclidean geometry: either there is *no* way of doing this, or there are infinitely many. We can say that the geodesics are straight, because they at least obey what were always recognized to be the operational constraints on the notion of a straight line; but they do not obey the old geometry. In short, either we say that the geodesics are what we always meant by 'straight line', or we say that there is nothing clear that we used to mean by that expression. (Or, perhaps, that what answered to the expression were not-too-long *pieces* of geodesics; that the old notion of 'straight line' cannot successfully be applied 'in the large'.) But in neither case is it correct to say: it is just a matter of shifting the 'label' 'straight line' from one set of paths to another. A good maxim in this connection would be: *Do not seek to have a revolution and minimize it too.*

On the other hand, one is not committed by the foregoing to denying that 'straight line' changes its meaning as one goes over to the GTR. Perhaps one wants to say it does, perhaps one does not want to say this. That is a question about the best way to use the expression 'change of meaning' in a very difficult class of cases. The important thing is that it does not 'change its meaning' in the trivial way that one might at first suspect. Once one appreciates that something that was formerly literally unimaginable has indeed happened, then one also appreciates that the usual 'linguistic' moves only help to distort the nature of the discovery, and not to clarify it.

III. The logic of quantum mechanics

In classical physics, the state of a system S consisting of N particles $o_1, \ldots, o_N$ is specified by giving the $3N$ position coordinates and the $3N$ momentum coordinates. Any reasonably well-behaved mathematical function of these $6N$ quantities represents a possible physical 'magnitude' $m(s)$. Statements of the form $m(s) = r$ – 'the magnitude m has the value r in the system S' – are the sorts of statements we shall call *basic physical propositions* here.

The basic mathematical idea of *quantum mechanics* is this: a certain infinite dimensional vector space $H(s)$ is coordinated to each physical system S, and each basic physical proposition is coordinated to a subspace of the vector space. In the case of what is called a 'nondegenerate' magnitude $m(s)$, the subspace corresponding to the basic physical proposition $m(s) = r$ is a one-dimensional one, say V_r, and the one-dimensional subspaces V_r 'span' the whole space. The reader may easily picture this situation as follows: pretend the space $H(s)$ is a finite dimensional space. (Indeed, nothing is lost if we pretend for now that all physical magnitudes have finitely many values, instead of continuously many, and in such a world the space $H(s)$ would be just an ordinary finite dimensional space.) Then to each number r which is a possible value of such a physical magnitude as position (remember that we are pretending that there are only *finitely* many such!), there corresponds a single *dimension* of the space $H(s)$ – i.e. V_r is simply a *straight line* in the space – and the lines V_r corresponding to all the possible values of, say, position form a possible coordinate system for the space. (In fact, they meet at right angles, as good coordinates should.) If we change from one physical magnitude $m_1(s)$ to a different magnitude $m_2(s)$ (say, from *position* to *momentum*) then the new coordinates V'_r will be inclined at an angle to the old, and will not coincide with the old. But each possible momentum r will correspond to a straight line V'_r, though not to a straight line which coincides with any one of the lines V_r corresponding to a possible *position* r.

So far we have said that there exists a 'prescription' for coordinating basic physical propositions to subspaces of the space $H(s)$. This mapping can be extended to compound statements by the following rules (here, let $S(p)$ be the space corresponding to a proposition p):

$$S(p \vee q) = \text{the } span \text{ of the spaces } S(p) \text{ and } S(q), \qquad (1)$$
$$S(pq) = \text{the } intersection \text{ of the spaces } S(p) \text{ and } S(q), \qquad (2)$$
$$S(-p) = \text{the orthocomplement of } S(p). \qquad (3)$$

These rules can also be extended to quantifiers, since, as is well known, the existential quantifier works like an extended disjunction and the universal quantifier works like an extended conjunction.

These rules, however, *conflict with classical logic*. To see this, let $r_1, r_2, \ldots, r_n$ be all the possible values of some 'non-degenerate' magnitude. Then the straight lines $V_{r_1}, \ldots, V_{r_n}$ 'span' the whole space $H(s)$ – i.e. the number n is also the number of dimensions of $H(s)$, and any vector in the space $H(s)$ can be expressed as a combination of vectors in these directions V_{r_i}. Since the 'span' of $V_{r_1}, \ldots, V_{r_n}$ is the whole space, the statement:

$$m(s) = r_1 \vee m(s) = r_2 \vee \cdots \vee m(s) = r_n \qquad (1)$$

(where m is the magnitude in question) is an always-true statement.

Now let m' be any magnitude such that the straight line V'_r representing the statement $m'(s) = r$ (where r is some real number) does not coincide with any of the n lines $V_{r_1}, V_{r_2}, \ldots, V_{r_n}$ (such a magnitude can always be found). The statement:

$$m'(s) = r \cdot [m(s) = r_1 \vee \cdots \vee m(s) = r_n] \qquad (2)$$

corresponds to the intersection of V'_r with the whole space, and this is just V'_r. Thus (2) is equivalent to:

$$m'(s) = r. \qquad (3)$$

On the other hand, consider the disjunction

$$[m'(s) = r \cdot m(s) = r_1] \vee [m'(s) = r \cdot m(s) = r_2] \vee \cdots$$
$$\vee [m'(s) = r \cdot m(s) = r_n] \qquad (4)$$

Each term in this disjunction corresponds to the o-dimensional subspace (the origin), which we consider to represent the always-false proposition. For a typical term $[m'(s) = r \cdot m(s) = r_i]$ corresponds to the intersection of the two one-dimensional subspaces V'_r and V_{r_i}, and this is just the origin. Thus (4) is the space spanned by n o-dimensional subspaces, and this is just the o-dimensional subspace. So two propositions which are *equivalent* according to classical logic, viz. (2) and (4), are mapped onto different subspaces of the space $H(s)$ representing all possible physical propositions about S. *Conclusion:* the mapping is nonsense – or, *we must change our logic.*

Suppose we are willing to adopt the heroic course of changing our logic. What then?

It turns out that there is a very natural way of doing this. Namely, *just read the logic off from the Hilbert space $H(S)$*. Two propositions are to be considered equivalent just in case they are mapped onto the same subspace of $H(S)$, and a proposition P_1 is to be considered as 'implying' a proposition P_2 just in case S_{P_1} is a subspace of S_{P_2}. This idea was first advanced by Birkhoff and von Neumann some years ago, and has been recently revived by a number of physicists. Perhaps the first physicist to fully appreciate that *all* so-called 'anomalies' in quantum mechanics come down to the *non-standardness of the logic* is David Finkelstein of Yeshiva University, who has defended this interpretation in a number of lectures.

IV. The peculiarities of quantum mechanics

It is instructive to examine some of the peculiarities of quantum mechanics in the light of the foregoing proposal. Consider, first of all, *complementarity*. Let S be a system consisting of a single electron with a definite position r (strictly speaking, position is a *vector* and not a real number, but we may consider a one-dimensional world). The subspace of $H(S)$ representing the statement 'the position of S is r' is, as already remarked, a one-dimensional subspace V_r. Let V'_r represent the statement 'the momentum of S is r'', where r' is any real number. Then the intersection $V_r \cap V'_r$ is o-dimensional; thus we have:

> For all r, r', the statement: (*the electron has position r. the electron* (1)
> *has momentum r'*) is logically false.

Or, in ordinary language 'an electron cannot have an exact position and an exact momentum at the same time'.

All cases of 'complementarity' are of this kind: they correspond to *logical incompatibilities* in quantum logic.

Next, consider the famous 'two-slit' experiment. Let A_1 be the statement 'the photon passes through slit 1' and let A_2 be the statement 'the photon passes through slit 2'. Let the probability that the photon hits a tiny region R on the photographic plate on the assumption A_1 be $P(A_1, R)$ and let the probability of hitting R on the assumption A_2 be $P(A_{,2} R)$. These probabilities may be computed from quantum mechanics or classical mechanics (they are the same, as long as only one slit is open), and checked experimentally by closing one slit, and leaving only A_1 (respectively A_2) open in the case of $P(A_1, R)$ (respectively, $P(A_2, R)$).

Now, it is often said that if both slits are open the probability that the particle hits R 'should be' $\frac{1}{2}P(A_1, R) + \frac{1}{2}P(A_2, R)$. This is the probability predicted by *classical* mechanics. However, it is *not* the *observed* probability (which is *correctly* predicted by *quantum* mechanics, and *not* by classical mechanics). How was this 'classical' prediction arrived at?

First of all, the probability that the photon hits the slit A_1 = the probability that it hits A_2. This can be tested experimentally, and also insured theoretically by symmetrizing the apparatus, rotating the apparatus periodically, etc. Since we count only experiments in which the photon gets through the barrier, and hence in which the disjunction $A_1 \vee A_2$ is true, we have

$$
\begin{aligned}
P(A_1 \vee A_2, R) &= P[(A_1 \vee A_2) \cdot R] / P(A_1 \vee A_2) \\
&= P(A_1 \cdot R \vee A_2 \cdot R) / P(A_1 \vee A_2) \\
&= P(A_1 \cdot R) / P(A_1 \vee A_2) + P(A_2 \cdot R) / P(A_1 \vee A_2)
\end{aligned}
$$

(here '$P[(A_1 \lor A_2) \cdot R]$' means 'the probability of $(A_1 \lor A_2) \cdot R$', and similarly for '$P(A_1 \lor A_2)$', etc.

Since
$$P(A_1) = P(A_2),$$

we have
$$P(A_1 \lor A_2) = 2P(A_1) = 2P(A_2).$$

Thus
$$P(A_1 \cdot R)/P(A_1 \lor A_2) = P(A_1 \cdot R)/2P(A_1) = \tfrac{1}{2}P(A_1, R)$$

and similarly

$$P(A_2 \cdot R)/P(A_1 \lor A_2) = P(A_2 \cdot R)/2P(A_2) = \tfrac{1}{2}P(A_2, R).$$

Substituting these expressions in the above equation yields:

$$P(A_1 \lor A_2, R) = \tfrac{1}{2}P(A_1, R) + \tfrac{1}{2}P(A_2, R). \tag{2}$$

Now a crucial step in this derivation was the expansion of $(A_1 \lor A_2) \cdot R$ into $A_1 \cdot R \lor A_2 \cdot R$. This expansion is *fallacious* in quantum logic; thus the above derivation also fails. Someone who believes classical logic must conclude from the failure of the classical law that one photon can somehow go through two slits (which would invalidate the above deduction, which relied at many points on the incompatibility of A_1 and A_2), or believe that the electron somehow 'prefers' one slit to the other (but only when no detector is placed in the slit to detect this mysterious preference), or believe that in some strange way the electron going through slit 1 'knows' that slit 2 is open and behaves differently than it would if slit 2 were closed; while someone who believes quantum logic would see no reason to predict $P(A_1 \lor A_2, R) = \tfrac{1}{2}P(A_1, R) + \tfrac{1}{2}P(A_2, R)$ in the first place.

For another example, imagine a population P consisting of hydrogen atoms, all at the same energy level **e**. Let D be the relative distance between the proton and the electron, and let E be the magnitude 'energy'. Then we are assuming that E has the same value, namely **e**, in the case of every atom belonging to P, whereas D may have different values $d_1, d_2, \ldots$ in the case of different atoms $A_1, A_2, \ldots$

The atom is, of course, a system consisting of two parts – the electron and the proton – and the proton exerts a central force on the electron. As an analogy, one may think of the proton as the earth and of the electron as a satellite in orbit around the earth. The satellite has a potential energy that depends upon its height above the earth, and that can be recovered as usable energy if the satellite is made to fall. It is clear from the analogy that this potential energy P associated with the electron (the satellite), can become large if the distance D is allowed to become sufficiently great. However, P cannot be greater than E (the total energy). So if E is known, as in the present case, we can compute a number d

such that D cannot exceed d, because if it did, P would exceed e (and hence P would be greater than E, which is absurd). Let us imagine a sphere with radius d and whose center is the proton. Then if all that we know about the particular hydrogen atom is that its energy E has the value e, we can still say that wherever the electron may be, it cannot be outside the sphere. The boundary of the sphere is a 'potential barrier' that the electron is unable to pass.

All this is correct in classical physics. In quantum physics we get

> Every atom in the population has the energy level e (3)

and we may *also* get

> 10% of the atoms in the population P have values of D which *exceed d*.   (4)

The 'resolution' of this paradox favored by many people is as follows. They claim that there is a mysterious 'disturbance by the measurement', and that (3) and (4) refer to values *after measurement*. Thus, in this view, (3) and (4) should be replaced by

> If an energy measurement is made on any atom in P, then the value e is obtained,   (3′)

and

> If a D-measurement is made on any atom in P, then in 10% of the cases a value greater than d will be obtained.   (4′)

These statements are consistent in view of

> An E-measurement and a D-measurement cannot be performed at the same time (i.e. the experimental arrangements are mutually incompatible).   (5)

Moreover, we do not have to accept (3′) and (4′) simply on the authority of quantum mechanics. These statements can easily be checked, not, indeed, by performing both a D-measurement and an E-measurement on each atom (that is impossible, in view of (5)), but by performing a D-measurement on every atom in one large, fair sample selected from P to check (3′) and an E-measurement on every atom in a different large, fair sample from P to check (4′). So (3′) and (4′) are both known to be true.

In view of (3′), it is natural to say of the atoms in P that they all 'have the energy level e'. But what (4′) indicates is that, paradoxically, some of the electrons will be found on the wrong side of the potential barrier. They have, so to speak, 'passed through' the potential barrier. In fact, quantum mechanics predicts that the distance D will assume arbitrarily large values even for a fixed energy **e**.

The trouble with the above 'resolution' of this paradox is twofold. In the first place, if distance measurement, or energy measurement (or both) disturb the very magnitude that they seek to measure, then there should be some *theory* of this disturbance. Such a theory is notoriously lacking, and it has been erected into an article of faith in the state of Denmark that there can be no such theory. Secondly, if a procedure distorts the very thing it seeks to measure, it is peculiar it should be accepted as a good measurement, and fantastic that a relatively simple theory should predict the *disturbed* values when it can say nothing about the *undisturbed* values.

The resolution provided by quantum logic is quite straightforward. The statement

(Such-and-such electrons in P [some specific 10%] have a D-value in excess of d). The energy level of every electron in P is **e**

is logically false in *both* classical and quantum logic.

Let $S_1, S_2, \ldots, S_R$ be all the statements of the form 'such-and-such electrons in P [some specific 10%] have a D-value in excess of d', i.e. let R be the number of subsets of P of size 10%, and let there be one S_i for each such subset. Then what was just said can be rephrased thus: for each *fixed i*, $i = 1, 2, \ldots, R$ the statement below is logically false:

$$S_i \cdot E = \mathbf{e}. \tag{6}$$

Also, in *both* classical and quantum logic, the following statement is likewise false:

$$S_1 \cdot (E = \mathbf{e}) \vee S_2 \cdot (E = \mathbf{e}) \vee \cdots \vee S_R \cdot (E = \mathbf{e}) \tag{7}$$

(since the o-space, any number of times, only spans the o-space).

However,

$$(E = \mathbf{e}) \cdot (S_1 \vee S_2 \vee \cdots \vee S_R) \tag{8}$$

is *not* logically false in quantum logic! In fact, the subspace of $(E = \mathbf{e})$ is included in the subspace of $(S_1 \vee S_2 \vee \cdots \vee S_R)$, so that (8) is *equivalent* to just

$$E = \mathbf{e} \tag{9}$$

which has a *consequence*

$$S_1 \vee S_2 \vee \cdots \vee S_R. \tag{10}$$

In words: the statement (9) (or (3)) is not incompatible with but *implies* the statement (10) (or (4)).

These examples should make the principle clear. The only laws of classical logic that are given up in quantum logic are distributive laws, e.g. $p \cdot (q \vee r) \equiv p \cdot q \vee q \cdot r$; and every single anomaly vanishes once we give these up.

V. The quantum mechanical view of the world

We must now ask: what is the nature of the world if the proposed interpretation of quantum mechanics is the correct one? The answer is both radical and simple. *Logic is as empirical as geometry.* It makes as much sense to speak of 'physical logic' as of 'physical geometry'. We live in a world with a non-classical logic. Certain statements – just the ones we encounter in daily life – *do* obey classical logic, but this is so because the corresponding subspaces of $H(S)$ form a very special lattice under the inclusion relation: a so-called 'Boolean lattice'. Quantum mechanics itself explains the *approximate* validity of *classical* logic 'in the large', just as non-Euclidean geometry explains the *approximate* validity of *Euclidean* geometry 'in the small'.

The world consists of particles, in this view,[†] and the laws of physics are 'deterministic' in a modified sense. In the classical physics there was (idealizing) *one* proposition P which was true of S and such that every physical proposition about S was implied by P. This P was simply the *complete description of the state of S*. The laws of classical physics say that if such a P is the state of S at t_0, then the state after the lapse of time t will be a certain $f(P)$. This is classical determinism.

In quantum mechanics, let us say that any P whose corresponding S_p is one-dimensional is a *state-description*. Let $S_1, S_2, \ldots, S_R$ be all the possible positions of a one-particle system S, and let $T_1, T_2, \ldots, T_R$ be all the possible momenta. Then

$$S_1 \vee S_2 \vee \cdots \vee S_R \tag{1}$$

is a valid statement in quantum logic, and so is:

$$T_1 \vee T_2 \vee \cdots \vee T_R \tag{2}$$

In words:

	Some S_i is a true state-description	(1′)
and	Some T_j is a true state-description.	(2′)

† This is so only because we are quantizing a particle theory. If we quantize a field theory, we will say 'the world consists of fields', etc.

However, as we have already noted, the conjunction $S_i \cdot T_j$ is *inconsistent* for all i, j. Thus the notion 'state' must be used with more-than-customary caution if quantum logic is accepted. A system has *no complete description* in quantum mechanics; such a thing is a *logical* impossibility, since it would have to imply one of the S_i, in view of (1), and also have to imply one of the T_j in view of (2). A system has a position-state *and* it has a momentum-state (which is not to say 'it has position r_i and it has momentum r_j', for any r_i, r_j, but to say [(it has position $r_i \vee \cdots \vee$ it has position r_R). (It has momentum $r_1 \vee \cdots \vee$ it has momentum r_R)], as already explained); and a system has many other 'states' besides (one for each 'non-degenerate' magnitude). These are 'states' in the sense of *logically strongest consistent statements*, but not in the sense of 'the statement which implies every true physical proposition about S'.

Once we understand this we understand the notion of 'determinism' appropriate to quantum mechanics; if the state at t_0 is, say, S_z, then quantum mechanics says that after time t has elapsed the state will be $U(S_z)$, where U is a certain 'unitary transformation'; and, similarly, quantum mechanics says that if the state at t_0 is T_j, then the state after t will be $U(T_j)$. So the state at t_0 determines the state after any period of time t, just as in classical physics. *But*, it may happen that I know the state after time t has elapsed to be $U(T_j) = T_z$, say, and I measure not momentum but position. In this case I cannot *predict* (except with probability) what the result will be, because the statement T_j does not imply any value of position. 'Indeterminacy' comes in not because the laws are indeterministic, but because the states themselves, although logically strongest factual statements, do not contain the answers to all physically meaningful questions. This illustrates how the conflict between 'determinacy and indeterminacy' is resolved in quantum mechanics from the standpoint of quantum logic.

Finally, it remains to say how probabilities enter in quantum mechanics under this interpretation. (This has been pointed out by David Finkelstein, whose account I follow.) Suppose I have a system S, and I wish to determine the probability of some magnitude M having a value in a certain interval, given some information T about S. I imagine a system P consisting of a large number N of non-interacting 'copies' of S, all in the same 'state' T. This new system P has a Hilbert space $H(P)$ of its own, i.e. $H(P)$ is a vector space representing all possible physical propositions about P. Let R_p be the statement that $R\%$ of the systems S in P have an M-value in the interval I am interested in. It turns out that, as N is allowed to approach infinity, the subspace corresponding to R_p either contains the subspace corresponding to 'all the "copies" are in state T' 'in the limit' or is orthogonal to it 'in the limit'. In other words, given that all of the systems in P are in state T and that they do not interact, it *follows with almost certainty* that $R\%$ of them have $M(S)$ in the interval or with *almost certainty* that some other percent have $M(S)$ in the interval, if the number of such systems in P is very large. But, if we can say 'if we had sufficiently many identical copies of S, $R\%$ of them would have the property we are interested in', then, on any reasonable theory of probability, we can say 'the probability that S has the property is R'. In short, *probability* (on this view) *enters in quantum mechanics just as it entered in classical physics, via considering large populations.* Whatever problems may remain in the analysis of probability, they have nothing special to do with quantum mechanics.

Lastly, we must say something about 'disturbance by measurement' in this interpretation. If I have a system in 'state' S_z (i.e. 'the position is r_z'), and I make a *momentum* measurement, I must 'disturb' S_z. This is so because whatever result T_j I get is going to be *incompatible* with S_z. Thus, when I get T_j, I will have to say that S_z is *no longer true*; but this is no paradox, since the *momentum* measurement disturbed the *position* even according to *classical* physics. Thus the only 'disturbance' on this interpretation is the classical disturbance; we do not have to adopt the strange view that *position* measurement 'disturbs' (or 'brings into being', etc.) *position*, or that *momentum* measurement disturbs (or 'brings into being', etc.) *momentum*, or anything of that kind.

The idea that momentum measurement 'brings into being' the value found arises very naturally, if one does not appreciate the logic being employed in quantum mechanics. If I know that S_z is true, then I know that for *each* T_j the conjunction $S_z \cdot T_j$ is false. It is natural to conclude ('smuggling in' classical logic) that $S_z \cdot (T_1 \vee T_2 \vee \ldots \vee T_R)$ is false, and hence that we must reject $(T_1 \vee T_2 \vee \ldots \vee T_R)$ – i.e. we must say 'the particle has no momentum'. Then one measures momentum, and one gets a momentum – say, one finds that T_M. Clearly, the particle *now* has a momentum – so the measurement must have 'brought it into being'. However, the error was in passing from the falsity of $S_z \cdot T_1 \vee S_z \cdot T_2 \vee \ldots \vee S_z \cdot T_R$ to the falsity of $S_z \cdot (T_1 \vee T_2 \vee \ldots \vee T_R)$. This latter statement is *true* (assuming S_z); so it is *true* that 'the particle has a momentum' (even if it is also true that 'the position is r_3'); and the momentum measurement merely *finds* this momentum (while disturbing the *position*); it does not create it, or disturb it in any way. It is as simple as that.

At this point let us return to the question of 'determinism' for the last time. Suppose I know a 'logically strongest factual statement' about S at t_0, and I deduce a similar statement about S after time t has elapsed

– say, S_3. Then I measure *momentum*. Why can I not predict the outcome? We already said: 'because S_3 does not *imply* T_j for any j'. But a stronger statement is true: S_3 is incompatible with T_j, for all j! But it does *not* follow that S_3 is incompatible with $(T_1 \vee T_2 \vee \ldots \vee T_R)$.

Thus it is still true, even assuming S_3, that 'the particle *has* a momentum'; and if I measure I shall find it. However, S_3 cannot tell me what I shall find, because whatever I find will be incompatible with S_3 (which will no longer be true, when I find T_j). *Quantum mechanics is more deterministic than indeterministic in that all inability to predict is due to ignorance.*

Let $U_1, U_2, \ldots, U_R$ be statements about S at t_0 such that 'U_i at t_0' is equivalent to 'T_i after time t has elapsed', for $i = 1, 2, \ldots, R$. Then it can be shown that

$$U_1 \vee U_2 \vee \ldots \vee U_R$$

is logically true – i.e. *there is a statement which is true of S at t_0 from which it follows what the momentum (or whatever) will be after the lapse of time t.* In this sense, my inability to say what momentum S has now is due to 'ignorance' – ignorance of what U_i was true at t_0. However, the situation is not due to *mere* ignorance; for I *could* not know which U_i was true at t_0, given that I knew something that implied that S_3 would be true now, *without knowing a logical contradiction.*

In sum:

(1) For any such question as 'what is the value of $M(S)$ now', where M is a physical magnitude, *there exists* a statement U_i which was true of S at t_0 such that *had I known U_i* was true at t_0, I could have predicted the value of $M(S)$ now; but

(2) It is *logically impossible* to possess a statement U_i which was true of S at t_0 from which one could have predicted the value of *every* magnitude M now.

You can predict any *one* magnitude, if you make an appropriate measurement, but you cannot predict them *all*.

VI. The 'change of meaning' issue

While many philosophers are willing to admit that we could adopt a different logic, frequently this is qualified by the assertion that of course to do so would merely amount to a 'change of language'. How seriously do we have to take this?

The philosophical doctrine most frequently associated with this 'change of language' move is conventionalism. In, say, Carnap's version this amounts to something like this:

(1) There are alleged to be 'rules of language' which *stipulate* that certain sentences are to be true, among them the axioms of logic and mathematics.

(2) Changing our logic and mathematics, if we ever do it, will be just an instance of the general phenomenon of *change of conventions*.

I have criticized this view at length elsewhere, and I shall not repeat the criticism here. Suffice it to say that if there *were* such conventions I do not see how they could be *justified*. To stipulate that certain sentences shall be immune from revision is *irrational* if that stipulation may lead one into serious difficulties, such as having to postulate either mysterious disturbances by the measurement (or to say that the measurement brings what it measures into existence) or 'hidden variables'. Moreover, if our aim is a true description of the world, then we should not adopt arbitrary linguistic stipulations concerning the form of our language unless there is an argument to show that these cannot interfere with that aim. If the rules of classical logic *were* really arbitrary linguistic stipulations (which I do not for a moment believe), then I have no idea how we are supposed to know that these stipulations are compatible with the aims of inquiry. And to say that they are nonarbitrary stipulations, that we are only free to adopt conventions whose *consequences* are consistent (in a non-syntactical sense of 'consequence') is to *presuppose* the notion of 'consequence', and hence of logic. In practice, as Quine has so well put it, the radical thesis that logic is true by language alone quickly becomes replaced by the harmless truism that 'logical truth is truth by virtue of language plus *logic*'. Those who begin by 'explaining' the truth of the principles of logic and mathematics in terms of some such notion as 'rule of language' end by smuggling in a quite old fashioned and unexplained notion of *a prioricity*.

Even if we reject the idea that a language literally has rules stipulating that the axioms of logic are immune from revision, the 'change of meaning' issue may come up in several ways. Perhaps the most sophisticated way in which the issue might be raised is this: it might be suggested that we identify the logical connectives by the logical principles they satisfy. To mean 'or' e.g. a connective must satisfy such principles as: 'p implies p or q' and 'q implies p or q', simply because these formulate the properties that we count as 'the meaning' of 'or'.

Even if this be true, little of interest to the philosophy of logic follows, however. From the fact that 'a language which does not have a word V which obeys such-and-such patterns of inference does not contain the concept *or* (or whatever) in its customary meaning' it does not follow either that a language which is adequate for the purpose of formulating true and significant statements about physical reality *must* contain a word V which obeys such-and-such patterns of inference, or that it *should* contain a word V which obeys such-and-such patterns of inference. Indeed, it does not even follow that an optimal scientific language *can* contain such a word V; it may be that having such a connective (and 'closing' under it, i.e. stipulating that for all sentences S_1, S_2 of the language there is to be a sentence $S_1 \vee S_2$) commits one to either changing the laws of physics one accepts (e.g. quantum mechanics), or accepting 'anomalies' of the kind we have discussed. If one does not believe (1) that the laws of quantum mechanics are false; nor (2) that there are 'hidden variables'; nor (3) that the mysterious 'cut between the observer and the observed system' exists; one perfectly possible option is this: to *deny* that there are *any* precise and meaningful operations on propositions which have the properties classically attributed to 'and' and 'or'. In other words, instead of arguing: 'classical logic *must* be right; so something is wrong with these features of quantum mechanics' (i.e. with complementarity and superposition of states), one may perfectly well decide 'quantum mechanics may not be right in all details; but complementarity and superposition of states are probably right. If these are right, and classical logic is also right, then either there are hidden variables, or there is a mysterious cut between the observer and the system, or something of that kind. But I think it is *more likely that classical logic is wrong* than that there are either hidden variables, or "cuts between the observer and the system", etc.' Notice that this completely *bypasses* the issue of whether adopting quantum logic is 'changing the meaning' of 'and', 'or', etc. If it is, so much the worse for 'the meaning'.

From the classical point of view, all this is nonsense, of course, since no empirical proposition could literally be *more likely* than that classical logic is right. But from the classical point of view, no empirical proposition could be *more likely* than that straight lines could not behave as depicted in Figure 1. What the classical point of view overlooks is that the *a prioricity* of logic and geometry vanishes as soon as *alternative logics* and *alternative geometries* begin to have serious physical application.

But *is* the adoption of quantum logic a 'change of meaning'? The following principles:

$$p \text{ implies } p \vee q. \qquad (1)$$
$$q \text{ implies } p \vee q. \qquad (2)$$
$$\text{if } p \text{ implies } r \text{ and } q \text{ implies } r, \text{ then } p \vee q \text{ implies } r. \qquad (3)$$

all *hold* in quantum logic, and these seem to be the basic properties of 'or'. Similarly

$$p, q \text{ together imply } p \cdot q. \qquad (4)$$

(Moreover, $p \cdot q$ *is the unique proposition that is implied by every proposition that implies both p and q*.)

$$p \cdot q \text{ implies } p. \tag{5}$$
$$p \cdot q \text{ implies } q. \tag{6}$$

all *hold* in quantum logic. And for *negation* we have

$$p \text{ and } -p \text{ never both hold. } (p \cdot -p \text{ *is a contradiction*}) \tag{7}$$
$$(p \vee -p) \text{ holds.} \tag{8}$$
$$-\ -p \text{ is equivalent to } p. \tag{9}$$

Thus a strong case could be made out for the view that adopting quantum logic is *not* changing the meaning of the logical connectives, but merely changing our minds about the law

$$p \cdot (q \vee r) \text{ is equivalent to } p \cdot q \vee p \cdot r \text{ (which fails in quantum logic).} \tag{10}$$

Only if it can be made out that (10) is 'part of the meaning' of 'or' and/or 'and' (which? and how does one decide?) can it be maintained that quantum mechanics involves a 'change in the meaning' of one or both of these connectives.

My own point of view, to state it summarily, is that we simply do not possess a notion of 'change of meaning' refined enough to handle this issue. Moreover, even if we were to develop one, that would be of interest only to philosophy of *linguistics* and not the philosophy of *logic*.

The important fact to keep in mind, however one may choose to phrase it, is that the whole 'change of meaning' issue is raised by philosophers only to *minimize* a conceptual revolution. But only a demonstration that a certain *kind* of change of meaning is involved, namely, *arbitrary linguistic change*, would successfully demolish the philosophical significance I am claiming for these conceptual revolutions. And *this* kind of 'change of meaning' is certainly *not* what is involved.

VII. The analogy between logic and geometry

It should now be clear that I regard the analogy between the epistemological situation in logic and the epistemological situation in geometry as a perfect one. In the remainder of this essay, I shall try to deal with two points of apparent disanalogy: (1) that geometrical notions such as 'straight line' have a kind of operational meaning that logical operations lack; and (2) that physical geometry is about something 'real', viz. physical space, while there is nothing 'real' in that sense for logic to be 'about'. But it is useful at this point to summarize how far the analogy has already been seen to extend.

We saw at the beginning of this essay that one *could* keep Euclidean geometry, but at a very high cost. If we choose paths which obey Euclid's axioms for 'straight line' in some arbitrary way (and there are infinitely many ways in which this could be done, at least in space *topologically* equivalent to Euclidean space), and we retain the law that $F = ma$, then we must explain the fact that bodies do not follow our 'straight lines', even in the absence of differential forces† by *postulating mysterious forces*. For, if we believe that such forces do not really exist, and that $F = ma$, then we have no choice but to reject Euclidean geometry. Now then, Reichenbach contended that the choice of any metric for physical space is a matter of 'definition'. If this is so, and, if '$F = ma$' is true even when we change the metric (as Reichenbach assumed), then *what forces exist is also a matter of definition*, at least in part. I submit that:

On the *customary* conception of 'force' it is *false* (and *not* a matter of a conventional option) that a body not being acted upon by differential forces is being pushed about by mysterious 'universal forces'; and that the whole significance of the revolution in geometry may be summarized in the following two propositions:

(A) *There is nothing in reality answering to the notion of a 'universal force'.*

(B) *There is something in reality approximately answering to the traditional notion of a 'straight line', namely a geodesic. If this is not a 'straight line', then nothing is.*

(Reichenbach's view is that (A) represents a *definition*; it is precisely this that I am denying, except in the trivial sense in which it is a 'definition' that 'force' refers to force and not to something else.)

Now then, the situation in quantum mechanics may be expressed thus: we *could* keep classical logic, but at a very high price. Just as we have to postulate mysterious 'universal forces' if we are to keep Euclidean geometry 'come what may', so we have to postulate equally mysterious and really very similar agencies – e.g. in their indetectability, their violation of all natural causal rules, their *ad hoc* character – if we are to reconcile quantum mechanics with classical logic *via* either the 'quantum potentials' of the hidden variable theorists, or the metaphysics of Bohr. Once again, anyone who really regards the choice of a *logic* as a 'matter of convention', will have to say that whether 'hidden variables exist', or

† By a 'differential' force what is meant is one that has a source, that affects different bodies differently (depending on their physical and chemical composition), etc. The 'forces' that one has to postulate to account for the behavior of rigid rods if one uses an unnatural metric for a space are called 'universal forces' by Reichenbach (who introduced the terminology *'differential/universal'*); these have *no* assignable source, affect *all* bodies the same way, etc.

whether, perhaps, a mysterious 'disturbance by the measurement exists'. or a fundamental difference between macro- and micro-cosm exists, etc., is likewise a matter of convention. And, once again, our standpoint can be summarized in two propositions:

(A') There is nothing really answering to the Copenhagen idea that two kinds of description (classical and quantum mechanical) must always be used for the description of physical reality (one kind for the 'observer' and the other for the 'system'), nor to the idea that measurement changes what is measured in an indescribable way (or even brings it into existence), nor to the 'quantum potential', 'pilot waves', etc. of the hidden variable theorists. These no more exist than Reichenbach's 'universal forces'.

(B') There are operations approximately answering to the classical logical operations, viz. the ∨, ·, and − of quantum logic. If these are not the operations of disjunction, conjunction, and negation, then no operations are.

VIII. The 'operational meaning' of the logical connectives

It is well known that operationalism has failed, at least if considered as a program of strict epistemological reduction. No nonobservational term of any importance in science is strictly definable in terms of 'observation vocabulary', 'measuring operations', or what not. In spite of this, the idea of an 'operational definition' retains a certain usefulness in science. What is usually meant by this – by scientists, not by philosophers – is simply a description of an idealized way of determining whether or not something is true, or what the value of a magnitude is, etc. For example, the 'operational meaning' of relativity is frequently expounded by imagining that we had errorless clocks at every space–time point.

Now the physicist who expounds the 'operational meaning' of a theory in terms of 'clocks at every space–time point' knows perfectly well that (1) all clocks have, in practice, some error; (2) no criterion can be formulated in purely observational language for determining the amount of error exactly and in all conceivable cases; (3) anyway, clocks have a certain minimum size, so one could not really have a clock at each point. Nevertheless, this kind of *Gedankenexperiment* is useful, because the situation it envisages can at least be approximated (i.e. one can have pretty accurate clocks which are pretty small and pretty close together), and because seeing exactly what happens in such situations according to the theory (which is made easier by the idealization – that is the whole reason for 'idealizing') gives one a better grasp of the theory. Provided one does not slip over from the view that 'operational definitions' are a useful heuristic device for getting a better grasp of a theory, to the view that they really tell one what theoretical terms *mean* (and that theoretical statements are then mere 'shorthand' for statements about measuring operations), no harm results. It has to be emphasized, however, that the idealized clocks, etc., involved in typical 'operational analyses' are themselves highly theoretical entities – indeed, their very designations 'clock', 'rigid rod', etc., presuppose the very theoretical notions they are used to clarify – so that operational analyses would be *circular* (in addition to other defects) if they really were intended as definitions in the strict sense of 'definition'.

What the 'operational meaning', in the loose sense just discussed, of the geometrical terms is, is well known. A 'straight line', for example, is the path of a light ray; or of a stretched thread. It is also the 'shortest distance' between any two of its points, as measured again by, say, a tape measure, or by laying off rigid rods, etc. The idealizations here are obvious. Light, for example, is really a wave phenomenon. A 'light ray', strictly speaking, is simply a normal to a wave front. And the notion of a 'normal' (i.e. a perpendicular) presupposes both the notions of straight line and angle.

If we avoid the wave nature of light by speaking of the 'path' of a single photon, we run into difficulties with complementarity: the photon can never be assigned a particular position r_i and a particular momentum r_j at the some time (in quantum logic the conjunction: the position of E is r_i and the momentum of E is r_j is even inconsistent), so the notion 'path of a photon' is operationally meaningless. And the idealized character of 'operational definitions' of 'straight line' in terms of stretched threads, rigid rods, etc., should be completely evident to anyone.

In spite of this, as we remarked before, such 'operational definitions' have a certain heuristic value. It enables us to grasp the idea of a non-Euclidean world somewhat better if we are able to picture exactly how light rays, stretched threads, etc., are going to behave in that world. But do the *logical connectives* have analogous 'operational definitions', even in this loose sense?

The answer 'yes' has been advanced in a provocative paper by David Finkelstein (Finkelstein, 1964). In order to explain the sense in which Finkelstein ascribes operational meaning to the logical vocabulary, it is necessary to first summarize a few well-known facts about logic. (These hold in both classical and quantum logic.)

First of all, propositions form what is called a *partial ordering* with respect to the relation of *implication*; that is, the relation of implication has the properties of being reflexive (p implies p), and transitive (if p

implies q and q implies r, then p implies r). Moreover, if we agree to count equivalent propositions as the same, then implication has the property: p implies q and q implies p both hold only when p and q are the same proposition.

The proposition $p \lor q$ is what is called an *upper bound* on both p and q: for it is implied by p (think of the implicandum as 'above' the implicans) and implied by q. Moreover, it is the *least upper bound*; for every proposition which is 'above' both p and q in the partial ordering of propositions by the implication relation is above their disjunction $p \lor q$. Similarly, the conjunction $p \cdot q$ is the *greatest lower bound* of p and q in the partial ordering.

In mathematics, a partial ordering in which there are for any two elements x, y a least upper bound $x \lor y$ and a greatest lower bound $x \cdot y$ is called a lattice; what we have just said is that propositions form a *lattice* with respect to implication.

The tautological proposition $p \lor -p$ is the greatest element in the whole lattice (we denote it by '1') and the inconsistent proposition is the least element, since every proposition implies a tautology and is implied by an inconsistency.

The proposition $-p$ has the property that its greatest lower bound with p (i.e. its conjunction with p) is o and its least upper bound with p is 1. A lattice in which there is for every x a complement $-x$ with these properties is called a *complemented lattice*. Thus propositions form a complemented lattice. (So do *sets*, under inclusion, and many other things.)

A lattice in which the laws:

$$x \cdot (y \lor z) = (x \cdot y) \lor (x \cdot z) \quad \text{and} \quad x \lor (y \cdot z) = (x \lor y) \cdot (x \lor z)$$

hold (where '$\cdot$' denotes the greatest lower bound and '$\lor$' denotes the least upper bound) is called a *distributive* lattice. The whole difference between classical and quantum logic lies in this: that propositions do not form a distributive lattice according to quantum logic, whereas according to classical logic they do.

In this paper we explained quantum logic in terms of the vector space representing the states of a system: however, one could equally well take the lattice of propositions as basic, for the subspaces of the vector space are in one-one correspondence to the propositions, and all the inclusion relations are the same.

Since conjunction and disjunction are simply greatest lower bound and least upper bound under implication, and negation is likewise characterized in terms of the implication lattice, it suffices to give an operational meaning to 'implication'.

This is accomplished by Finkelstein with the aid of the notion of a *test*. Let us pretend that to every physical property P there corresponds a test T such that something has P just in case it 'passes' T (i.e. it *would* pass T, if T were performed). This is no more than the idealization which is always made in all operational analysis. Then we define the following natural 'inclusion' relation among tests:

$$T_1 \subset T_2 \text{ just in case everything that 'passes' } T_1 \text{ 'passes' } T_2. \quad (\text{1})$$

This inclusion relation may be operationally tested in the following way: take a large population of things which are supposed to all pass T_1 (say, on the basis of theory). Take out a large fair sample S_1, and apply test T_1 to every member of the sample. If they all pass, then our hypothesis that P consists of things which pass T_1 is operationally confirmed. Otherwise, we have to look for a different P. Now (assuming they all pass), we take a *different* fair sample S_2 from P and apply T_2. If all the elements of S_2 pass T_2, the hypothesis that 'all things that pass T_1 also pass T_2' has been confirmed. (Note that we do *not* test $T_1 \subset T_2$ by applying T_2 to the things to which T_1 was applied; this would be bad practice because those things may no longer have P *after* T_1 has been performed – i.e. they might not still pass T_1 – if T_1 were repeated.)

Now then, if quantum mechanics is true, then it turns out that there is an idealized test $T_1 \lor T_2$ which is passed by everything which passes T_1 and by everything which passes T_2 and which is such that the things that pass this test pass *every* test T such that $T_1 \subset T$ and $T_2 \subset T$. This test $T_1 \lor T_2$ is the *least upper bound* on T_1 and T_2 in the lattice of tests. Similarly, there is a greatest lower bound $T_1 \cdot T_2$, with the properties $T_1 \cdot T_2 \subset T_1$, $T_1 \cdot T_2 \subset T_2$, and $T \subset T_1 \cdot T_2$ for all T such that $T \subset T_1$ and $T \subset T_2$, and a test $- T$ which is a 'complement to T' in the sense that $T \cdot - T = 0$ (the impossible test) and $T \lor - T = 1$ (the vacuous test).

Quantum mechanically these tests may be described as follows. Let p be a proposition corresponding to a subspace S_p and q be a proposition corresponding to a subspace S_q of the vector space $H(S)$. Let S_t be the subspace spanned by S_p and S_q – i.e. the smallest subspace containing both S_p and S_q. Then there is always a physical magnitude m, and a set Σ of values of m, such that S_t is exactly the subspace corresponding to the proposition that $m(S) \in \Sigma$. Let T be the test which consists in measuring m, and 'passing' if a value in Σ is obtained. Then it follows from quantum mechanics that everything that passes the test corresponding (in the same sense) to the subspace S_p passes T, i.e. $T_p \subset T$, and similarly $T_q \subset T$. Moreover, as Finkelstein shows, any possible test T' such that $T_p \subset T'$ and $T_q \subset T'$ is such that $T \subset T'$; so T is indeed a 'least upper bound' on T_p and T_q.

Now then, suppose the proposition '$P \vee Q$' has any operational meaning at all (i.e. that there is any test T at all which is passed by all and only the things which have either property P or property Q).

Since the things that have property P all pass the corresponding test T_p, and everything that has P certainly has $P \vee Q$, it must be that $T_p \subset T$. Similarly, it must be that $T_q \subset T$. On the other hand, let T' be any test such that $T_p \subset T'$ and $T_q \subset T'$. Since everything that has P passes T_p, by hypothesis, it follows that the things with property P are a subset of the things which pass T', and similarly the things with property Q are a subset of the things which pass T'. So the things with $P \vee Q$ are a subset of the things which pass T', and since the things with $P \vee Q$ are assumed to be just the things which pass T, it follows that $T \subset T'$. Thus if there is *any* test at all (even 'idealizing', as we have been) which corresponds to the *disjunction* $P \vee Q$, it must have the property of being a *least upper bound* on T_p and T_q.

But, by Finkelstein's result, the only tests which have this property are the ones which are equivalent to the test T corresponding to the subspace spanned by S_p and S_q. Similarly, if conjunction is to correspond to any test at all, it must be the test determined by the intersection of the spaces S_p and S_q, and negation must correspond to the orthocomplement of S_p. Thus we are led directly to *quantum logic* and not to classical logic!

In sum: if we seek to preserve the (approximate) 'operational meaning' that the logical connectives *always* had, then we have to *change* our logic; if we insist on the old logic, then *no* operational meaning at all can be found for the logical connectives that will work in all cases. (Of course, for *macroscopic* propositions the lattice is distributive; so we may keep the classical logic *and* the classical tests for these cases. But as soon as one tries to extend this to an operational meaning for *microscopic* propositions in any consistent way, we will be in trouble in view of Finkelstein's result.)

Two points may now be made with regard to the philosophical significance of Finkelstein's work.

First, the operational analysis of the logical connectives has the same heuristic value that the operational analysis of the geometrical notions does. If we interpret '$P \vee Q$' (as applied to a system S) as meaning 'S passes the test T which is the least upper bound on T_p and T_q' – this is equivalent to 'S passes *every* test which is passed by everything that passes T_p and passed by everything that passes T_q' – and similarly interpret '$P \cdot Q$' as 'S passes the test T which is such that $T \subset T_p$ and $T \subset T_q$ and T is passed by everything that passes any test T' such that $T' \subset T_p$ and $T' \subset T_q$', and '$-P$' as 'S passes the test T which is such that nothing that passes T passes T_p and everything passes any test which is passed by everything that passes T_p and everything that passes T', then we can figure out exactly what to expect in a quantum logical world. Reichenbach once replied to the neo-Kantian claim that Euclidean geometry is the 'only geometry that can be visualized' roughly as follows: 'If to imagine a non-Euclidean world is to imagine a world upon which a Euclidean description *cannot* be forced, even by introducing "universal forces", then of course no such world can be imagined: for *any* geometry can be imposed on a world, if we are willing to adopt enough *ad hoc* hypotheses. But if to imagine a non-Euclidean world is to imagine a world which conforms to the standard operational significance of *non*-Euclidean geometry, then of course such a world can be imagined.' (Poincaré had earlier made a similar point: to imagine a non-Euclidean world, imagine the experiences you would have if you lived in such a world. – Assuming, of course, the standard 'operational definitions'.)

In exactly the same way, one can say: if to imagine a world which does not obey classical logic is to imagine a world upon which a description presupposing classical logic cannot be *forced*, even by introducing 'hidden variables' or Copenhagen double-think, then of course this cannot be done. But one can imagine a world which conforms to the *operational significance* of quantum logic. Or, adapting Poincaré: 'to imagine a quantum logical world, imagine the experiences you would have in such a world'. (*You live in one.*) Assuming, of course, the 'operational definitions' of the logical connectives, as we just analyzed them.

II

Time and physical geometry*

I think that if we attempted to set out the 'man on the street's' view of the nature of time, we would find that the main principle underlying his convictions on this subject might be stated somewhat as follows:

(1) All (and only) things that exist *now* are real.

Future things (which do not already exist) are not real (on this view); although, of course they *will* be real when the appropriate time has come to be the present time. Similarly, past things (which have ceased to exist) are not real, although they *were* real in the past.

Obviously, we shall have to make some assumptions about the concept *real* if we are to discuss the 'man on the street's' view at all. The assumptions I shall make are as follows:

I. I-now am real. (Of course, this assumption changes each time I announce that I am making it, since 'I-now' refers to a different instantaneous 'me'.)

II. At least one other observer is real, and it is possible for this other observer to be in motion relative to me.

And, the most important assumption, which will be referred to (when properly understood) as the principle that There Are No Privileged Observers:

III. If it is the case that all and only the things that stand in a certain relation R to me-now are real, and you-now are also real, then it is also the case that all and only the things that stand in the relation R to you-now are real.

If the assumption III is not to be vacuous, it is necessary to understand the expression 'the things' as referring not just to present things, but to all things, past, present, and future, whether we regard all these things as 'real' or not. (That is, III is to be understood as stated in 'tenseless' language, except for the indexical terms 'me-now' and 'you-now'.) Secondly, R must be restricted to physical relations that are supposed to be independent of the choice of a coordinate system (as simultaneity was in *classical* physics) and to be definable in a 'tenseless' way in terms

* First published in *The Journal of Philosophy*, LXIV, 8 (27 April 1967).

of the fundamental notions of physics. And, lastly, it must not depend on anything *accidental* (physically speaking) that all and only the things that stand in the relation R to me-now are real.

We note that, if we assume classical physics and take the relation R to be the relation of simultaneity, then, on the view (1), it is true that all and only the things that stand in the relation R to me-now are real, and the principle III is satisfied because the relation of simultaneity is transitive.

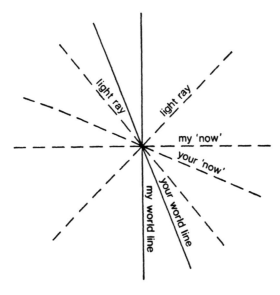

Figure 1

Finally, I shall assume Special Relativity.

We now discover something really remarkable. Namely, on every natural choice of the relation R, it turns out that *future* things (or events) are already real! For example, suppose I take the relation R to be the relation of simultaneity (as is suggested by (1)). For the remainder of this paper I shall assume the following (perfectly possible) physical situation to be actual: you-now and I-now are at the same place now, but moving with relative velocities which are very large (relative to the speed of light, which I take to be = 1). Thus, our world-lines look as shown in Figure 1 (I have also drawn our 'light-cone' for the purpose of the later discussion):

Now then, we cannot take the relation of simultaneity-in-my-coordinate-system to be R without violating the way in which the principle that There Are No Privileged Observers is intended to be understood.

Rather, we have to take R to be the relation of simultaneity-in-the-*observer's*-coordinate-system. Then, if I assume that all and only those things which stand in *this* relation R to me-now are real, I find that you-now are also real (since I-now and you-now are simultaneous-at-a-point and, hence, simultaneous in *every* coordinate system). Hence, everything that stands in the relation R to you – everything that is simultaneous to you-now in *your* coordinate system – is also real. But it is well known that, as a consequence of Special Relativity, there are events which lie in 'the future' according to *my* coordinate system and which lie in the 'present' of you-now according to *your* coordinate system. Since these things stand in the relation R to you-now, and you-now are real, and it was assumed that all and only the things that stand in the relation R to me-now are real, the principle III requires that I call these future things and events real! (But actually, I now have a contradiction: for these future things do not stand in the relation R to me-now, and so my assumption that all and only the things that stand in *this* relation R to me-now are real was already inconsistent with the principle that There Are No Privileged Observers.)

The difficulty is obvious: what the principle that There Are No Privileged Observers requires is simply that the relation R be *transitive*; i.e. that it have the property that from xRy and yRz it follow that xRz. Simultaneity-in-*my*-coordinate-system has this property, since if x is simultaneous with y in my coordinate system, and y is simultaneous with z in my coordinate system, then x is also simultaneous with z in my coordinate system; but simultaneity-in-*my*-coordinate system is not admissible as a choice of R, because it depends on the coordinate system. And the relation 'x is simultaneous with y in the coordinate system of x' (which is essentially the relation we just considered), while admissible, is not transitive, since, if I-now am simultaneous with you-now in the coordinate system of me-now, and you-now are simultaneous with event X in the coordinate system of you-now, it does *not* follow that I-now am simultaneous with event X in the coordinate system of me-now.

Now then, if we combine the fact that the relation R is required by III to be transitive with our desire to preserve the following principle, which is one-half of (1):

(2) All things that exist now are real.

— then we quickly see that future things must be real.

For, if the relation R satisfies (2) – and I take (2) to mean (at least when I assert it) that all things that exist *now* according to *my* coordinate system are real – and you-now are as in Figure 1, then you-now must stand in the relation R to me-now, since you exist both *now* and *here*.

But, if the relation R always holds between all the events that are on some one 'simultaneity line' in my coordinate system and me-at-the-appropriate-time, then (since the laws of nature are invariant under Lorentz transformation, by the principle of Special Relativity), the relation R must also hold between all the events on some one 'simultaneity-line' in *any* observer's coordinate system and that-observer-at-the-appropriate-time. Hence, all the events that are simultaneous with you-now in *your* coordinate system must *also* bear the relation R to you-now. Let event X be one such event which is 'in the future' according to *my* coordinate system (if our velocities are as shown in Figure 1, then such an event X must always exist). Then, since the event X bears the relation R to you-now, and you-now bear the relation R to me-now, the event X bears the relation R to me-now. But we chose R to be such that all and only those events which bear R to me-now are real. So the event X, which is a *future* event according to my coordinate system, is already real!

The space-fight tomorrow

A different problem connected with time was studied by Aristotle. Aristotle was what we would today call an 'indeterminist'. He did not think that the outcome of certain future events – his example was, Who will win the sea-fight tomorrow? – was determined at the present time. Given this assumption, he found it difficult to believe that the statement 'The *A*s will win the sea-fight' *already* has a truth value, that it is already true or already false. Rather he suggested that such 'future contingent' statements have *no* truth value.

Some philosophers would say that this belief rested upon a simple *mistake* on Aristotle's part, that 'is true' and 'is false' are *tenseless* predicates, and that statements are 'true' and 'false' *simpliciter*, not 'already true' or 'now true' or 'going to be true', etc. But this is begging the question. If any statement that is true at some time is true at every time, then indeed there is some point to dropping this way of speaking and making 'true' and 'false' tenseless (which they clearly are *not* in ordinary language, by the way). But the correctness of this view – whether you express it by saying that truth is eternal or by saying that truth is 'tenseless' does not matter – was what Aristotle was challenging. It is unfair to assume a form of language which presupposes that Aristotle was wrong and then to use the assumed correctness of this linguistic formalism as an argument against Aristotle.

It would be easy to write down a list of predicates $P_1, \ldots, P_n$ such that, on Aristotle's view, the statement that any past or (completed) present event has one of the properties $P_1, \ldots, P_n$ always has a truth value, but the statement that a future event has one of the properties P_i does not have a truth value. Thus Aristotle's view is a sort of supplement to what we called the 'man on the street's' view of time. The principle (1) asserts a fundamental difference between the present and both the past and the future; but it does not assert any difference between the past and the future; neither past things nor future things can now be called real. Aristotle would perhaps have accepted this, but he would have added that there is a fundamental difference, nonetheless, between the past and the future, viz., that past events are now determined, the relevant statements about them have now acquired truth values which will 'stick' for all time; but future events are undetermined, and at least some statements about them are not yet either true or false. (The fact that you can change the future but not the past would have been explained by Aristotle in terms of this fundamental asymmetry between the two.)

Aristotle was wrong. At least he was wrong if Relativity is right; and there is today better reason to believe Relativity than to believe Aristotle, on this point at least. To see this, let us revert to the situation depicted in Figure 1, and let there be a space-fight which is 'in the future' in my coordinate system (i.e. above the 'now' of 'me-now') but 'in the past' in your coordinate system (i.e. below the 'now' of 'you-now'). Then, since the space-fight is 'in the future' (for me), I must say that the statement 'the As will win or have already won' has no truth value. But, if you say that this same statement *has* a truth value – as you must, if Aristotle is right, since it lies 'in the past' for you – then you and I cannot both be right. Nor can just one of us be right, without becoming a Privileged Observer!†

† Nor can one avoid the difficulty by abandoning the objectivity of truth, and speaking of 'truth-for-me' and 'truth-for-you'. For, if it is even true-for-me that the statement in question *has a truth-value-for-you*, then either it is true-for-me that the statement in question is true-for-you, or it is true-for-me that the statement in question is false-for-you (since 'true-for-you' and 'false-for-you' are the only two truth-values–for-you). But if it is true-for-me, that 'the As have won or will win the space-fight' is true-for-you, then, to all intents and purposes, 'the As have won or will win the space-fight' is true-for-me too, since both facts – the outcome of the space-fight and the truth-value–for-you of this statement – have exactly the same status relative to me; and, likewise, if it is true-for-me that 'the As have won or will win the space-fight' is false-for-you, then, to all intents and purposes, 'the As have won or will win the space-fight-is false-for-me too. So I must not regard it as true that 'the As have won or will win the space-fight' is either true-for-you or false-for-you, if I wish to maintain that this future contingent has no truth value. And hence, if it is true that:

   For every contingent statement S and every observer O, (S is true-for-O or S is false-for-O) if and only if S refers to an event in the lower half of O's light-cone.
– then this 'truth' cannot itself be stated by any observer on pain of contradiction!
   (Here I assume this much concerning the logic of 'true-for-me': that, if $p \vee q$ is

The point becomes even clearer when viewed a bit more abstractly. Let us assume that Aristotle was right. Then I shall show that we can define a relation of Absolute Simultaneity, contradicting Relativity. Namely, I define an event to be Absolutely Future if the statement that it has the property P_i (where P_i is one of the properties in the list alluded to before) has no truth value. I define an event to be Absolutely Present if it is *not* Absolutely Future and if every event that is in its proper future (i.e. occurs later on its world-line – this is relativistically invariant) *is* Absolutely Future. Then two events are Absolutely Simultaneous if and only if there is a time (in the coordinate system of *any* observer) at which they are both Absolutely Present. (Also, one can define *that* observer to be at Absolute Rest.)

It is no criticism of Aristotle's ability to say that he was mistaken, any more than it is a criticism of Newton that he failed to foresee Relativity. Had the world been different than it is, these sorts of objections could not be brought against Aristotle's view. But it is important to see that Aristotle's view depends upon an *absolute* 'pastness' and 'futurity' just as much as Newtonian physics does, and that it is obsolete for the same reason.

I might try to save the view that 'future contingents' have no truth value by saying that even *present-tense* statements have no truth value if they refer to the outcome of events that are so far away that a causal signal informing me of the outcome could not have reached me-now without traveling faster than light. In other words, I might attempt saying that statements about events that are in neither the upper half nor the lower half of my light-cone have no truth value. In addition, statements about events in the upper half of my light-cone have no truth value, since they are in my future according to every coordinate system. So only statements about events in the lower half of my light-cone have a truth value; only events that are in 'my past' according to *all* observers are determined.

This last move, however, flagrantly violates the idea that there are no Privileged Observers. Why should a statement's having or not having a truth value depend upon the relation of the events referred to in the statement to just one special human being, *me*? Moreover, the following highly undesirable consequence flows from this last view: let Oscar be a person whose whole world-line is outside of the light-cone of me-now. Let me-future be a future 'stage' of me such that Oscar is in the lower half of the light-cone of me-future. Then, when that future becomes the

true-for-me, then either p is true-for-me or else q is true-for-me. Of course, we might try still more radical alterations of logic. But it seems considerably preferable – to me at least – *not* to do this, and rather to concede the objectivity of future contingents.)

present, it will become true that Oscar *existed*, although it will never have had a truth value to say in the present tense 'Oscar exists now'. Things could come to *have been*, without its ever having been true that they *are*!

Morals and conclusions

What moral should we draw from all this? In the first part of this chapter I showed that, if we couple the assumption III with the principle (2), we must say that future things are real, even if they do not exist yet. Strictly speaking, I showed only that *some* future things are real, and that only on the assumption that the situation shown in Figure 1 obtains. But the argument can obviously be extended (bringing in certain further assumptions with which virtually no one would argue) to show that *all* future things are real ('things' here includes 'events'), and likewise that all *past* things are real, even though they do not exist *now*. To spell this out: if I accept (2), then I must say that all those things are real (but not *only* those things, as we saw) which bear to me the following relation: the relation y bears to x just in case y is simultaneous with x in the coordinate system of x. I must describe the relation this way, and not as 'the relation y bears to x just in case y is simultaneous with x *in the coordinate system of me-now*,' because otherwise I would be committed to the view that the real things bear to me-now a relation which is not Lorentz-invariant and, hence, that they define an absolute simultaneity. But principle III then requires that I also count every thing and event which bears the *transitive closure* of R to me (i.e. which bears R to me, or which bears R to something that bears R to me, or which bears R to something that bears R to something that bears R to me, or....) as real. But every thing and event in space–time bears the transitive closure of the above relation R to me, at least if there are enough observers. And, if we allow all physical systems (even electromagnetic fields, etc.) as 'observers' (as why should we not?) and allow observers to use coordinate systems in which they are not at rest, then there are certainly 'enough observers'.

In the second part of this chapter I showed that contingent statements about future events already have a truth value. In consequence, the 'tenseless' notion of existence (i.e. the notion that amounts to 'will exist, or has existed, or exists right now') is perfectly well-defined. This is fortunate, since the upshot of the first part of the chapter could also have been stated by saying that the notion of being 'real' turns out to be coextensive with the *tenseless* notion of existence.

I conclude that the problem of the reality and the determinateness of future events is now solved. Moreover, it is solved by physics and not by philosophy. We have learned that we live in a four-dimensional and not a three-dimensional world, and that space and time – or, better, space-like separations and time-like separations – are just two aspects of a single four-dimensional continuum with a peculiar metric which sometimes permits distance$(y, x) = 0$ even when $x \neq y$. Indeed, I do not believe that there are any longer any *philosophical* problems about Time; there is only the physical problem of determining the exact physical geometry of the four-dimensional continuum that we inhabit.

In this paper I have talked only about the relativistic aspects of the problem of physical time: there is, of course, also the problem of thermodynamics, and whether the Second Law does or does not explain the existence of 'irreversible' processes (the so-called 'problem of the direction of time'), and the problem of the existence or nonexistence of true irreversibilities in quantum mechanics, which, I gather, is currently under hot discussion. I have not talked about these problems.

12
Memo on 'conventionalism'*

(1) Grünbaum's last memorandum in defense of 'conventionalism' with respect to geometry seemed to me to mark a definite shift in his position since our Princeton conference. I hope that this shift was conscious and intentional, and that 'polemical' considerations will not lead Grünbaum to repudiate this 'shift' if I point it out!

The shift I have in mind is this: at the Princeton conference Grünbaum emphasized that his 'conventionalism' is a logical (or perhaps metaphysical?) thesis, but certainly not an empirical one. I specifically put to him the following question, which I regard as of central importance for this area: what he would say if the world were 'Newtonian' (e.g. the Michelson–Morley experiment had the 'other' result, the velocity of light relative to the 'ether' were measurable, etc.); and he replied that his position would be unchanged. That is, in a Newtonian world, even one in which the notion of 'absolute velocity' has obvious physical meaning, length is still not an *intrinsic* (Grünbaum's word, not mine!) property of physical objects, and simultaneity is still not an intrinsic relation between physical events. The choice of a metric is still a matter of convention; although (in such a world) one convention† is immensely superior to all rivals, from a practical standpoint. (This was Reichenbach's standpoint in his Space–Time book as I pointed out; and it is precisely this notion of a 'convention' which cannot be told from a bona-fide physical law‡ that I find so puzzling.) Now, however, it

* First published by the Minnesota Center for the Philosophy of Science, University of Minnesota (22 April 1959) and circulated privately.
† Here 'conventions' are counted as the same if they differ only in the choice of a unit. The conventional character of the choice of a unit is, of course, not at issue.
‡ The 'conventions' mentioned by Reichenbach are most naturally stated as laws: 'Transportation of rigid bodies does not change their length'; 'the velocity of light (in a vacuum) is a constant, no matter what rest system is employed', etc. It would be virtually impossible to reformulate these statements as explicit definitions. Yet if they are stated as laws, then all the objections of my 'Analytic-Synthetic' paper weigh against giving them a status any different from the status of other mechanical or optical laws. (The first of these statements could be reworded as an explicit definition of 'rigid'; but then it is no longer a 'coordinating definition' in Reichenbach's sense. Rather, 'iron bars are rigid' (or some such statement) now becomes the 'congruence-definition'.)

appears that the strongest arguments for conventionalism are from its alleged empirical import; and this empirical import of 'conventionalism' is linked by Grünbaum to the special theory of relativity. I conclude, therefore, (I hope not rashly!) that Grünbaum's position is *now* this: that in *our* world, by virtue of certain physical facts, there is a sense in which the choice of a metric is to a certain extent arbitrary. With this I agree (although I have reservations about Grünbaum's arguments); but this is very different from the claim that even in a Newtonian world the length of a body is not an intrinsic property of it, whereas its temperature is. The difference is just the difference between the traditional 'philosophical' relativity (i.e. the claim that distance is 'really' a relation to a material standard, and that a world of just empty space – without material clocks and measuring rods – is 'really' impossible) and scientific Relativity. The difference, I take it, is no less than that between a metaphysical position and a scientific theory.

(2) The argument that Grünbaum uses, and that I am dubious about, is this: that the conventionalist does not maintain that we are free to use geometric words in any way we wish. He points out that there are certain ways in which we *cannot* consistently use them (given certain physical facts about our world). In doing this he indirectly points to certain important physical laws. Thus conventionalism with respect to geometry has empirical content, and cannot be assimilated to the more trivial observation that the assignment of referents to linguistic expressions is, in a certain sense, arbitrary.

It is easy to show that this argument does *not* show that the assignment of meanings to geometric terms ('length', 'congruent', etc.), is arbitrary in a sense more significant than the trivial sense in which one might point out that it is 'arbitrary' that the sound 'red' is used to denote the color red rather than the color green, or the city of London, or anything else. In fact, we need only to point out that there are ways in which one *cannot* consistently use the sound 'red' (given certain physical facts about our world). For instance, if we wish 'red' to denote a color that human beings are able to see, then it cannot be used to denote ultraviolet. For instance, if we wish 'red' to denote a property of visible physical objects, then it cannot be used to refer to just those objects whose size is greater than one ten-thousandth and smaller than one-thousandth of an inch. The first of these uses of 'red' is not possible (consistently with the intention stated) because of the highly significant empirical fact that humans cannot see ultraviolet. The second use is impossible for a similar reason: human beings cannot see objects in the size range mentioned, at least not accurately enough to classify them with respect to size.

Then, in analogy with Grünbaum's argument, we can conclude that 'conventionalism with respect to colors' is highly significant because it entails certain physical laws, and hence (incorrectly) that it is not a special case of the Principle of Conventionality† (that, except for a few isolated cases of onomatopoeia like 'bow-wow', etc., there is no 'natural' link between a given thing and the sound-sequence used to refer to it in a given language). The mistake should be obvious: the empirical content is not in the statement that the choice between certain possible uses is a matter of convention, but in the exclusion of certain uses as not possible for empirical reasons. Such exclusions can be made in the case of any words; that we can make them in the case of geometric words is the last thing that could possibly show that the choice of a metric is 'arbitrary' in a way that the choice of a referent for the sound-sequence 'red' is not.

(3) The relevance of the special theory of relativity seems to me to be much more direct (in the mathematician's sense of 'direct' and 'indirect', almost) than Grünbaum makes it to be. I maintain (i) that the word 'length' refers, in English, to a certain magnitude which was incorrectly believed (before Einstein) to be a function of one argument (the physical object whose length is being talked about), but which is *in fact* a function of two arguments: the physical object whose length is being talked about, and the inertial system which is selected to be the 'rest system'. This dependence on the second argument was able to go unnoticed for so long simply because the relative velocities of different rest systems are, in practice, always so small as not to make any difference. (ii) The special theory of relativity asserts, among other things, that the laws of nature have a certain symmetry with respect to all inertial systems. (iii) The magnitude 'length' can be *transformed* into a function of one argument by artificially fixing the second argument (choosing a fixed 'rest system' by fiat). One obtains in this way a family of one-place magnitudes each of which corresponds to intuitive 'length' for an English-speaking observer in a suitable rest system. But it is a *direct* consequence of (ii) that the choice of one such concept of 'length' over any other is physically arbitrary.

There are a number of points 'behind' the above account which cannot be gone into fully here. One is this: I have described the English language (or rather the word 'length' in the English language) by saying that the word 'length' has a fixed referent which did not change when the Einstein theory was adopted (although the nature of that referent became better understood). In part, this account is suggested by

---

† This principle, in a more precise form, is stated by Paul Ziff in a forthcoming book. The informal statement that 'the choice of a referent for a given word is arbitrary' is objectionable since, literally speaking, the meaning of words like 'red' is not product of *choice* at all, but of slow evolution.

plausible reasoning from a number of other cases: we know very well, in general, that people may be able to measure a magnitude pretty well without being able to measure it exactly in all cases, and without being clear as to its real nature. Hence, it is not implausible that people may be talking about the same thing when they use the word 'length', even if they do not always agree in their measurements, or always use the same measuring devices, and even if their theory of what 'length' is has recently undergone some substantial modifications. But there is also a definite linguistic Principle† behind my view: namely, the methodological principle that one should prefer that account of a language that splits words into different 'senses' in the smallest number of cases (subject to certain constraints, to insure that a description does not lump together uses of 'homonyms' even in cases where there is no statable definition that covers the joint distribution). In other words, one cannot say that a word has the same sense in two different contexts unless one can formulate a definition (in the sense of 'dictionary definition' not '*Principia Mathematica* definition') that covers both cases reasonably well; but, by the same token, whenever one can cover a whole distributional area by a single definition, one should prefer a description that does this to a description which splits the area into several 'senses', or, even worse, into several 'homonyms'. (Paul Ziff has put this rather well in a maxim he calls 'Occam's Eraser': 'Dictionary entries are not to be multiplied without necessity'.) It should be observed that this principle has direct relevance to the Philosophy of Science. For example, Bridgman's suggestion, that one should count as many meanings of 'length' as there are ways of measuring length, would be immediately ruled out as violating sound linguistic methodology.

Applying 'Occam's Eraser' to the present case: if D_1 is a description of English according to which the word 'length' had *no* referent before

---

† This is called the 'Principle of Composition' in Ziff's book (Ziff, 1960). It is closely related to Frege's principle that 'the sense of a sentence is a function of the sense of its parts'. In my opinion this principle (properly formulated) is the most important of all characteristics of all human languages (natural or constructed). The only 'languages' that are exceptions to it are completely finite codes: and the most important fact about real language is that it is *not* a finite code. A real language consists of an infinite (or potentially infinite) set of utterances (sentences, well formed formulas, etc.) built up out of a finite set of building blocks in such a way that the meaning will be *predictable*. This already insures that (1) there must be a certain parallelism between syntax and semantics; and (2) that differences and similarities in meaning must in general go along with differences and similarities in utterance-structure. This last fact underlies the methodological principle alluded to in the text: we try to give the same word (morpheme, construction, etc.) the same meaning wherever possible in our analysis of a language so as to maximize the extent to which meaning (of the whole utterance) will be determined by *composition* plus the meanings associated with the basic units.

Einstein (although people mistakenly thought it had), and D_2 is a description according to which 'length' had the same referent before and after Einstein (although people's views about that referent have changed), and D_1 and D_2 meet the obvious empirical constraints equally well, then, other things being equal, D_2 is to be preferred to D_1. This reflects (A) the conception of linguistics as a theoretical, rather than a purely 'descriptive' science; and (B) the view that general linguistic theory is concerned with criteria for *choosing between* descriptions, not with discovery methods. (For a brilliant discussion of both points, see Chomsky (1957)).

Assuming that my description of the word 'length' in English is accepted, then the rest follows easily: that the special theory of relativity asserts that length is a two-place function, not a one-place function, and that the theory also asserts that, while 'length' can be transformed into a one-place function by picking a rest system R_0 (i.e. one can agree to mean λ_x (length (x, R_0)) and not $\lambda_x\lambda_y$ (length (x, y)) by 'length'†), the choice of an R_0 to serve this purpose is physically arbitrary. This can be fairly well expressed by saying

(i) According to the special theory of relativity, the choice of a rest system makes no difference to the form of the physical laws.

or, with more vagueness, by saying:

(ii) According to the special theory of relativity, the choice of a metric (from a certain class) is a matter of arbitrary convention.

It will be seen then, that a certain 'conventionalism' with respect to the choice of a metric is a part of the special theory of relativity. On the other hand, this is so for physical reasons: in a Newtonian world there would be no reason to assert that length is not a one-place function. Hence, (i) and (ii) have no analogues in a Newtonian world.

(Even in a Newtonian world, the *measurement* of length depends on the use of a standard body; but so does the measurement of anything. This does not make length, as a theoretical magnitude used by the Newtonian physicist to describe his world, a 'relation to a standard object' any more

† Instead of saying that the word 'length' (in isolation) refers to the two place predicate length (x, y) (the length of x in the system y), it would perhaps be clearer to say that the expression 'the length of x' (in English) means length $(x$, some contextually definite y); and that the contribution of Special Relativity was to make the context-dependence explicit. (I am not saying that the context-dependence is a linguistic matter; it is not, it is something connected with the physical nature of the aspect of the world we are talking about when we use the word 'length'. On the other hand, the context dependence would speedily become noticeable at the linguistic (behavioral) level if people travelled about the universe at very high relative velocities.)

than temperature is a 'relation to a thermometer'. Also the choice of a unit depends on a standard, in both cases: but the existence of such magnitudes as temperature and length does not depend on human beings having selected corresponding units.)

(4) Let me note in passing that general relativity does not offer a clear verdict (as Grünbaum himself has pointed out) for the strong forms of 'relativity' associated with the names of Leibnitz and Mach. The analogue of (ii) indeed holds (with the 'certain class' very much widened); however, the claim that talk about distance is disguised talk about *material bodies* (inconceivability of 'empty space') is by no means upheld. On the contrary, 'space' itself (space-time) has certain invariant properties (given by the g_{ik} tensor). The g_{ik} tensor could itself be regarded as a disguised description of relations between material bodies only if it could be explicitly defined in terms of (non-spatial) predicates of material objects. But such a definition is not possible in general relativity; and in fact more than one g_{ik} tensor is formally possible even in an 'empty' universe (i.e. one in which the T_{ik} tensor vanishes). It is for this reason, I assume, that Grünbaum adopted a sort of 'halfway' conventionalism at Princeton, according to which the metrical properties of space are imposed by arbitrary choice (subject to the limitations fixed by the topology) but the topology has an absolute (= 'intrinsic') significance. Such a 'half-way' position seems obviously unstable!

Moreover, it should be remarked that the tendency of the later Einstein was not towards a closer conformity to the Mach-Leibnitz ideal (elimination of 'space', the 'ether', etc.) but in just the opposite direction: statements about space were not to be construed as statements about material bodies: rather, material bodies were to be regarded as 'funny' places in the field, and the field itself was to be identified with *space-time* as characterized by appropriate tensors (g_{ik}, T_{ik}, etc.). In short, the tendency was to translate all statements about matter into statements about one object ('the field' = space-time). Einstein himself remarked that this was a kind of rehabilitation of the ether concept; and it is hard to see how either Leibnitz or Mach would have taken much comfort from the ideal of a completed Unified Field Theory.

(5) Since I have alluded here to the problem of the alleged inconceivability of 'empty space', and since many philosophers today seem to share the feeling that there are *logical* difficulties associated with the idea of a universe with nothing in it, it may be worth while presenting my reasons for rejecting the view that a material 'standard' is necessary for the existence of even space-time.

The most striking difficulty with the view that a material 'standard' is necessary is the vagueness of the notion of a 'material object'. Will one electron do? Or must there be two electrons, at a distance from each other? (Must they be moving in different directions?) Is a whole hydrogen atom necessary? Is a macro-object necessary? etc.

For a Verificationist there is of course no difficulty. A Verificationist would have to demand the existence of measuring rods – superimposable macro-objects. But a consistent Verificationist should also regard the notion of a world with no living things in it as meaningless: for the existence of a world which *never* had living things in it (at any time, past, present, or future) would be unverifiable (because only living things can 'verify' anything). I assume, however, that none of us are 'Verificationists' in this sense. We are all willing to admit (I assume) that a world with only hydrogen atoms in it is perfectly conceivable (although one cannot conceive of verifying that the universe has (tenselessly) nothing but hydrogen atoms in it). What I cannot see is, once one has given up the demand for 'verification', why even 'hydrogen atoms' should be necessary.

Suppose 'hydrogen atoms' are enough (i.e. a universe with nothing in it but hydrogen atoms is admitted to be conceivable). Why not a world with nothing in it but electrons? And if this is conceivable, why not a world with nothing in it but electromagnetic fields? Indeed, the electromagnetic field is just the photon field (the field whose quanta are photons) as the Dirac field is the electron field (the field whose quanta are positive and negative electrons). Finally, if a universe with only electromagnetic fields (and photons) is admitted to be conceivable, why not a universe with only gravitational fields?

Up to the last few months one might have replied: because gravitational fields have no corpuscular character; but this reply is no longer available! Dirac and Peter Bergmann have (independently) shown that gravitational waves can carry energy. If Dirac's theoretical prediction of the existence of 'gravitrons' is borne out (as no doubt it will be), then the gravitational field has the same dual nature as the other fields.

This last point bears directly on Grünbaum's remark (in the Philosophical Review) that the fact that general relativity admits of solutions even when T_{ik} vanishes (empty universes) shows a certain incompatibility with 'Mach's Principle' (I would prefer to say: with the relativistic philosophy of space associated with the principle). First, one or two technicalities (unavoidably): (i) Vanishing of the energy is *not* the same thing (quantum mechanically!) as vanishing of the number of particles. Indeed, these two quantities (energy and number of particles) do not *commute*: hence, if the number of particles is zero, then the energy can only have zero as *average* value. This means that there is electro-magnetic radiation *even in a hard vacuum*: this is called the 'vacuum field' in quantum mechanics. Conversely, if the energy vanishes, then the number of particles (in positive levels of the energy) will have zero as average value (sometimes the number will be 'negative', i.e. more particles will have negative than positive energy levels). (ii) This means that one cannot have a universe with *both* no particles and no fields: if there are no particles, there will be the 'vacuum field'; and if the energy is zero, there will be particles in both positive and negative energy-levels. (In fact, you can't have a universe with no fields; since every particle has a field associated with it.) Now the point: why is a universe with no field *except* the 'vacuum' field not perfectly conceivable?

A computationalist might reply that the 'vacuum' field is a mere theoretical fiction. For a realist, this position is a bit difficult to sustain, since the 'vacuum' field has perfectly detectable experimental effects!

To sum up: if you say that space is inconceivable without matter, I feel entitled to reply: 'What is matter?' 'And are you sure it's not the same thing as space?' Is a universe with 'only' a g_{ik} field in it really the same thing as an 'empty' universe? Is it the same thing to say that the g_{ik} field has the value zero everywhere (Euclidean space) and that the g_{ik} field does not exist? (Remember that O is a 'sharp' value, and that if *any* entity has a sharp value, then *other* entities, that do not commute with the entity in question, will not have sharp values.)

To put it bluntly: if spatial notions make sense only in terms of such ideas as 'superimposing' two measuring-rods, then spatial notions don't make sense unless there are bodies large enough for the notion of 'touching' to make sense. In particular, spatial notions should not make sense in a world which contained nothing but *photons* (although the T_{ik} tensor would not vanish). Similarly, if periodicity of some kind is necessary for temporal notions to make sense, then one must require the existence of at least hydrogen atoms before one can speak of time. If, on the other hand, the periodicities and the distance-relationships required to make sense of spatial and temporal notions (on the relativist view) are already provided by the existence of *waves* (wave periodicity, superposition of waves), then these notions make sense even in a hard vacuum (the vacuum field has a perfectly good wave character).

Against the foregoing, it might be replied that what is held to be inconceivable is an empty *Newtonian* universe. But this claim seems to be in no better shape (except contingently: Newtonian physics is false, so any kind of Newtonian world, 'empty' or 'full', is *physically* impossible). Why should corpuscles be necessary in a Newtonian world? Why not just a Newtonian ether with waves in it? (Of course the ether doesn't exist: but this is a *physical* fact. The ether was a perfectly good – and extremely fruitful – 'hypothetical construct'.) And why not an ether with *no* waves in it? If the reply is that *time* makes no sense without motion, I can only reply that the idea of a universe with no *relative* motion (one billiard ball) seems perfectly consistent. Of course one could not *measure* time in such a universe; but that is not the issue. The issue is rather: can we, having introduced time (or spatio-temporal distance) as a theoretical magnitude in terms of which we describe our world, sensibly speak of *other*, logically or physically possible worlds, in which that magnitude would not be *measurable* (*ex hypothesu*) but on which it would have definite values?

CONJECTURE: All arguments against the foregoing have the following form: time *must* be defined in terms of (relative) motion; therefore, time without relative motion is inconceivable. REPLY: 'time' isn't *defined* at all. Time can only be *measured* by utilizing relative motion; but time cannot be defined in terms of non-temporal notions.† Statements about empty universes are consistent precisely because there are no logical (= formal, definitional) relations between temporal and non-temporal concepts. Such statements are meaningful in the only relevant sense; they have a use in physical theory. Their use is not as descriptions of the actual world (nobody claims that nothing exists) but as descriptions of theoretically interesting limiting cases.

† In particular, I regard Grünbaum's attempt to explicitly define temporal in terms of causal notions in the forthcoming Carnap-volume essay as quite unsuccessful.

13
What theories are not*

In this paper I consider the role of *theories* in empirical science, and attack what may be called the 'received view' – that theories are to be thought of as 'partially interpreted calculi' in which only the 'observation terms' are 'directly interpreted' (the theoretical terms being only 'partially interpreted', or, some people even say, 'partially understood').

To begin, let us review this received view. The view divides the non-logical vocabulary of science into two parts:

OBSERVATION TERMS	THEORETICAL TERMS
such terms as	such terms as
'red'	'electron'
'touches'	'dream'
'stick', etc.	'gene', etc.

The basis for the division appears to be as follows: the observation terms apply to what may be called publicly observable things and signify observable qualities of these things, while the theoretical terms correspond to the remaining unobservable qualities and things.

This division of terms into two classes is then allowed to generate a division of statements into two† classes as follows:

OBSERVATIONAL STATEMENTS	THEORETICAL STATEMENTS
statements containing only observation terms and logical vocabulary	statements containing theoretical terms

* First published in E. Nagel, P. Suppes, and A. Tarski (eds.), *Logic, Methodology and Philosophy of Science* and reprinted with the permission of the publishers, Stanford University Press. © 1962 by the Board of Trustees of the Leland Stanford Junior University.

† Sometimes a *tripartite* division is used: observation statements, theoretical statements (containing *only* theoretical terms), and 'mixed' statements (containing both kinds of terms). This refinement is not considered here, because it avoids none of the objections presented below.

Lastly, a scientific theory is conceived of as an axiomatic system which may be thought of as initially uninterpreted, and which gains 'empirical meaning' as a result of a specification of meaning *for the observation terms alone*. A kind of partial meaning is then thought of as drawn up to the theoretical terms, by osmosis, as it were.

The observational–theoretical dichotomy

One can think of many distinctions that are crying out to be made ('new' terms vs. 'old' terms, technical terms vs. non-technical ones, terms more-or-less peculiar to one science vs. terms common to many, for a start). My contention here is simply:

(1) The *problem* for which this dichotomy was invented ('how is it possible to interpret theoretical terms?') does not exist.

(2) A basic reason some people have given for introducing the dichotomy is false: namely, justification in science does *not* proceed 'down' in the direction of observation terms. In fact, justification in science proceeds in any direction that may be handy – more observational assertions sometimes being justified with the aid of more theoretical ones, and vice versa. Moreover, as we shall see, while the notion of an *observation report* has some importance in the philosophy of science, such reports cannot be identified on the basis of the vocabulary they do or do not contain.

(3) In any case, whether the reasons for introducing the dichotomy were good ones or bad ones, the double distinction (observation terms–theoretical terms, observation statements–theoretical statements) presented above is, in fact, completely broken-backed. This I shall try to establish now.

In the first place, it should be noted that the dichotomy under discussion was intended as an explicative and not merely a stipulative one. That is, the words 'observational' and 'theoretical' are not having arbitrary new meanings bestowed upon them; rather, pre-existing uses of these words (especially in the philosophy of science) are presumably being sharpened and made clear. And, in the second place, it should be recalled that we are dealing with a double, not just a single, distinction. That is to say, part of the contention I am criticizing is that, once the distinction between observational and theoretical *terms* has been drawn as above, the distinction between theoretical statements and observational reports or assertions (in something like the sense usual in methodological discussions) can be drawn in terms of it. What I mean when I say that the dichotomy is 'completely broken-backed' is this:

(A) If an 'observation term' is one that cannot apply to an unobservable, then there are no observation terms.†
(B) Many terms that refer primarily to what Carnap would class as 'unobservables' are not theoretical terms; and at least some theoretical terms refer primarily to observables.
(C) Observational reports can and frequently do contain theoretical terms.
(D) A scientific theory, properly so-called, may refer only to observables. (Darwin's theory of evolution, as originally put forward, is one example.)

To start with the notion of an 'observation term': Carnap's formulation in *Testability and Meaning* (Carnap, 1955) was that for a term to be an observation term not only must it correspond to an observable quality, but the determination whether the quality is present or not must be able to be made by the observer in a relatively short time, and with a high degree of confirmation. In his most recent authoritative publication (Carnap, 1956), Carnap is rather brief. He writes, 'the terms of V_O [the 'observation vocabulary'] are predicates designating observable properties of events or things (e.g. 'blue', 'hot', 'large', etc.) or observable relations between them (e.g. 'x is warmer than y', 'x is contiguous to y', etc.) (Carnap, 1956, p. 41). The only other clarifying remarks I could find are the following: 'The name "observation language" may be understood in a narrower or in a wider sense; the observation language in the wider sense includes the disposition terms. In this article I take the observation language L_O in the narrower sense' (Carnap, 1956, p. 63). 'An observable property may be regarded as a simple special case of a testable disposition: for example, the operation for finding out whether a thing is blue or hissing or cold, consists simply in looking or listening or touching the thing, respectively. Nevertheless, *in the reconstruction of the language* [italics mine] it seems convenient to take some properties for which the test procedure is extremely simple (as in the examples given) as directly observable, and use them as primitives in L_O' (Carnap, 1956, p. 63).

These paragraphs reveal that Carnap, at least, thinks of observation terms as corresponding to qualities that can be detected without the aid of instruments. But always so detected? Or can an observation term refer sometimes to an observable thing and sometimes to an unobservable? While I have not been able to find any explicit statement on this point,

† I neglect the possibility of trivially constructing terms that refer only to observables: namely, by conjoining 'and is an observable thing' to terms that would otherwise apply to some unobservables. 'Being an observable thing' is, in a sense, highly theoretical and yet applies only to observables!

217

it seems to me that writers like Carnap must be *neglecting* the fact that *all* terms – including the 'observation terms' – have at least the possibility of applying to unobservables. Thus their problem has sometimes been formulated in quasi-historical terms – 'How could theoretical terms have been introduced into the language?' And the usual discussion strongly suggests that the following puzzle is meant: if we imagine a time at which people could only talk about observables (had not available any theoretical terms), how did they ever manage to *start* talking about unobservables?

It is possible that I am here doing Carnap and his followers an injustice. However, polemics aside, the following points must be emphasized:

(1) Terms referring to unobservables are *invariably* explained, in the actual history of science, with the aid of already present locutions referring to unobservables. There never was a stage of language at which it was impossible to talk about unobservables. Even a three-year-old child can understand a story about 'people too little to see'† and not a single 'theoretical term' occurs in this phrase.

(2) There is not even a single *term* of which it is true to say that it *could not* (without changing or extending its meaning) be used to refer to unobservables. 'Red', for example, was so used by Newton when he postulated that red light consists of *red corpuscles*.‡

In short: if an 'observation term' is a term which *can*, in principle, only be used to refer to observable things, then *there are no observation terms*. If, on the other hand, it is granted that locutions consisting of just observation terms can refer to unobservables, there is no longer any reason to maintain *either* that theories and speculations about the unobservable parts of the world must contain 'theoretical (= non-observation) terms' *or* that there is any general problem as to how one

† Von Wright has suggested (in conversation) that this is an *extended* use of language (because we first learn words like 'people' in connection with people we *can* see). This argument from 'the way we learn to use the word' appears to be unsound, however (cf. Fodor, 1961).

‡ Some authors (although not Carnap) explain the intelligibility of such discourse in terms of logically possible submicroscopic observers. But (a) such observers could not see single photons (or light corpuscles) even on Newton's theory; and (b) once such physically impossible (though logically possible) 'observers' are introduced, why not go further and have observers with sense organs for electric charge, or the curvature of space, etc.! Presumably because *we* can see *red*, but not *charge*. But then, this just makes the point that we *understand* 'red' even when applied outside our normal 'range', even though we learn it ostensively, without *explaining* that fact. (The explanation lies in this: that understanding any term – even 'red' – involves at least two elements: internalizing the syntax of a natural language, and acquiring a background of ideas. Overemphasis on the way 'red' is *taught* has led some philosophers to misunderstand how it is *learned*.)

can introduce terms referring to unobservables. Those philosophers who find difficulty in how we understand theoretical terms should find an equal difficulty in how we understand 'red' and 'smaller than'.

So much for the notion of an 'observation term'. Of course, one may recognize the point just made – that the 'observation terms' also apply, in some contexts, to unobservables – and retain the class (with a suitable warning as to how the label 'observational term' is to be understood). But can we agree that the complementary class – what should be called the 'nonobservable terms' – is to be labelled 'theoretical terms'? No, for the identification of 'theoretical term' with 'term (other than the "disposition terms", which are given a special place in Carnap's scheme) designating an unobservable quality' is unnatural and misleading. On the one hand, it is clearly an enormous (and, I believe, insufficiently motivated) extension of common usage to classify such terms as 'angry', 'loves', and so forth, as 'theoretical terms' simply because they allegedly do not refer to public observables. A theoretical term, properly so-called, is one which comes from a scientific *theory* (and the almost untouched problem, in thirty years of writing about 'theoretical terms' is what is *really* distinctive about such terms). In this sense (and I think it the sense important for discussions of science) 'satellite' is, for example, a theoretical term (although the things it refers to are quite observable†) and 'dislikes' clearly is not.

Our criticisms so far might be met by re-labelling the first dichotomy (the dichotomy of terms) 'observation vs. non-observation', and suitably 'hedging' the notion of 'observation'. But more serious difficulties are connected with the identification upon which the second dichotomy is based – the identification of 'theoretical statements' with statements containing nonobservation ('theoretical' terms) and 'observation statements' with 'statements in the observational vocabulary'.

That observation statements may contain theoretical terms is easy to establish. For example, it is easy to imagine a situation in which the following sentence might occur: 'We also *observed* the creation of two electron-positron pairs'.

This objection is sometimes dealt with by proposing to 'relativize' the observation-theoretical dichotomy to the context. (Carnap, however, rejects this way out in the article we have been citing.) This proposal

---

† Carnap might exclude 'satellite' as an observation term, on the ground that it takes a comparatively long time to verify that something is a satellite with the naked eye, even if the satellite is close to the parent body (although this could be debated). However, 'satellite' cannot be excluded on the quite different ground that many satellites are too far away to see (which is the ground that first comes to mind) since the same is true of the huge majority of all *red* things.

to 'relativize' the dichotomy does not seem to me to be very helpful. In the first place, one can easily imagine a context in which 'electron' would occur, in the same text, in *both* observational reports and in theoretical conclusions from those reports. (So that one would have distortions if one tried to put the term in either the 'observational term' box or in the 'theoretical term' box.) In the second place, for what philosophical problem or point does one require even the relativized dichotomy?

The usual answer is that sometimes a statement A (observational) is offered in support of a statement B (theoretical). Then, in order to explain why A is not itself questioned in the context, we need to be able to say that A is functioning, in that context, as an observation report. But this misses the point I have been making! I do not deny the need for some such notion as 'observation report'. What I deny is that the distinction between observation reports and, among other things, theoretical statements, can or should be drawn on the basis of vocabulary. In addition, a relativized dichotomy will not serve Carnap's purposes. One can hardly maintain that theoretical terms are only partially interpreted, whereas observation terms are completely interpreted, if no sharp line exists between the two classes. (Recall that Carnap takes his problem to be 'reconstruction of the language', not of some isolated scientific context.)

Partial interpretation

The notion of 'partial interpretation' has a rather strange history – the term certainly has a technical ring to it, and someone encountering it in Carnap's writings, or Hempel's, or mine† certainly would be justified in supposing that it was a term from mathematical logic whose exact definition was supposed to be too well known to need repetition. The sad fact is that this is not so! In fact, the term was introduced by Carnap in a section of his monograph (Carnap, 1939), without definition (Carnap *asserted* that to interpret the observation terms of a calculus is automatically to 'partially interpret' the theoretical primitives, without explanation), and has been subsequently used by Carnap and other

---

† I used this notion uncritically in 'Mathematics and the existence of abstract entities' (Putnam, 1957). From the discussion, I seem to have had concept (2) (below) of 'partial interpretation' in mind, or a related concept. (I no longer think *either* that set theory is helpfully thought of as a 'partially interpreted calculus' in which only the 'nominalistic language' is directly interpreted, *or* that mathematics is best identified with set theory for the purposes of philosophical discussion.)

authors including myself) with copious cross references, but with no further explanation.

One can think of (at least) three things that 'partial interpretation' could mean. I will attempt to show that none of these three meanings is of any use in connection with the 'interpretation of scientific theories'. My discussion has been influenced by a remark of Ruth Anna Putnam to the effect that not only has the concept been used without any *definition*, but it has been applied indiscriminately to *terms, theories,* and *languages*.

(1) One might give the term a meaning from mathematical logic as follows (I assume familiarity here with the notion of a 'model' of a formalized theory): to 'partially interpret' a theory is to specify a non-empty class of intended models. If the specified class has one member, the interpretation is *complete*; if more than one, properly *partial*.

(2) To partially interpret a *term P* could mean (for a Verificationist, as Carnap is) to specify a verification–refutation procedure. If $\bar{a}_1$ is an individual constant designating an individual a_1 (Carnap frequently takes space–time points as the individuals, assuming a 'field' language for physics), and it is possible to verify $P(\bar{a}_1)$, then the individual a_1 is in the extension of the term P; if $P(\bar{a}_1)$ is refutable, then a_1 is in the extension of $\bar{P}$, the negation of P; and if the existing test procedures do not apply to a_1 (e.g. if a_1 fails to satisfy the antecedent conditions specified in the test procedures) then it is *undefined* if a_1 is in or out of the extension of P.

This notion of partial interpretation of terms immediately applies to terms introduced by reduction sentences† (Carnap calls these 'pure disposition terms'). In this case the individual a_1 is either in the extension of P or in the extension of $\bar{P}$, provided the antecedent of at least one reduction sentence 'introducing' the term P is true of a_1, and otherwise it is *undefined* whether $P(a_1)$ is true or not. But it can be extended to theoretical primitives in a *theory* as follows: if $P(\bar{a}_1)$ follows from the postulates and definitions of the theory and/or the set of all true observation sentences, then a_1 is in the extension of P; if $\bar{P}(\bar{a}_1)$ follows from the postulates and definitions of the theory and/or the set of all true observation sentences, then a_1 is in the extension of $\bar{P}$; in all other cases, $P(\bar{a}_1)$ has an *undefined* truth-value.

(3) Most simply, one might say that to partially interpret a formal *language* is to *interpret part* of the language (e.g. to provide translations into common language for some terms and leave the others mere dummy symbols).

† For the definition of this concept, see Carnap (1955).

Of these three notions, the first will not serve Carnap's purposes, because it is necessary to use theoretical terms in order to specify even a *class* of intended models for the usual scientific theories. Thus, consider the problem of specifying the intended values of the individual variables. If the language is a 'particle' language, then the individual variables range over 'things' – but things in a *theoretical* sense, including mass points and systems of mass points. It is surely odd to take the notion of a 'physical object' as either an observational or a purely logical one when it becomes wide enough to include point-electrons, at one extreme, and galaxies at the other. On the other hand, if the language is a 'field' language, then it is necessary to say that the individual variables range over *space–time points* – and the difficulty is the same as with the notion of a 'physical object'.

Moving to the predicate and function-symbol vocabulary: consider, for example, the problem of specifying either a unique intended interpretation or a suitable class of models for Maxwell's equations. We must say, at least, that E and H are intended to have as values vector-valued functions of *space–time points*, and that the norms of these vectors are to measure, roughly speaking, the velocity-independent force on a small test particle per unit charge, and the velocity-dependent force per unit charge. One might identify force with mass times (a suitable component of) acceleration, and handle the reference to an (idealized) test particle via 'reduction sentences', but we are still left with 'mass', 'charge', and, of course, 'space–time point'. ('Charge' and 'mass' have as values a real-valued function and a non-negative real-valued function, respectively, of space–time points; and the values of these functions are supposed to measure the intensities with which certain *physical magnitudes* are present at these points – where the last clause is necessary to rule out flagrantly unintended interpretations that can never be eliminated otherwise.)

(One qualification: I said that *theoretical* terms are necessary to specify even a *class* of intended models – or of models that a realistically minded scientist could accept as what he has in mind. But 'physical object', 'physical magnitude' and 'space–time point' are not – except for the last – 'theoretical terms', in any idiomatic sense, any more than they are 'observation terms'. Let us call them for the nonce simply 'broad spectrum terms' – noting that they pose much the same problems as do certain meta-scientific terms, for example, 'science' itself. Of them we might say, as Quine does of the latter term (Quine, 1957), that they are not defined in advance – rather science itself tells us (with many changes of opinion) what the scope of 'science' is, or of an individual science, for example, chemistry, what an 'object' is, what 'physical magnitudes' are. In this way, these terms, although not theoretical terms, tend eventually to acquire technical senses via theoretical definitions.)

A further difficulty with the first notion of 'partial interpretation' is that theories with false observational consequences have *no* interpretation (since they have no model that is 'standard' with respect to the observation terms). This certainly flies in the face of our usual notion of an interpretation, according to which such a theory is *wrong*, not *senseless*.

The second notion of partial interpretation that we listed appears to me to be totally inadequate even for the so-called 'pure disposition terms', for example, 'soluble'. Thus, let us suppose, for the sake of a simplified example, that there were only one known test for *solubility* – immersing the object in water. Can we really accept the conclusion that it has a *totally undefined* truth-value to say of something that is never immersed in water that it is soluble?

Let us suppose now that we notice that all the sugar cubes that we immerse in water dissolve. On the basis of this *evidence* we *conclude* that all sugar is soluble – even cubes that are never immersed. On the view we are criticizing, this has to be described as 'linguistic stipulation' rather than as 'discovery'! Namely, according to this concept of partial interpretation, what we do is *give* the term 'soluble' the *new* meaning 'soluble-in-the-old-sense-or-sugar'; and what we ordinarily describe as evidence that the unimmersed sugar cubes are soluble should rather be described as evidence that our new meaning of the term 'soluble' is compatible with the original 'bilateral reduction sentence'.

Also, although it will now be true to say 'sugar is soluble', it will still have a totally undefined truth-value to say of many, say, lumps of *salt* that *they* are soluble.

Ordinarily, 'change of meaning' refers to the sort of thing that happened to the word 'knave' (which once meant 'boy'), and 'extension of meaning' to the sort of thing that happened in Portugal to the word for family ('familhia'), which has come to include the servants. In these senses, which also seem to be the only ones useful for linguistic theory, it is simply *false* to say that in the case described (concluding that sugar is soluble) the word 'soluble' underwent either a change or an extension of meaning. The *method* of verification may have been extended by the discovery, but this is only evidence that method of verification is not meaning.

In any case, there does not seem to be any reason why we cannot agree with the customary account. What we meant all along by 'it's soluble' was, of course, 'if it *were* in water, it would dissolve'; and the case we described can *properly* be described as one of drawing an inductive inference—to the conclusion that all these objects (lumps of sugar, whether immersed or not) are soluble in *this* sense. Also, there is no reason to reject the view, which is certainly built into our use of the term 'soluble', that it has a definite (but sometimes unknown) truth-value to say of anything (of a suitable size and consistency) that it is soluble, whether it satisfies a presently-known test condition or not. Usually the objection is raised that it is 'not clear what it means' to say 'if it *were* in water, it would dissolve'; but there is no *linguistic* evidence of this unclarity. (Do people construe it in different ways? Do they ask for a paraphrase? Of course, there is a philosophical problem having to do with 'necessary connection', but one must not confuse a word's being connected with a philosophical problem with its possessing an unclear meaning.)

Coming now to theoretical terms (for the sake of simplicity, I assume that our world is nonquantum mechanical): if we want to preserve the ordinary world-picture at all, we certainly want to say it has a definite truth-value to say that there is a helium atom inside any not-too-tiny region X. But in fact, our test conditions – even if we allow tests implied by a theory, as outlined above under (2) – will not apply to small regions X in the interior of the sun, for example (or in the interior of many bodies at many times). Thus we get the following anomalous result: it is *true* to say that there are helium atoms in the sun, but it is neither true nor false that one of these is inside any given tiny subregion X! Similar things will happen in connection with theoretical statements about the very large, for example, it may be 'neither true nor false' that the average curvature of space is positive, or that the universe is finite. And once again, perfectly ordinary scientific discoveries will constantly have to be described as 'linguistic stipulations', 'extensions of meaning', and so forth.

Finally, the third sense of 'partial interpretation' leads to the view that theoretical terms have *no meaning at all*, that they are mere computing devices, and is thus unacceptable.

To sum up: we have seen that of the three notions of 'partial interpretation' discussed, each is either unsuitable for Carnap's purposes (starting with observation terms), or incompatible with a rather minimal scientific realism; and, in addition, the second notion depends upon gross and misleading changes in our use of language. Thus in *none* of these senses is 'a partially interpreted calculus in which only the observation terms are directly interpreted' an acceptable model for a scientific theory.

Introducing theoretical terms

We have been discussing a proposed solution to a philosophical problem. But what *is* the problem?

The problem is sometimes referred to as the problem of 'interpreting', that is, giving the meaning of theoretical terms in science. But this cannot be much of a *general* problem (it may, of course, be a problem in specific cases). Why should one not be able to give the meaning of a theoretical term? (Using, if necessary, *other* theoretical terms, 'broad-spectrum' terms, etc.) The problem might be restated – to give the meaning of theoretical terms, *using only observation terms*. But then, why should we suppose that this is or ought to be possible?

Something like this may be said: suppose we make a 'dictionary' of theoretical terms. If we allow theoretical terms to appear both as 'entries' and in the *definitions*, then there will be 'circles' in our dictionary. But there are circles in every dictionary!

We perhaps come closer to the problem if we observe that, while dictionaries are useful, they are useful only to speakers who already know a good deal of the language. One cannot learn one's native language to begin with from a dictionary. This suggests that the problem is really to give an account of how the use of theoretical terms is *learned* (in the life-history of an individual speaker); or, perhaps, of how theoretical terms are 'introduced' (in the history of the language).

To take the first form of the problem (the language-learning of the individual speaker): it appears that theoretical terms are learned in essentially the way most words are learned. Sometimes we are given lexical definitions (e.g. 'a *tigon* is a cross between a tiger and a lion'); more often, we simply imitate other speakers; many times we combine these (e.g. we are given a lexical definition, from which we obtain a rough idea of the use, and then we bring our linguistic behavior more closely into line with that of other speakers via imitation).

The story in connection with the introduction of a new technical term into the *language* is roughly similar. Usually, the scientist introduces the term via some kind of paraphrase. For example, one might explain 'mass' as 'that physical magnitude which determines how strongly a body resists being accelerated, e.g. if a body has twice the mass it will be twice as hard to accelerate'. (Instead of 'physical magnitude' one might say, in ordinary language, 'that property of the body', of 'that *in* the body which...' such 'broad-spectrum' notions occur in every natural language; and our present notion of a 'physical magnitude' is already an extreme refinement.) Frequently, as in the case of 'force' and 'mass', the term will be a common-language term whose new technical use is in some respects quite continuous with the ordinary use. In such cases, a lexical definition is frequently omitted, and in its place one has merely a statement of some of the differences between the usual use and the technical use being introduced. Usually one gains only a rough idea of the use of a technical term from these explicit metalinguistic statements, and this rough idea is then refined by reading the theory or text in which the term is employed. However, the role of the explicit metalinguistic statement should not be overlooked: one could hardly read the text or technical paper with understanding if one had neither explicit metalinguistic statements or previous and related uses of the technical words to go by.

It is instructive to compare here the situation with respect to the logical connectives, in their modern technical employment. We introduce the precise and technical sense of 'or', 'not', 'if-then', and so forth, using the imprecise 'ordinary language' *or, and, not* and so forth. For example, we say '$A \vee B$ shall be true *if A* is true, *and* true *if B* is true, *and A \vee B* shall be false *if A* is false *and B* is false. In particular, $A \vee B$ shall be true *even if A and B* are *both* true'. Notice that no one has proposed to say that '$\vee$' is only 'partially interpreted' because we use 'and', 'if', etc., in the ordinary imprecise way when we 'introduce' it.

In short, we can and do perform the feat of using imprecise language to introduce more precise language. This is like all use of tools – we use less-refined tools to manufacture more-refined ones. Secondly, there are even ideas that can be expressed in the more precise language that could not be intelligibly expressed in the original language. Thus, to borrow an example due to Alonzo Church, a statement of the form $\{[(A \supset B) \supset B] \supset B\}$ can probably not be intelligibly rendered in ordinary language – although one can understand it once one has had an explanation of '$\supset$' in ordinary language.

It may be, however, that the problem is supposed to be *this*: to *formalize* the process of introducing technical terms. Let us try this problem out on our last example (the logical connectives). Clearly we *could* formalize the process of introducing the usual truth-functional connectives. We would only have to take as primitives the 'ordinary language' *and, or, not* in their usual (imprecise) meanings, and then we could straightforwardly write down such characterizations as the one given above for the connective '$\vee$'. But if someone said: 'I want you to introduce the logical connectives, quantifiers, and so forth, without having any *imprecise* primitives (because using imprecise notions is not 'rational reconstruction') and also without having any *precise* logical symbols as primitives (because that would be 'circular')', we should just have to say that the task was impossible.

The case appears to me to be much the same with the 'theoretical terms'. If we take as primitives not only the 'observation terms' and the 'logical terms', but also the 'broad-spectrum' terms referred to before ('thing', 'physical magnitude', etc.), and, perhaps, certain imprecise but useful notions from common language – for example, *'harder* to accelerate', 'determines' – then we can introduce theoretical terms without difficulty:

(1) Some theoretical terms can actually be explicitly defined in Carnap's 'observation language'. Thus, suppose we had a theory according to which everything consisted of 'classical' elementary particles – little extended individual particles; and suppose no two of these were supposed to touch. Then 'elementary particle', which is a 'theoretical term' if anything is, would be explicitly definable: X is an elementary particle $\equiv$ X cannot be decomposed into parts Y and Z which are not contiguous – and the above definition requires only the notions 'X is a part of Y' and 'X is contiguous to Y'. (If we take *contiguity* as a reflexive relation, then 'part of' is definable in terms of it: X is a part of Y $\equiv$ everything that is contiguous to X is contiguous to Y. Also, Y and Z constitute a 'decomposition' of X if (i) nothing is a part of both Y and Z; (ii) X has no part which contains no part in common with either Y or Z. However, it would be perfectly reasonable, in my opinion, to take 'part of' as a *logical* primitive, along with 'is a member of' – although Carnap would probably disagree.)

We note that the, at first blush surprising, possibility of defining the obviously theoretical term 'elementary particle' in Carnap's 'observation language' rests on the fact that the notion of a *physical object* is smuggled into the language in the very interpretation of the individual variables.

(2) The kind of characterization we gave above for 'mass' (using the notion 'harder to accelerate') could be formalized. Again a broad-spectrum notion ('physical magnitude') plays a role in the definition.

But once again, no one would normally want to formalize such obviously informal definitions of theoretical terms. And once again, if someone says: 'I want you to introduce the theoretical terms using *only* Carnap's *observation terms*', we have to say, apart from special cases (like that of the 'classical' notion of an elementary particle), that this seems impossible. But why should it be possible? And what philosophic moral should we draw from the impossibility? – Perhaps only this: that we are able to have as rich a theoretical vocabulary as we do have because, thank goodness, we were never in the position of having *only* Carnap's observation vocabulary at our disposal.

14
Craig's theorem*

Craig's observation

'Craig's theorem' (Craig, 1953), as philosophers call it, is actually a corollary to an observation. The observation is that (I) *Every theory that admits a recursively enumerable set of axioms can be recursively axiomatized.*

Some explanations are in order here: (1) A *theory* is an infinite set of wffs (well-formed formulas) which is closed under the usual rules of deduction. One way of giving a theory T is to specify a set of sentences S (called the *axioms of T*) and to define T to consist of the sentences S together with all sentences that can be derived from (one or more) sentences in S by means of logic. (2) If T is a theory with axioms S, and S' is a subset of T such that every member of S can be deduced from sentences in S', then S' is called an *alternative set of axioms for T*. Every theory admits of infinitely many alternative axiomatizations – including the trivial axiomatization, in which every member of T is taken as an axiom (i.e. $S = T$). (3) A set S is called *recursive* if and only if it is decidable – i.e. there exists an effective procedure for telling whether or not an arbitrary wff belongs to S. (This is not the mathematical definition of 'recursive', of course, but the intuitive concept which the mathematical definition captures.) For 'effective procedure' one can also write 'Turing machine' (cf. Davis, 1958.).

A theory is *recursively axiomatizable* (often simply 'axiomatizable', in the literature) if it has at least one set of axioms that is recursive. Every finite set is recursive; thus all theories that can be finitely axiomatized are recursively axiomatizable. An example of a theory that can be recursively axiomatized but not finitely axiomatized is Peano arithmetic. The primitive predicates are $E(x, y)$ (also written $x = y$), $S(x, y, z)$ (also written $x+y = z$), $T(x, y, z)$ (also written $xy = z$), and $F(x, y)$ (also written $y = x'$ or $y = x+1$). The axioms are Peano's axioms for number theory plus the four formulas that recursively define addition and multiplication $[(x) (x+\text{o} = x), (x) (y) (x+y' = (x+y)'), (x) (x \cdot \text{o} = \text{o}),$

* First published in *The Journal of Philosophy*, LXII, 10 (13 May 1965).

$(x)\,(y)\,(x\cdot y' = xy+x)$, in slightly abbreviated notation]. The 'axiom' of mathematical induction says:

(II) $\qquad S_0\cdot(x)\,(y)\,[(S_x\cdot(y = x').\to S_y]:\to.(x)\,S_x$

where S_x is any wff not containing 'y', S_y contains 'y' where and only where S_x contains free 'x', and S_0 contains the individual constant 'o' wherever S_x contains free 'x'. Thus (II) 'says': 'if o satisfies the formula S_x and, for every x, when x satisfies S_x so does $y = x+1$, then every number satisfies S_x'.

Although Peano would have considered this a single 'axiom', to write it down we have to write down an infinite set of wffs – one instance of (II) for each wff S_x that can be built up out of the symbols o, E, S, T, F and logical symbols. Thus Peano arithmetic has an infinite set of axioms (and it has been proved that no finite alternative set of axioms exists). However, the usual set of axioms is *recursive*. To decide whether or not a wff is an axiom we see if it is one of the axioms that are not of the form (II) (there are only seven of these to check), and, if it is not, we then see whether or not the wff in question has the form (II) (which can be effectively decided). Thus theories with an infinite set of axioms play an important role in actual mathematics; however, it is always required in practice that the set of axioms be recursive. For, if there were no procedure for telling whether or not a wff was an axiom, then we could not tell whether or not an arbitrary sequence of wffs was a proof!

(4) A set is *recursively enumerable* if the members of the set are also the elements $S_1, S_2, S_3, \ldots$ of some sequence that can be effectively produced (i.e. produced by a Turing machine that is allowed to go on 'spinning out' the sequence forever). For example, the numbers, such as 159, whose digits occur successively in the decimal expansion of $\pi = 3\cdot14159\ldots$ are a recursively enumerable set, and they can be arranged in the sequence 3, 1, 31, 14, 4, 314, 141, 5, 41, 3141,... It is not known whether or not this set of integers is recursive, however. In fact, no one knows whether 7777 occurs in the decimal expansion of π or not.

The set of theorems of T, where T is a finitely axiomatized theory, is also a recursively enumerable set, and the theorems can be arranged in the sequence: (axioms of T, in lexicographic order), (theorems that can be obtained by one application of a rule of inference), (theorems that can be obtained by two applications of a rule of inference),... (If rules of inference such as the 'TF' of Quine's *Methods of Logic* are permitted, which can lead to infinitely many different results from a single finite set of wffs, then at the nth stage we write down only formulas of length less than 10^n which satisfy the condition given).

The set of theorems of T, where T is any *recursively* axiomatized theory, is also a recursively enumerable set. The idea of the proof is to arrange all the *proofs* in T in an effectively produced sequence (say, in order of increasing number of symbols). If one replaces the ith proof in the resulting sequence Proof$_1$, Proof$_2$, ... by the wff that is proved by Proof$_i$, one obtains a listing of all and only the theorems of T (with infinitely many repetitions, of course – however, these can be deleted if one wishes).

Is every recursively enumerable set recursive? According to a fundamental theorem of recursive-function theory, the answer is 'no'. There is a recursively enumerable set D of positive integers that is not recursive. In other words, there is a sequence a_1, a_2, ... of numbers that can be effectively continued as long as we wish, but such that there is *no method in principle* that will always tell whether or not an arbitrary integer eventually occurs in the sequence.

The set of theorems of quantification theory (first-order logic) is another example of a recursively enumerable nonrecursive set. The theorems can be effectively produced in a single infinite sequence; but there does not exist *in principle* an algorithm by means of which one can tell in a finite number of steps whether a wff will or will not eventually occur in the sequence – i.e. the 'Decision Problem' for pure logic is not solvable.

We now see what the observation (I) comes to. It says that all recursively enumerable theories can be recursively axiomatized. If the theory T is recursively enumerable (this is equivalent to having a recursively enumerable set of axioms), then a *recursive* set S can be found which is a set of axioms for T.

Craig's proof of (I)

Craig's proof of (I) is so remarkably simple that we shall give it in full. Let T be a theory with a recursively enumerable set S of axioms, and let an effectively produced sequence consisting of these axioms be S_1, S_2, ... We shall construct a new set S' which is an alternative set of axioms for T. Namely, for each positive integer i, S' contains the wff $S_i \cdot [S_i \cdot (\ldots)]$, with i conjuncts S_i.

Clearly, each S_i can be deduced from the corresponding axiom in S' by the rule $A \cdot B$ *implies* A. Also, each axiom in S' can be deduced from the corresponding S_i by repeated use of the rules: A implies $A \cdot A$, and A, B imply $A \cdot B$. It remains to show that S' is recursive.

Let A be a wff, and consider the problem of deciding whether or not A belongs to S'. Clearly, if $A \neq S_1$ and A is not of the form $[B \cdot (B \ldots)]$,

A is *not* in S'. If A is of the form $[B \cdot (B \ldots)]$ with k Bs, then A belongs to S' if and only if $B = S_k$. So we just continue the sequence $S_1, S_2, \ldots$ until we get to S_k and compare B with S_k. If $B = S_k$, A is in S'; otherwise A is not in S'. The proof is complete!

Notice that, although we have given a method for deciding whether or not a wff A is in S', we still have no method for deciding whether or not an arbitrary wff C is in S, even though S and S' are trivially equivalent sets of sentences, logically speaking. For we don't know how long an initial segment $S_1, \ldots, S_k$ we must produce before we can say that if C is not in the segment it is not in S at all. The fact that S' is decidable even if S is not constitutes an extremely instructive example.

'Craig's theorem' can now be stated: (III) Let T be a recursively enumerable theory, and consider any division of predicate letters of T into two disjoint sets, say $V_A = T_1, T_2, \ldots$ and $V_B = O_1, O_2, \ldots$ Let T_B consist of those theorems of T which contain only predicate letters from V_B. Then T_B is a recursively axiomatizable theory. Proof: Let $S_1, S_2, \ldots$ be an effectively produced sequence consisting of the theorems of T. By leaving out all wffs which are not in the subvocabulary V_B, we obtain the members of T_B, say as $V_1, V_2, \ldots$ Thus T_B is a recursively enumerable theory, and possesses a recursively enumerable axiomatization (take T_B itself as the set S of axioms). Then, by observation (I), T_B is recursively axiomatizable. Q.E.D.

The reader will observe that the proof assumes that the sets of predicate letters V_A and V_B are themselves recursive; strictly speaking we should have stated this. In practice these sets are usually finite sets, and thus trivially recursive.

The alleged philosophical significance of Craig's theorem

Imagine the entire 'language of science' to be formalized, and divide the primitive predicates into two classes (see the previous chapter for a criticism of this dichotomy): the so-called 'theoretical terms', T_1, $T_2, \ldots$, and the so-called 'observation terms', $O_1, O_2, \ldots$ Let us also imagine the *assertions* of science to be formalized in this language, so as to comprise a single theory T. T would, of course, be a very big theory, including everything from psychology to paleontology and from quantum mechanics to galactic astronomy. The 'predictions' of T are, presumably, to be found among those theorems of T which are in the vocabulary $V_O = O_1, O_2, \ldots$

Let T_O be the subtheory consisting of all those theorems of T which are expressible in the observational vocabulary V_O. Thus a statement belongs to T_O just in case it meets two conditions: (1) the statement must contain no theoretical terms; (2) the statement must be a consequence of the axioms of T. Then 'Craig's theorem' asserts that T_O is itself a recursively axiomatizable theory – and clearly T_O contains all the 'predictions' that T does!

This has led some authors to advance the argument that, since the purpose of science is successful prediction, *theoretical terms are in principle unnecessary*. For, the argument runs, we could (in principle) dispense with T altogether and just rely on T_O, since T_O implies all the predictions that T does. And T_O contains no theoretical terms.

Thus, Hempel, discussing Craig's method (Hempel, 1963), writes: 'Craig's result shows that no matter how we select from the total vocabulary V_T of an interpreted theory T a subset V_B of experiential or observational terms, the balance of V_T, constituting the "theoretical terms", can always be avoided in sense (c)' (p. 699). This sense Hempel calls 'functional replaceability' and defines as follows: 'The terms of T might be said to be avoidable if there exists another theory T_B couched in terms of V_B which is "functionally equivalent" to T in the sense of establishing exactly the same deductive connections between V_B sentences as T' (pp. 696–7).

It must be emphasized that Hempel does not rest content with this conclusion. He advances the argument we are considering only in order to reply to it. Hempel's objections are twofold: (1) the axioms of T_O (i.e. Hempel's 'T_B') are infinite in number, although effectively specified, and are 'practically unmanageable'. (2) The *inductive* connections among sentences of T_B may be altered when T_B is considered in the light of T. Purely observational sentences may bear probabilistic confirmation relations to one another by virtue of a theory T containing 'theoretical terms' which they would hardly bear to one another if they were considered simply in themselves, by someone ignorant of T. And science, Hempel argues, aims at systematizing observational sentences not only deductively but also *inductively*.

Speaking to Hempel's first objection, Israel Scheffler writes (Scheffler, 1960):

It seems to me, however, that if we take these programs [i.e. 'specific empiricist programs'] as requiring simply the reflection of non-transcendental assertions into replacing systems without transcendental terms, then we do not distort traditional notions of empiricism, and we have to acknowledge that Craig's result does the trick;... further cited problems remain but they are independent of empiricism as above formulated (p. 170).

The discussion of Craig's theorem in Nagel (1961) rests upon the misconception that 'the axioms A^* of L^* [our T_O.]...are specified by an effective procedure...upon the true observation statements W_O of L' (p. 136). It is not the totality of *true* sentences of L in the vocabulary V_O that has to be generated as an effectively produced sequence to obtain the recursive axiomatization of T_O, but only the totality of sentences of L in the vocabulary V_O that are *theorems of T*. Thus Nagel's criticism: 'Moreover, in order to specify the axioms for L^* we would have to know, *in advance* of any deductions made from them, *all* the true statements of L^* – in other words, Craig's method shows us how to construct the language L^* only *after* every possible inquiry into the subject matter of L^* has been completed' (p. 137), is simply incorrect. We do not have to know 'all the true statements' in the vocabulary V_O in order to construct a T_O with the 'same empirical content' as T – indeed, we need only be given the theory T. Craig's method can be applied whether the predictions of T are in fact true or false, and is a purely formal method – thus it is not necessary to complete *any* 'inquiry into the subject matter of L^*' in order to 'construct the language L^*'.

The argument reconsidered

To us there is something curiously unsatisfactory about the entire body of argumentation just reviewed. What all the participants in the debate appear to accept is the major premise (inherited from Mach and Comte, perhaps) that 'the aim of science' is successful prediction or deductive and inductive systematization of observation sentences or something of that kind. Given this orientation, the only possible reply to the fantastic proposal to employ T_O and scrap T (thereby leaving out of science all references to such 'theoretical entities' as viruses, radio stars, elementary particles, and unconscious drives, to mention only a small sample) is the kind of thing we have just cited – that T_O is 'unmanageable' or not 'heuristically fruitful and suggestive' – or that T_O does not in fact lead to the same predictions as T when probability (confirmation) relations are considered as well as absolute (deductively implied) predictions. But surely the most important thing has not been mentioned – that leaving out of science all mention of what cannot be seen with the naked eye would be leaving out just about *all* of science.

Let us spell this out a little bit. The use of such expressions as 'the aim of science', 'the function of scientific theories', 'the purpose of the terms and principles of a theory', is already extremely apt to be misleading. For there is no *one* 'aim of science', no one 'function of scientific theories', no one 'purpose of the terms and general principles' of a scientific theory. Different scientists have different purposes. Some

233

scientists are, indeed, primarily interested in prediction and control of human experience; but most scientists are interested in such objects as viruses and radio stars in their own right. Describing the behavior of viruses, radio stars, etc., may not be THE 'aim of science', but it is certainly an aim of scientists. And in terms of this aim one can give a very short answer to the question, Why theoretical terms? Why such terms as 'radio star', 'virus', and 'elementary particle'? *Because* without such terms we could not speak of radio stars, viruses, and elementary particles, for example – and we *wish* to speak of them, to learn more about them to explain their behavior and properties better.

Is the 'short answer' too short?

We fear that a great many philosophers of science would regard our 'short answer' to the question just posed as *too* short for two reasons: (1) it presupposes the existence of theoretical entities; and (2) it presupposes the intelligibility of theoretical terms. The second reason is the crucial one. Hardly any philosopher who felt no doubts about the intelligibility of such notions as 'radio star', 'virus', and 'elementary particle', would question the existence of these things. For surely the existence of radio stars, viruses, and elementary particles is well established; indeed, one can even *observe* viruses, with the electron microscope, observe radio stars, by means of radio telescopes, observe the tracks left by elementary particles in a cloud- or bubble-chamber. None of this amounts to *deductive proof*, to be sure; but this is an *uninteresting* objection (those of us who accept the scientific method are not unhappy because we cannot prove deductively what Hume showed to be incapable of deductive proof). Don't we have just about the best possible reasons for believing in the existence of radio stars, viruses, and elementary particles? If we don't, what would be *better* reasons?

This is not a cogent line to take, however, if the position is that no 'reasons' *could be* any good because we can't believe what we can't *understand*, and so-called 'theoretical discourse' is really unintelligible. That theoretical terms may be unintelligible is suggested by Scheffler's use of the expression 'transcendental term' (as if radio stars first made their appearance in Kantian metaphysics!), and his discussion shows that he takes this 'possibility' quite seriously:

To what extent is the pragmatist position in favor of a broader notion of significance positively supported by the arguments it presents? Its strong point is obviously its congruence with the *de facto* scientific use of transcendental theories and with the interdependence of parts of a scientific system undergoing test. These facts are, however, not in themselves *conclusive* evidence for significance, inasmuch as many kinds of things are used in science with no implication of cognitive significance, i.e. truth-or-falsity; and many things are interdependent under scientific test without our feeling that they are therefore included in the cognitive system of our assertions. Clearly 'is useful', 'is fruitful', 'is subject to modification under test', etc., are applicable also to nonlinguistic entities, e.g. telescopes and electronic computers. On the other hand, even linguistic units judged useful and controllable via empirical test may conceivably be construed as non-scientific machinery, and such construction is not definitely ruled out by pragmatist arguments (Scheffler, 1960, p. 163).

If it is really possible, however, that the term 'virus' is *meaningless* ('nonsignificant'), all appearance to the contrary notwithstanding, why is it not also possible that observation terms, such as 'chair' and 'red' are really meaningless? Indeed, traditional empiricists sometimes suggested that even such terms ('observation terms in thing language') might also be 'nonsignificant', and that the only *really* significant terms might be such as terms of 'pain', referring to sensations and states of mind. The traditional empiricist reason for regarding it as beyond question that *these* terms, at least, are 'significant' is that (allegedly) in the case of these terms the meaning is the referent, and I know the referent from my own case. However, this argument is self-undercutting: if 'pain' means what *I* have, then (on the traditional view) it is logically possible that no one else means by 'pain' what I do. Indeed, there may be *no* such thing as the *public* meaning of the word 'pain' at all! We believe that this traditional theory of meaning is in fact bankrupt and that it cannot account for the meaning of *any* term, least of all for sensation words.

In point of fact, a term is 'meaningful' – i.e. has meaning in the language – if it belongs to the common language or has been explained by means of terms already in the common language. The fact that we can and do have theoretical terms in our language rests upon the fact that there was never a 'pretheoretical' stage of language; the possibility of talking about unobservables is present in language from the beginning. 'Radio star', 'virus', and 'elementary particle' are perfectly good (meaningful) expressions in English. If you have doubts, consult your nearest dictionary! 'But maybe these words are *really* meaningless'. Well, they are defined in the dictionary. 'Maybe the definitions are meaningless too'. What sort of a doubt is this?

We are not urging that every word in the dictionary has a place in science or that the concept of 'meaning' in linguistic theory does not require further investigation (cf. Ziff 1960, for a pioneer investigation).

But we are urging that the concept of meaning used in everyday life and in linguistic theory is also the one appropriate to philosophy and that every term that has an established place in English syntax and an established use has meaning – that's how we *use* the word 'meaning'. The word 'God', for example, certainly has meaning. It does not follow that it belongs in science – but it does not belong *not* because it is 'nonsignificant', but just for the obvious reason – that there are no scientific procedures for determining whether or not God exists or what properties He has if He does exist. Whether or not the word *should* be employed as it is in *non*scientific contexts is another question. However, in the case of 'radio star' one could hardly argue 'there are no scientific procedures for detecting radio stars'. Thus 'radio star' is not only an expression with a meaning, but one which is usable in science. To suggest that it might be unintelligible is to promulgate an *unintelligible concept of intelligibility*.

Scheffler tries to make clear the notion of significance he has in mind by writing 'cognitive significance, i.e. *truth-or-falsity*', but this is no help at all. For '"Radio stars exist" is neither true nor false' means '"Radio star" is not a (sufficiently) clear expression' if it means anything. And the reply to a charge that 'radio star' was unclear would surely be, 'In what *way* is it unclear?' Theoretical terms can't just be 'unclear'; there has to be some relevant respect in which they might be sharpened and have not been.

We conclude that (a) theoretical terms are intelligible, in any ordinary sense of that term; and (b) theoretical entities have been established to exist (up to a sophomorish kind of skeptical doubt). Thus our 'short answer' stands: theoretical terms are not eliminable, 'Craig's theorem' notwithstanding, if one wishes to talk about theoretical entities; and we *do* wish to talk about theoretical entities.

15
It ain't necessarily so*

The two statements that Donnellan considered in his paper (Donnellan, 1962) are both more or less analytic in character. By that I mean that they are the sort of statement that most people would consider to be true by definition, if they considered them to be necessary truths at all. One might quarrel about whether 'all whales are mammals' is a necessary truth at all. But if one considers it to be a necessary truth, then one would consider it to be true by definition. And, similarly, most people would say that 'all cats are animals' is true by definition, notwithstanding the fact that they would be hard put to answer the question, 'true by *what* definition?'

I like what Donnellan had to say about these statements, and I liked especially the remark that occurs toward the end of his paper, that there are situations in which we are confronted by a question about how to talk, but in which it is not possible to describe one of the available decisions as deciding to retain our old way of talking (or 'not to change the meaning') and the other as deciding to adopt a 'new' way of talking (or to 'change the meaning').

In this paper I want to concentrate mostly on statements that look necessary, but that are not analytic; on 'synthetic necessary truths', so to speak. This is not to say that there are not serious problems connected with analyticity. On the contrary, there certainly are. The general area of necessary truths might be broken up, at least for the sake of preliminary exploration (and we philosophers are still in the stage of preliminary exploration even after thousands of years) into three main subareas: the area of analytic truths, the area of logical and mathematical truths, and the area of 'synthetic *a priori*' truths. I don't mean to beg any questions by this division. Thus, separating logical and mathematical truths from analytic truths is not meant to prejudge the question whether they are or are not ultimately analytic.

I. Analytic truths

The 'analyticity' of 'all cats are animals' or, what is closely related, the redundancy of the expression 'a cat which is an animal' seems, to

* First published in *The Journal of Philosophy*, LIX, 22, (25 October 1962).

depend on the fact that the word 'animal' is the name of a semantic category (the notion of a 'semantic category' is taken from Fodor and Katz, 1963) and the word 'cat' is a member of that category.

In a later chapter (chapter 2, volume 2), I call words that have an analytic definition 'one-criterion words'. Many words that are not one-criterion words fall into semantic categories – in fact, all nouns fall into semantic categories. 'House', for example, falls into the semantic category *material object*, 'red' falls into the semantic category *color*, and so on. Thus, for any noun one can find an analytic or quasi-analytic truth of the sort 'a cat is an animal', 'a house is a material object', 'red is a color', and so forth. But it hardly follows that all nouns are one-criterion words – in fact, there are only a few hundred one-criterion words in English.

It is important to distinguish 'analytic' truths of the sort 'all cats are animals' from analytic truths of the sort 'all bachelors are unmarried', in part because the former tend to be *less* necessary than the latter. It might not be the case that all cats are animals; they might be automata!

There are, in fact, several possibilities. If *some* cats are animals in every sense of the word, while others are automata, then there is no problem. I think we would all agree that these others were neither animals nor cats but only fake cats – very realistic and very clever fakes to be sure, but fakes nonetheless. Suppose, however, that all cats on earth are automata. In that case the situation is more complex. We should ask the question, 'Were there *ever* living cats?' If, let us say, up to fifty years ago there were living cats and the Martians killed all of them and replaced them all overnight with robots that look exactly like cats and can't be told from cats by present-day biologists (although, let us say, biologists will be able to detect the fake in fifty years more), then I think we should again say that the animals we all call cats are not in fact cats, and also not in fact animals, but robots. It is clear how we should talk in this case: 'there were cats up to fifty years ago; there aren't any any longer. Because of a very exceptional combination of circumstance we did not discover this fact until this time'.

Suppose, however, that there never have been cats, i.e. genuine non-fake cats. Suppose evolution has produced many things that come close to the cat but that it never actually produced the cat, and that the cat as we know it is and always was an artifact. Every movement of a cat, every twitch of a muscle, every meow, every flicker of an eyelid is thought out by a man in a control center on Mars and is then executed by the cat's body as the result of signals that emanate not from the cat's 'brain' but from a highly miniaturized radio receiver located, let us say, in the cat's pineal gland. It seems to me that in this last case, once we discovered the fake, we should continue to call these robots that we have mistaken for animals and that we have employed as house pets 'cats' but not 'animals'.

This is the sort of problem that Donnellan discussed in connection with the 'all whales are mammals' case. Once we find out that cats were created from the beginning by Martians, that they are not self-directed, that they are automata, and so on, then it is clear that we have a problem of how to speak. What is not clear is which of the available decisions should be described as the decision to keep the meaning of either word ('cat' or 'animal') unchanged, and which decision should be described as the decision to change the meaning. I agree with Donnellan that this question has no clear sense. My own feeling is that to say that cats turned out not to be animals is to keep the meaning of both words unchanged. Someone else may feel that the correct thing is to say is, 'It's turned out that there aren't and never were any cats'. Someone else may feel that the correct thing to say is, 'It's turned out that some animals are robots'. Today it doesn't seem to make much difference what we say; while in the context of a developed linguistic theory it may make a difference whether we say that talking in one of these ways is changing the meaning and talking in another of these ways is keeping the meaning unchanged. But that is hardly relevant here and now; when linguistic theory becomes *that* developed, then 'meaning' will itself have become a technical term, and presumably our question now is not which decision is changing the meaning in some future technical sense of 'meaning', but what we can say in our present language.

II. Synthetic necessary statements

Later I shall make a few remarks about the truths of logic and mathematics. For the moment, let me turn to statements of quite a different kind, for example, the statement that *if one did X and then Y, X must have been done at an earlier time than Y*, or the statement that *space has three dimensions*, or the statement that *the relation* earlier than *is transitive*. All of these statements are still classified as necessary by philosophers today. Those who feel that the first statement is 'conceptually necessary' reject time travel as a 'conceptual impossibility'.

Once again I will beg your pardon for engaging in philosophical science fiction. I want to imagine that something has happened (which is in fact a possibility) namely, that modern physics has *definitely* come to the conclusion that space is Riemannian. Now, with this assumption in force, let us discuss the status of the statement that *one cannot reach the place from which one came by traveling away from it in a straight line and*

239

continuing to move in a constant sense. This is a geometrical statement. I want to understand it, however, not as a statement of pure geometry, not as a statement about space 'in the abstract', but as a statement about physical space, about the space in which we live and move and have our being. It may be claimed that in space, in the space of actual physical experience, the notion of a 'straight line' has no application, because straight lines are supposed to have no thickness at all and not the slightest variation in curvature, and we cannot identify by any physical means paths in space with these properties (although we can approximate them as closely as we desire). However, this is not relevant in the present context. Approximate straight lines will do. In fact, it is possible to state one of the differences between a Riemannian and a Euclidean world *without using* the notion of 'straightness' at all. Suppose that by a 'place' we agree to mean, for the nonce, a cubical spatial region whose volume is one thousand cubic feet. Then, in a Riemannian world, there are only a finite number of disjoint 'places'. There is no path curved *or* straight, on which one can 'go sight-seeing' in a Riemannian world and hope to see more than a certain number N of disjoint 'places'. Thus, the statement that *one can see as many distinct disjoint places as one likes if one travels far enough along a suitable path* is one which is true in any Euclidean world, and not in any Riemannian world.

Now, I think it is intuitively clear that the two propositions just mentioned: that one cannot return to one's starting point by traveling on a straight line unless one reverses the sense of one's motion at some point, and that one can visit an arbitrary number of distinct and disjoint 'places' by continuing far enough on a suitable path – that is, that there is no finite upper bound on the total number of 'places' – had the *status of* necessary truths before the nineteenth century.

Let me say, of a statement that enjoys the status with respect to a body of knowledge that these statements enjoyed before the nineteenth century and that the other statements alluded to enjoy today, that it is 'necessary *relative to* the appropriate body of knowledge'. This notion of *necessity relative to a body of knowledge* is a technical notion being introduced here for special purposes and to which we give special properties. In particular, when we say that a statement is necessary relative to a body of knowledge, we imply that it is included in that body of knowledge and that it enjoys a special role in that body of knowledge. For example, one is not expected to give much of a reason for that kind of statement. But we do not imply that the statement is necessarily *true*, although, of course, it is thought to be true by someone whose knowledge that body of knowledge is.

We are now confronted with this problem: a statement that was necessary relative to a body of knowledge later came to be declared false in science. What can we say about this?

Many philosophers have found it tempting to say nothing significant happened, that we simply changed the meaning of words. This is to assimilate the case to Donnellan's first case, the whales–mammals case. Just as 'whale' may, perhaps, have two meanings – one for laymen and one for scientists – so 'straight line' may have two meanings, one for laymen and one for scientists. Paths that are straight in the layman's sense may not be straight in the scientist's sense, and paths that are straight in the scientist's sense may not be straight in the layman's sense. One cannot come back to one's starting point by proceeding indefinitely on a path that is straight in the old Euclidean sense; when the scientist says that one can come back to one's starting point by continuing long enough on a straight line, there is no paradox and no contradiction with common sense. What he means by a straight line may be a curved line in the layman's sense, and we all *know* you can come back to your starting point by continuing long enough on a closed curve.

This account is not tenable, however. To see that it isn't, let us shift to the second statement. Here we are in immediate difficulties because there seems to be *no* difference, even today, between the layman's sense of 'path' and the scientist's sense of 'path'. Anything that the layman could trace out if we gave him but world enough and time, the scientists would accept as a 'path', and anything that the scientist would trace out as a 'path', the layman would accept as a 'path'. (Here I do not count microscopic 'paths' as paths). Similarly with 'place'. To put it another way: if Euclidean geometry is only apparently false owing to a change in the meaning of words, then if we keep the meanings of the words *unchanged*, if we use the words in the old way, Euclidean geometry must *still* be true. In that case, in addition to the N 'places' to which one can get by following the various paths in our Riemannian space, there must be infinitely many additional 'places' to which one can get by following other paths that somehow the scientist has overlooked. Where are these places? Where are these other paths? In fact, they don't exist. If someone believes that they exist, then he must invent special physical laws to explain why, try as we may, we never succeed in seeing one of these other places or in sticking to one of these other paths. If someone did accept such laws and insisted on holding on to Euclidean geometry in the face of all present scientific experience, it is clear that he would not have simply 'made a decision to keep the meanings of words unchanged'; he would have adopted a metaphysical theory.

The statement that *there are only finitely many disjoint 'places' to get to, travel as you may* expresses a downright 'conceptual impossibility' within the framework of Euclidean geometry. And one cannot say that *all* that has happened is that we have changed the meaning of the word 'path', because in that case one would be committed to the metaphysical hypothesis that, in addition to the 'paths' that are *still* so called, there exist others which are somehow physically inaccessible and additional 'places' which are somehow physically inaccessible and which, together with what the physicists presently recognize as places and paths, fill out a Euclidean space.

Insofar as the terms 'place', 'path', and 'straight line' have any application at all in physical space, they still have the application that they always had; something that was literally inconceivable has turned out to be true.

Incidentally, although modern physics does not *yet* say that space is Riemannian, it does say that our space has variable curvature. This means that if two light rays stay a constant distance apart for a long time and then come closer together after passing the sun, we do *not* say that these two light rays are following curved paths through space, but we say rather that they follow straight paths and that two straight paths may have a constant distance from each other for a long time and then later have a decreasing distance from each other. Once again, if anyone wishes to say, 'Well, those paths aren't straight in the old sense of "straight",' then I invite him to tell me *which* paths in the space near the sun are 'really straight'. And I guarantee that, first, no matter which paths he chooses as the straight ones, I will be able to embarrass him acutely. I will be able to show, for example, not only that light rays refuse to travel along the paths he claims to be really straight, but that they are not the shortest paths by any method of measurement he may elect; one cannot even travel along those paths in a rocket ship without accelerations, decelerations, twists, turns, etc. In short, the paths he claims are 'really straight' will look crooked, act crooked, and feel crooked. Moreover, if anyone does say that certain nongeodesics are the really straight paths in the space near the sun, then his decision will have to be a quite arbitrary one; and the theory that more or less *arbitrarily* selected curved paths near the sun are 'really straight' (because they obey the laws of Euclidean geometry and the geodesics do not) would again be a metaphysical theory, and the decision to accept it would certainly *not* be a mere decision to 'keep the meaning of words unchanged'.

III. The cluster character of geometric notions

Distance cannot be 'operationally defined' as distance according to an aluminum ruler, nor as distance according to a wooden ruler, nor as distance according to an iron ruler, nor as distance according to an optical measuring instrument, nor in terms of any other *one* operational criterion. The criteria we have for distance define the notion collectively, not individually, and the connection between any one criterion and the rest of the bundle may be viewed as completely synthetic. For example, there is no contradiction involved in supposing that light sometimes travels in curved lines. It is because of the cluster character of geometrical concepts that the methods usually suggested by operationists for demonstrating the falsity of Euclidean geometry by isolated experiments would not have succeeded before the development of non-Euclidean geometry. If someone had been able to construct a very large light-ray triangle and had shown that the sum of the angles exceeded 180°, even allowing for measuring errors, he would not have shown the ancient Greek that Euclidean geometry was false, but only that light did not travel in straight lines.

What accounted for the necessity of the principles of Euclidean geometry relative to pre-nineteenth-century knowledge? An answer would be difficult to give in detail, but I believe that the general outlines of an answer are not hard to see. Spatial locations play an obviously fundamental role in all of our scientific knowledge and in many of the operations of daily life. The use of spatial locations requires, however, the acceptance of some systematic body of geometrical theory. To abandon Euclidean geometry before non-Euclidean geometry was invented would be to 'let our concepts crumble'.

IV. Time travel

I believe that an attempt to describe in ordinary language what time travel would be like can easily lead to absurdities and even downright contradictions. But if one has a mathematical technique of representing all the phenomena subsumed under some particular notion of 'time travel', then it is easy to work out a way of speaking, and even a way of thinking, corresponding to the mathematical technique. A mathematical technique for representing at least one set of occurrences that might be meant by the term 'time travel' already exists. This is the technique of *world lines* and Minkowski space–time diagrams. Thus, suppose, for example, that a time traveler – we'll call him Oscar Smith – and his apparatus have world lines as shown in the diagram.

From the diagram we can at once read off what an observer sees at various times. At t_0, for example, he sees Oscar Smith not yet a time traveler. At time t_1 he still sees Oscar Smith at place A, but also he sees something else at place B. At place B he sees, namely, an event of

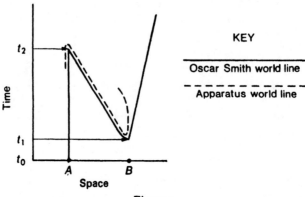

KEY

———————

Oscar Smith world line

– – – – – – – – – –

Apparatus world line

Figure 1

'creation' – not 'particle-antiparticle creation', but the creation of two macro-objects which separate. One of these macro-objects is easily described. It is simply an older Oscar Smith, or an individual resembling in all possible ways an older version of Oscar Smith, together with the apparatus of a time machine. The world-line diagram (Figure 1) shows that the older Oscar – let's call him Oscar₃ – leaves his time machine. The other object that was created in the same event is a somewhat peculiar object. It is a system consisting of a third Oscar Smith, or, more precisely, of a body like that of Oscar Smith, seated in a time machine. But this system consisting of the Oscar Smith body and the time machine is a very exceptional physical system. If we take a moving picture of this physical system during its entire period of existence, we will find that if the movie is played backward then the events in it are all normal. In short, this is a system running backward in time – entropy in the system is decreasing instead of increasing, cigarette butts are growing into whole cigarettes, the Oscar Smith body is emitting noises that resemble speech sounds played backward, and so forth. The system that is running backward in time continues to exist until the time t_2, when it merges with Oscar Smith, and we see annihilation – not 'particle-antiparticle annihilation', but the annihilation of the real Oscar Smith and the running-backward system. During a certain period of time, there are three Oscar Smiths: Oscar Smith₁, Oscar Smith₃, and the Oscar Smith who is living backward in time (Oscar Smith₂, we shall call him). We can even predict subjective phenomena from the world-line diagram. We can say, for example, what sort of memories Oscar Smith₃ has at the moment of his 'creation'. He has, namely, all the memories that someone would have if he had all the experiences of Oscar₁ up to the moment of his annihilation and then all the experiences shown as occurring to the living-backward Oscar, Oscar$_2$, on a movie film, provided the movie film is reversed so that these experiences are shown as happening in the normal order. I have no doubt whatsoever as to how any reasonable scientist would describe these events, if they actually transpired. He would interpret them in terms of the world-line diagram; i.e. he would say: 'There are not really three Oscar Smiths; there is only one Oscar Smith. The Oscar Smith you call Oscar Smith$_1$ lives forward in time until the time t_2; at t_2 his world line for some reason bends backwards in time, and he lives backward in time from t_2 back to the time t_1. At t_1 he starts living forward in time again and continues living forward in time for the rest of his life.'

I remember having a discussion concerning time travel with a philosopher friend a number of years ago. I told him the story I have just told you. My friend's view, in a nutshell, was that time travel was a conceptual impossibility. The phenomena I described can, of course, be imagined; but they are correctly described in the way I first described them; i.e. Oscar Smith$_1$ lives until t_2, at which time he collides with the strange system. When the two systems merge, they are both annihilated. At the time t_1 this strange physical system was created, as was also another individual very much resembling Oscar Smith$_1$, but with an entirely fictitious set of memories including, incidentally, memories of Oscar Smith$_1$'s existence up to the time t_2.

Let us ask ourselves what makes us so sure that there is here a consistently imaginable set of circumstances to be described. The answer is that it is the mathematical representation, i.e. the world-line diagram itself, that gives us this assurance. Similarly, in the case of space with variable curvature, near the sun, the only thing that makes us sure that there is a consistently imaginable set of phenomena to be described is their representation in terms of the mathematics of non-Euclidean geometry.

The present case also exhibits *dis*analogies to the geometric case. In the geometric case we could not go on using language in the old way – if to preserve Euclidean geometry is to go on using language in the old way – without finding ourselves committed to ghost places and ghost paths. In the present case, we can go on using language in the old way – if my friend's way of describing the situation *is* the one which corresponds to 'using language in the old way' – without having to countenance any ghost entities. But there are a host of difficulties which make us doubt whether to speak in this way is to go on using language without any change of usage or meaning. First of all, consider the sort of system that a physicist would describe as a 'human being living backward in time'. The same system would be described by my friend not as a *person*

245

at all, but as a human *body* going through a rather nauseating succession of physical states. Thus, on my friend's account, $Oscar_2$ is not a person at all, and $Oscar_1$ and $Oscar_3$ are two quite different persons. $Oscar_1$ is a person who had a normal life up to the time t_2 when something very abnormal happened to him, *namely, he vanished,* and $Oscar_3$ is a person who had a normal life from the time t_1 on, but who came into existence in a very abnormal way: *he appeared out of thin air.* Consider now the legal problems that might arise and whose resolution might depend on whether we accepted the physicist's account or the account of my friend. Suppose $Oscar_1$ murders someone but is not apprehended before the time at which he vanishes. Can we or can we not punish $Oscar_3$ for $Oscar_1$'s crime? On the physicist's account, $Oscar_3$ *is* $Oscar_1$, only grown older, and can hence be held responsible for all the actions of $Oscar_1$. On my friend's account, $Oscar_3$ is only a person under the unfortunate delusion that he is $Oscar_1$ grown older, and should be treated with appropriate kindness rather than punishment. And, of course, no one is responsible for $Oscar_2$'s actions on this view, since they are not really *actions* at all. And if $Oscar_1$'s wife lives with $Oscar_3$ after t_2, she is guilty of unlawful cohabitation, while if she lives with $Oscar_3$ prior to t_2, the lady is guilty of adultery. In this kind of case, to go into court and tell the story as my friend would tell it would be to use language in a most *extraordinary* way.

This case differs importantly from Donnellan's cases in that, although our problem can be described as a problem of how to speak, it is not merely a problem of how to speak, since moral and social questions depend on how we decide to speak.

V. Conclusions

In the last few years I have been amused and irritated by the spate of articles proving that time travel is a 'conceptual impossibility'. All these articles make the same mistake. They take it to be enough to show that, if we start talking about time travel, things go wrong with ordinary language in countless places and in countless ways. For example, it makes no sense to *prevent an occurrence that is in the past,* yet a time traveler about to visit the Ice Age in his time machine may well take an overcoat to keep from freezing to death several million years *ago.* Exactly similar objections could have been raised against the notion of their being only finitely many 'places' prior to the development of Riemannian geometry. It is precisely the existence of the world-line language that makes us sure that all these apparently insurmountable difficulties about 'preventing', 'expecting', etc., can be met. For example, the proper way to reformulate the principle that it makes no sense to prevent an occurrence in the past is to introduce the relativistic notion of *proper time*, i.e. time along the world line, and to say that it makes no sense to prevent an occurrence not in one's proper future. Also, even if an event is in one's proper future, but one already knows its outcome, say, because it is in the objective past and has been recorded, then it cannot be prevented (although one can try). For example if reliable records show that an older self of you is going to freeze to death two million years ago, then, try as you may, you will not succeed in preventing this event. But this actually introduces nothing new into human life; it is the analogue of the principle that, if you know with certainty that something is going to happen to you in the future, then, try as you may, you won't succeed in forestalling it. It is just that there is a new way of knowing with certainty that something is going to happen to you in your proper future – namely, if your proper future happens to be also your present past, then you may be able to know with certainty what will happen to you *by using records*.

The principle of the transistivity of the relation *earlier than* involves similar considerations. If the world line of the universe as a whole happened to be a closed curve, then we should have to abandon that principle altogether. As Gödel has pointed out (Gödel, 1951), there is no contradiction with General Relativity in the supposition that the universe may have a closed world line.

The history of the causal principle is yet another case in point. Before quantum mechanics, if we found that an event A was sometimes succeeded by an event B and sometimes by a different event B', this was taken as *conclusive* proof that there were factors, say C and C', differing in the two situations. Physicists abandoned the principle that in such cases one should *always* postulate such unknown factors only because a worked-out mathematical language of an acausal character appeared on the scene, the mathematical language of unitary transformation and projections that is used in quantum mechanics. This is a mathematical way of representing phenomena, but it is not *just* a mathematical way of representing phenomena. Once again, it influences the way in which ordinary men decide questions. When cases arise of the sort Donnellan foresaw in his paper, then ordinary men may choose one way of speaking over another precisely because one way of speaking links up with a scientific way of talking and thinking whereas the other way of speaking does not link up with *any* coherent conceptual system.

The causality case is analogous to the geometry case in that the decision to preserve the older way of speaking – that is, to say, whenever an event A appears to produce either of two different events B and B', that there must be some hidden difference in the attendant circumstances – involves postulating ghost entities. The ghost entities in question are called 'hidden variables' in the literature of quantum mechanics.

I am inclined to think that the situation is not substantially different in logic and mathematics. I believe that if I had the time I could describe for you a case in which we would have a choice between accepting a physical theory based upon a non-standard logic, on the one hand, and retaining standard logic and postulating hidden variables on the other. In this case, too, the decision to retain the old logic is not merely the decision to keep the meaning of certain words unchanged, for it has physical and perhaps metaphysical consequences. In quantum mechanics, for example, the customary interpretation says that an electron does not have a definite position prior to a position measurement; the position measurement causes the electron to take on suddenly the property that we call its 'position' (this is the so-called 'quantum jump'). Attempts to work out a theory of quantum jumps and of measurement in quantum mechanics have been notoriously unsuccessful to date. It has been pointed out† that it is entirely unnecessary to postulate the absence of sharp values prior to measurement and the occurrence of quantum jumps, if we are willing to regard quantum mechanics as a theory formalized within a certain nonstandard logic, the modular logic proposed in 1935 by Birkhoff and von Neumann, for precisely the purpose of formalizing quantum mechanics.

There seems to be only one conclusion to come to, and I herewith come to it. The distinction between statements necessary relative to a body of knowledge and statements contingent relative to that body of knowledge is an important methodological distinction and should not be jettisoned. But the traditional philosophical distinction between statements necessary in some eternal sense and statements contingent in some eternal sense is not workable. The rescuing move which consists in saying that if a statement which appears to be necessary relative to a body of knowledge at one time is not necessary relative to the body of knowledge at a later time, then it is not really the same statement that is involved, that words have changed their meaning, and that the old statement would still be a necessary truth if the meanings of the words had been kept unchanged, is unsuccessful. The rescuing move which consists in saying that such statements were only mistaken to be necessary truths, that they were contingent statements all along, and that their 'necessity' was 'merely psychological' is just the other side of the same

---

† This was pointed out by David Finkelstein in his lecture to the informal subgroup on Measurement in Quantum Mechanics at the Boulder Symposium on Mathematical Physics, sponsored by the American Mathematical Society in the summer of 1960.

blunder. For the difference between statements that can be overthrown by merely conceiving of suitable experiments and statements that can be overthrown only by conceiving of whole new theoretical structures – sometimes structures, like Relativity and Quantum Mechanics, that change our whole way of reasoning about nature – is of logical and methodological significance, and not just of psychological interest.

16

The 'corroboration' of theories*

Sir Karl Popper is a philosopher whose work has influenced and stimu-
lated that of virtually every student in the philosophy of science. In part
this influence is explainable on the basis of the healthy-mindedness of
some of Sir Karl's fundamental attitudes: 'There is no method peculiar
to philosophy'. 'The growth of knowledge can be studied best by
studying the growth of scientific knowledge.'

Philosophers should not be specialists. For myself, I am interested in science
and in philosophy only because I want to learn something about the riddle of
the world in which we live, and the riddle of man's knowledge of that world.
And I believe that only a revival of interest in these riddles can save the
sciences and philosophy from an obscurantist faith in the expert's special skill
and in his personal knowledge and authority.

These attitudes are perhaps a little narrow (can the growth of knowledge
be studied without also studying nonscientific knowledge? Are the
problems Popper mentioned of merely theoretical interest – just
'riddles'?), but much less narrow than those of many philosophers; and
the 'obscurantist faith' Popper warns against is a real danger. In part
this influence stems from Popper's realism, his refusal to accept the
peculiar meaning theories of the positivists, and his separation of the
problems of scientific methodology from the various problems about
the 'interpretation of scientific theories' which are internal to the
meaning theories of the positivists and which positivistic philosophers
of science have continued to wrangle about (I have discussed positivistic
meaning theory in chapter 14 and also in chapter 5, volume 2).

In this paper I want to examine his views about scientific methodology
– about what is generally called 'induction', although Popper rejects
the concept – and, in particular, to criticize assumptions that Popper
has in common with received philosophy of science, rather than assumptions that are peculiar to Popper. For I think that there are a number of
such common assumptions, and that they represent a mistaken way of
looking at science.

* First published in P. A. Schilpp (ed.), *The Philosophy of Karl Popper*, II (La Salle,
Ill., The Open Court Publishing Co., 1974).

1. Popper's view of 'induction'

Popper himself uses the term 'induction' to refer to any method for verifying or showing to be true (or even probable) general laws on the basis of observational or experimental data (what he calls 'basic statements'). His views are radically Humean: no such method exists or can exist. A principle of induction would have to be either synthetic *a priori* (a possibility that Popper rejects) or justified by a higher level principle. But the latter course necessarily leads to an infinite regress.

What is novel is that Popper concludes neither that empirical science is impossible nor that empirical science rests upon principles that are themselves incapable of justification. Rather, his position is that empirical science does not really rely upon a principle of induction!

Popper does not deny that scientists state general laws, nor that they test these general laws against observational data. What he says is that when a scientists 'corroborates' a general law, that scientist does not thereby assert that law to be true or even probable. 'I have corroborated this law to a high degree' only means 'I have subjected this law to severe tests and it has withstood them'. Scientific laws are *falsifiable* not verifiable. Since scientists are not even trying to *verify* laws, but only to falsify them, Hume's problem does not arise for empirical scientists.

2. A brief criticism of Popper's view

It is a remarkable fact about Popper's book, *The Logic of Scientific Discovery* that it contains but a half-dozen brief references to the *application* of scientific theories and laws; and then all that is said is that application is yet another *test* of the laws. 'My view is that...the theorist is interested in explanations as such, that is to say, in testable explanatory theories: applications and predictions interest him only for theoretical reasons – because they may be used as *tests* of theories' (Popper, 1959, p. 59).

When a scientist accepts a law, he is recommending to other men that they rely on it – rely on it, often, in practical contexts. Only by wrenching science altogether out of the context in which it really arises – the context of men trying to change and control the world – can Popper even put forward his peculiar view on induction. Ideas are not *just* ideas; they are guides to action. Our notions of 'knowledge', 'probability', 'certainty', etc., are all linked to and frequently used in contexts in which action is at issue: may I confidently rely upon a certain idea? Shall I rely upon it tentatively, with a certain caution? Is it necessary to check on it?

If 'this law is highly corroborated', 'this law is scientifically accepted', and like locutions merely meant 'this law has withstood severe tests' – and there were no suggestion at all that a law which has withstood severe tests is likely to withstand further tests, such as the tests involved in an application or attempted application, then Popper would be right; but then science would be a wholly unimportant activity. It would be practically unimportant, because scientists would never tell us that any law or theory is safe to rely upon for practical purposes; and it would be unimportant for the purpose of understanding, since in Popper's view, scientists never tell us that any law or theory is true or even probable. Knowing that certain 'conjectures' (according to Popper all scientific laws are 'provisional conjectures') have not yet been refuted is *not understanding anything*.

Since the application of scientific laws does involve the anticipation of future successes, Popper is not right in maintaining that induction is unnecessary. Even if scientists do not inductively anticipate the future (and, of course, they do), men who apply scientific laws and theories do so. And 'don't make inductions' is hardly reasonable advice to give these men.

The advice to regard all knowledge as 'provisional conjectures' is also not reasonable. Consider men striking against sweatshop conditions. Should they say 'it is only a provisional conjecture that the boss is a bastard. Let us call off our strike and try appealing to his better nature'. The distinction between *knowledge* and *conjecture* does real work in our lives; Popper can maintain his extreme skepticism only because of his extreme tendency to regard theory as an end for itself.

3. Popper's view of corroboration

Although scientists, on Popper's view, do not make inductions, they do 'corroborate' scientific theories. And although the statement that a theory is highly corroborated does not mean, according to Popper, that the theory may be accepted as true, or even as approximately true,† or even as probably approximately true, still, there is no doubt that most readers of Popper read his account of corroboration as an account of something like the verification of theories, in spite of his protests. In this sense, Popper has, *contre lui* a theory of induction. And it is this theory, or certain presuppositions of this theory, that I shall criticize in the body of this paper.

Popper's reaction to this way of reading him is as follows:

† For a discussion of 'approximate truth', see chapter 5, volume 2.

My reaction to this reply would be regret at my continued failure to explain my main point with sufficient clarity. For the sole purpose of the elimination advocated by all these inductivists was to *establish as firmly as possible the surviving theory* which, they thought must be the *true* one (or, perhaps, only a *highly probable* one, in so far as we may not have fully succeeded in eliminating every theory except the true one).

As against this, I do not think that we can ever seriously reduce by elimination, the number of the competing theories, since this number remains always infinite. What we do – or should do – is to *hold on, for the time being, to the most improbable of the surviving theories* or, more precisely, to the one that can be most severely tested. We tentatively '*accept*' this theory – but only in the sense that we select it as worthy to be subjected to further criticism, and to the severest tests we can design.

On the positive side, we may be entitled to add that the surviving theory is the best theory – and the best tested theory – of which we know. (Popper, 1959, p. 419).

If we leave out the last sentence, we have the doctrine we have been criticizing in pure form: when a scientist 'accepts' a theory, he does not assert that it is probable. In fact, he 'selects' it as most improbable! In the last sentence, however, am I mistaken, or do I detect an inductivist quaver? What does 'best theory' mean? Surely Popper cannot mean 'most likely'?

4. The scientific method – The received schema

Standard 'inductivist' accounts of the confirmation† of scientific theories go somewhat like this: theory implies prediction (basic sentence, or observation sentence); if prediction is false, theory is falsified; if sufficiently many predictions are true, theory is confirmed. For all his attack on inductivism, Popper's schema is not *so* different: theory implies prediction (basic sentence); if prediction is false, theory is falsified; if sufficiently many predictions are true, and certain further conditions are fulfilled, theory is highly corroborated.

Moreover, this reading of Popper does have certain support. Popper does say that the 'surviving theory' is *accepted* – his account is, therefore, an account of the logic of accepting theories. We must separate two questions: is Popper right about what the scientist means – or should mean – when he speaks of a theory as 'accepted'; and is Popper right about the methodology involved in according a theory that status?

† 'Confirmation' is the term in standard use for *support* a positive experimental or observational result gives to a hypothesis; Popper uses the term 'corroboration' instead, as a rule, because he objects to the connotations of 'showing to be true' (or at least probable) which he sees as attaching to the former term.

What I am urging is that his account of that methodology fits the received schema, even if his interpretation of the status is very different.

To be sure there are some important conditions that Popper adds. Predictions that one could have made on the basis of background knowledge do not test a theory; it is only predictions that are *improbable* relative to background knowledge that test a theory. And a theory is not corroborated, according to Popper, unless we make sincere attempts to derive false predictions from it. Popper regards these further conditions as anti-Bayesian;† but this seems to me to be a confusion, at least in part. A theory which implies an improbable prediction is improbable, that is true, but it may be the most probable of all theories which imply that prediction. If so, and the prediction turns out true, then Bayes's theorem itself explains why the theory receives a high probability. Popper says that we select the most improbable of the *surviving* theories – i.e. the accepted theory is most improbable even *after* the prediction has turned out true; but, of course, this depends on using 'probable' in a way no other philosopher of science would accept. And a Bayesian is not committed to the view that *any* true prediction significantly confirms a theory. I share Popper's view that quantitative measures of the probability of theories are not a hopeful venture in the philosophy of science (cf. chapter 18); but that does not mean that Bayes's theorem does not have a certain *qualitative* rightness, at least in many situations.

Be all this as it may, the heart of Popper's schema is the theory-prediction link. It is because theories imply basic sentences in the sense of 'imply' associated with deductive logic – because basic sentences are DEDUCIBLE from theories – that, according to Popper, theories and general laws can be falsifiable by basic sentences. And this same link is the heart of the 'inductivist' schema. Both schemes say: *look at the predictions that a theory implies; see if those predictions are true.*

My criticism is going to be a criticism of this link, of this one point on which Popper and the 'inductivists' agree. I claim: in a great many important cases, scientific theories do not imply predictions at all. In the remainder of this paper I want to elaborate this point, and show its significance for the philosophy of science.

† *Bayes's theorem* asserts, roughly, that the probability of a hypothesis H on given evidence E is directly proportional to the probability of E on the hypothesis H, and also directly proportional to the antecedent probability of H – i.e. the probability of H if one does not know that E. The theorem also asserts that the probability of H on the evidence E is less, other things being equal, if the probability of E on the assumption $\bar{H}$ (*not -H*) is greater. Today probability theorists are divided between those who accept the notion of 'antecedent probability of a hypothesis', which is crucial to the theorem, and those who reject this notion, and therefore the notion of the probability of a hypothesis on given evidence. The former school are called 'Bayesians'; the latter 'anti-Bayesians'.

5. The theory of universal gravitation

The theory that I will use to illustrate my points is one that the reader will be familiar with: it is Newton's theory of universal gravitation. The theory consists of the law that every body a exerts on every other body b a force F_{ab} whose direction is towards a and whose magnitude is a universal constant g times $M_a M_b / d^2$, together with Newton's three laws. The choice of this particular theory is not essential to my case: Maxwell's theory, or Mendel's, or Darwin's would have done just as well. But this one has the advantage of familiarity.

Note that this theory does not imply a single basic sentence! Indeed, any motions whatsoever are compatible with this theory, since the theory says nothing about what forces other than gravitation may be present. The forces F_{ab} are not themselves directly measurable; consequently not a single *prediction* can be deduced from the theory.

What do we do, then, when we apply this theory to an astronomical situation? Typically we make certain simplifying assumptions. For example, if we are deducing the orbit of the earth we might assume as a first approximation:

(I) No bodies exist except the sun and the earth.
(II) The sun and the earth exist in a hard vacuum.
(III) The sun and the earth are subject to no forces except mutually induced gravitational forces.

From the conjunction of the theory of universal gravitation (UG) and these auxiliary statements (AS) we can, indeed, deduce certain predictions – e.g. Kepler's laws. By making (I), (II), (III) more 'realistic' – i.e. incorporating further bodies in our model solar system – we can obtain better predictions. But it is important to note that these predictions do not come from the theory alone, but from the conjunction of the theory with AS. As scientists actually use the term 'theory', the statements AS are hardly part of the 'theory' of gravitation.

6. Is the point terminological?

I am not interested in making a merely *terminological* point, however. The point is not just that scientists don't use the term 'theory' to refer to the conjunction of UG with AS, but that such a usage would obscure profound methodological issues. A *theory*, as the term is actually used, is a set of *laws*. Laws are statements that we hope to be *true*; they are supposed to be true by the nature of things, and not just by accident. None of the statements (I), (II), (III) has this character. We do not really believe that *no* bodies except the sun and the earth exist, for example, but only that all other bodies exert forces small enough to be neglected. This statement is not supposed to be a law of nature: it is a statement about the 'boundary conditions' which obtain as a matter of fact in a particular system. To blur the difference between AS and UG is to blur the difference between *laws* and *accidental statements*, between statements the scientist wishes to establish as *true* (the laws), and statements he already knows to be false (the oversimplifications (I), (II), (III)).

7. Uranus, Mercury, 'dark companions'

Although the statements AS *could* be more carefully worded to avoid the objection that they are known to be false, it is striking that they are not in practice. In fact, they are not 'worded' at all. Newton's calculation of Kepler's laws makes the assumptions (I), (II), (III) without more than a casual indication that this is what is done. One of the most striking indications of the difference between a theory (such as UG) and a set of AS is the great care which scientists use in stating the theory, as contrasted with the careless way in which they introduce the various assumptions which make up AS.

The AS are also far more subject to revision than the theory. For over two hundred years the law of universal gravitation was accepted as unquestionably true, and used as a premise in countless scientific arguments. If the standard kind of AS had not led to successful prediction in that period, they would have been modified, not the theory. In fact, we have an example of this. When the predictions about the orbit of Uranus that were made on the basis of the theory of universal gravitation and the assumption that the known planets were all there were turned out to be wrong, Leverrier in France and Adams in England simultaneously predicted that there must be another planet. In fact, this planet was discovered – it was Neptune. Had this modification of the AS not been successful, still others might have been tried – e.g. postulating a medium through which the planets are moving, instead of a hard vacuum, or postulating significant nongravitational forces.

It may be argued that it was crucial that the new planet should itself be observable. But this is not so. Certain stars, for example, exhibit irregular behavior. This has been explained by postulating companions. When those companions are not visible through a telescope, this is handled by suggesting that the stars have *dark companions* – companions which cannot be seen through a telescope. The fact is that many of the assumptions made in the sciences cannot be directly tested – there are many 'dark companions' in scientific theory.

Lastly, of course, there is the case of Mercury. The orbit of this planet can almost but not quite be successfully explained by Newton's theory. Does this show that Newton's theory is wrong? *In the light of an alternative theory*, say the General Theory of Relativity, one answers 'yes'. But, in the absence of such a theory, the orbit of Mercury is just a slight anomaly, cause: unknown.

What I am urging is that all this is perfectly good scientific practice. The fact that any one of the statements AS may be false – indeed, they are false, as stated, and even more careful and guarded statements might well be false – is important. We do not know for sure all the bodies in the solar system; we do not know for sure that the medium through which they move is (to a sufficiently high degree of approximation in all cases) a hard vacuum; we do not know that nongravitational forces can be neglected in all cases. Given the overwhelming success of the Law of Universal Gravitation in almost all cases, one or two anomalies are not reason to reject it. It is more *likely* that the AS are false than that the theory is false, at least when no alternative theory has seriously been put forward.

8. The effect on Popper's doctrine

The effect of this fact on Popper's doctrine is immediate. The Law of Universal Gravitation is *not* strongly falsifiable at all; yet it is surely a paradigm of a scientific theory. Scientists for over two hundred years did not falsify UG; they derived predictions from UG in order to explain various astronomical facts. If a fact proved recalcitrant to this sort of explanation it was put aside as an anomaly (the case of Mercury). Popper's doctrine gives a correct account of neither the nature of the scientific theory nor of the practice of the scientific community in this case.

Popper might reply that he is not describing what scientists do, but what they *should* do. Should scientists then not have put forward UG? Was Newton a bad scientist? Scientists did not try to falsify UG because they could not try to falsify it; laboratory tests were excluded by the technology of the time and the weakness of the gravitational interactions. Scientists were thus limited to astronomical data for a long time. And, even in the astronomical cases, the problem arises that one cannot be absolutely sure that no nongravitational force is relevant in a given situation (or that one has summed *all* the gravitational forces). It is for this reason that astronomical data can *support* UG, but they can hardly *falsify* it. It would have been incorrect to reject UG because of the deviancy of the orbit of Mercury; given that UG predicted the other

257

orbits, to the limits of measurement error, the possibility could not be excluded that the deviancy in this one case was due to an unknown force, gravitational or nongravitational, and in putting the case aside as one they could neither explain nor attach systematic significance to, scientists *were* acting as they 'should'.†

So far we have said that (1) theories do not imply predictions; it is only the conjunction of a theory with certain 'auxiliary statements' (AS) that, in general, implies a prediction. (2) The AS are frequently suppositions about boundary conditions (including initial conditions as a special case of 'boundary conditions'), and highly risky suppositions at that. (3) Since we are very unsure of the AS, we cannot regard a false prediction as definitively falsifying a theory; theories are *not* strongly falsifiable.

All this is not to deny that scientists do sometimes derive predictions from theories and AS in order to test the theories. If Newton had not been able to derive Kepler's laws, for example, he would not have even put forward UG. But even if the predictions Newton had obtained from UG had been wildly wrong, UG might still have been true: the AS might have been wrong. Thus, even if a theory is 'knocked out' by an experimental test, the theory may still be right, and the theory may come back in at a later stage when it is discovered the AS were not useful approximations to the true situation. As has previously been pointed out,‡ falsification in science is no more conclusive than verification.

All this refutes Popper's view that what the scientist does is to put forward 'highly falsifiable' theories, derive predictions from them, and then attempt to falsify the theories by falsifying the predictions. But it does not refute the standard view (what Popper calls the 'inductivist' view) that scientists try to *confirm* theories *and* AS by deriving predictions from them and verifying the predictions. There is the objection that (in the case of UG) the AS were known to be false, so scientists could hardly have been trying to confirm them; but this could be met by saying that the AS could, in principle, have been formulated in a more guarded way, and would not have been false if sufficiently guarded.§

† Popper's reply to this sort of criticism is discussed in 'Kuhn versus Popper' in Schilpp, 1974.
‡ This point is made by many authors. The point that is often missed is that, in cases such as the one discussed, the auxiliary statements are much less certain than the theory under test; without this remark, the criticism that one *might* preserve a theory by revising the AS looks like a bit of formal logic, without real relation to scientific practice. (See 'Kuhn versus Popper' in Schilpp, 1974.)
§ I have in mind saying 'the planets exert forces on each other which are more than o·999 (or whatever) gravitational', rather than 'the planets exert *no* non-gravitational forces on each other'. Similar changes in the other AS could presumably turn them into true statements – though it is not methodologically unimportant that no scientist,

I think that, in fact, there is some truth in the 'inductivist' view: scientific theories are shown to be correct by their successes, just as all human ideas are shown to be correct, to the extent that they are, by their successes in practice. But the inductivist schema is still inadequate, except as a picture of one aspect of scientific procedure. In the next sections, I shall try to show that scientific activity cannot, in general, be thought of as a matter of deriving predictions from the conjunction of theories and AS, whether for the purpose of confirmation or for the purpose of falsification.

9. Kuhn's view of science

Recently a number of philosophers have begun to put forward a rather new view of scientific activity. I believe that I anticipated this view about ten years ago when I urged that some scientific theories cannot be over-thrown by experiments and observations *alone*, but only by alternative theories (Putnam, 1962). The view is also anticipated by Hanson (Hanson, 1958), but it reaches its sharpest expression in the writings of Thomas Kuhn (Kuhn, 1962) and Louis Althusser (Althusser, 1965). I believe that both of these philosophers commit errors; but I also believe that the tendency they represent (and that I also represent, for that matter) is a needed corrective to the deductivism we have been examining. In this section, I shall present some of Kuhn's views, and then try to advance on them in the direction of a sharper formulation.

The heart of Kuhn's account is the notion of a *paradigm*. Kuhn has been legitimately criticized for some inconsistencies and unclarities in the use of this notion; but at least one of his explanations of the notion seems to me to be quite clear and suitable for his purposes. On this explanation, a paradigm is simply a scientific theory together with an example of a successful and striking application. It is important that the application – say, a successful explanation of some fact, or a successful and novel prediction – be *striking*; what this means is that the success is sufficiently impressive that scientists – especially young scientists choosing a career – are led to try to emulate that success by seeking further explanations, predictions, or whatever on the same model. For example, once UG had been put forward and one had the example of Newton's derivation of Kepler's laws together with the example of the derivation of, say, a planetary orbit or two, then one had a paradigm. The most important paradigms are the ones that generate scientific fields; the field generated by the Newtonian paradigm was, in the first

to my knowledge, has bothered to calculate exactly what changes in the AS would render them true while preserving their usefulness.

instance, the entire field of Celestial Mechanics. (Of course, this field was only a part of the larger field of Newtonian mechanics, and the paradigm on which Celestial Mechanics is based is only one of a number of paradigms which collectively structure Newtonian mechanics.)

Kuhn maintains that the paradigm that structures a field is highly immune to falsification – in particular, it can only be overthrown by a new paradigm. In one sense, this is an exaggeration: Newtonian physics would probably have been abandoned, even in the absence of a new paradigm, if the world had started to act in a markedly non-Newtonian way. (Although even then – would we have concluded that Newtonian physics was false, or just that we didn't know what the devil was going on?) But then even the old successes, the successes which were paradigmatic for Newtonian physics, would have ceased to be available. What is true, I believe, is that in the absence of such a drastic and unprecedented change in the world, and in the absence of its turning out that the paradigmatic successes had something 'phony' about them (e.g. the data were faked, or there was a mistake in the deductions), a theory which is paradigmatic is not given up because of observational and experimental results by themselves, but only because and when a better theory is available.

Once a paradigm has been set up, and a scientific field has grown up around that paradigm, we get an interval of what Kuhn calls 'normal science'. The activity of scientists during such an interval is described by Kuhn as 'puzzle solving' – a notion I shall return to.

In general, the interval of normal science continues even though not all the puzzles of the field can be successfully solved (after all, it is only human experience that some problems are too hard to solve), and even though some of the solutions may look *ad hoc*. What finally terminates the interval is the introduction of a new paradigm which manages to supersede the old.

Kuhn's most controversial assertions have to do with the process whereby a new paradigm supplants an older paradigm. Here he tends to be radically subjectivistic (overly so, in my opinion): data, in the usual sense, cannot establish the superiority of one paradigm over another because data themselves are perceived through the spectacles of one paradigm or another. Changing from one paradigm to another requires a 'Gestalt switch'. The history and methodology of science get rewritten when there are major paradigm changes; so there are no 'neutral' historical and methodological canons to which to appeal. Kuhn also holds views on meaning and truth which are relativistic and, in my view, incorrect; but I do not wish to discuss these here.

What I want to explore is the interval which Kuhn calls 'normal science'. The term 'puzzle solving' is unfortunately trivializing; searching for explanations of phenomena and for ways to harness nature is too important a part of human life to be demeaned (here Kuhn shows the same tendency that leads Popper to call the problem of the nature of knowledge a 'riddle'). But the term is also striking: clearly, Kuhn sees normal science as neither an activity of trying to falsify one's paradigm nor as an activity of trying to confirm it, but as something else. I want to try to advance on Kuhn by presenting a schema for normal science, or rather for one aspect of normal science; a schema which may indicate why a major philosopher and historian of science would use the metaphor of solving puzzles in the way Kuhn does.

10. Schemata for scientific problems

Consider the following two schemata:

Schema I
_____
Theory
Auxiliary Statements
_____
Prediction – True or false?

Schema II
_____
Theory
???
_____
Fact to be explained

These are both schemata for scientific problems. In the first type of problem we have a theory, we have some AS, we have derived a prediction, and our problem is to see if the prediction is true or false: the situation emphasized by standard philosophy of science. The second type of problem is quite different. In this type of problem we have a theory, we have a fact to be explained, but the AS are missing: the problem is to find AS if we can, which are true, or approximately true (i.e. useful oversimplifications of the truth), and which have to be conjoined to the theory to get an explanation of the fact.

We might, in passing, mention also a third schema which is neglected by standard philosophy of science:

Schema III
_____
Theory
Auxiliary Statements
???

261

This represents the type of problem in which we have a theory, we have some AS, and we want to know what consequences we can derive. This type of problem is neglected because the problem is 'purely mathematical'. But knowing whether a set of statements has testable consequences at all depends upon the solution to this type of problem, and the problem is frequently of great difficulty – e.g. little is known to this day concerning just what the physical consequences of Einstein's 'unified field theory' are, precisely because the mathematical problem of deriving those consequences is too difficult. Philosophers of science frequently write as if it is *clear*, given a set of statements, just what consequences those statements do and do not have.

Let us, however, return to Schema II. Given the known facts concerning the orbit of Uranus, and given the known facts (prior to 1846) concerning what bodies make up the solar system, and the standard AS that those bodies are moving in a hard vacuum, subject only to mutual gravitational forces, etc., it was clear that there was a problem: the orbit of Uranus could not be successfully calculated if we assumed that Mercury, Venus, Earth, Mars, Saturn, Jupiter, and Uranus were all the planets there are, and that these planets together with the sun make up the whole solar system. Let S_1 be the conjunction of the various AS we just mentioned, including the statement that the solar system consists of at least, but not necessarily of only, the bodies mentioned. Then we have the following problem:

Theory: UG
AS: S_1
Further AS: ???

Explanandum: The orbit of Uranus

– note that the problem is not to find further explanatory laws (although sometimes it may be, in a problem of the form of Schema II); it is to find further assumptions about the initial and boundary conditions governing the solar system which, together with the Law of Universal Gravitation and the other laws which make up UG (i.e. the laws of Newtonian mechanics) will enable one to explain the orbit of Uranus. If one does not require that the missing statements be true, or approximately true, then there are an infinite number of solutions, mathematically speaking. Even if one includes in S_1 that no nongravitational forces are acting on the planets or the sun, there are still an infinite number of solutions. But one tries first the simplest assumption, namely:

(S_2) There is one and only one planet in the solar system in addition to the planets mentioned in S_1.

Now one considers the following problem:

Theory: UG
AS: S_1, S_2

Consequence ??? – turns out to be that the unknown planet must have a certain orbit O.

This problem is a mathematical problem – the one Leverrier and Adams both solved (an instance of Schema III). Now one considers the following empirical problem:

Theory: UG
AS: S_1, S_2

Prediction: a planet exists moving in orbit O – True or False?

– this problem is an instance of Schema I – an instance one would not normally consider, because one of the AS, namely the statement S_2 is not at all known to be true. S_2 is, in fact, functioning as a low-level hypothesis which we wish to test. But the test is not an inductive one in the usual sense, because a verification of the prediction is also a verification of S_2 – or rather, of the approximate truth of S_2 (which is all that is of interest in this context) – Neptune was not the only planet unknown in 1846; there was also Pluto to be later discovered. The fact is that we are interested in the above problem in 1846, because we know that if the prediction turns out to be true, then that prediction is precisely the statement S_3 that we need for the following deduction;

Theory: UG
AS: S_1, S_2, S_3

Explanandum: the orbit of Uranus

– i.e. the statement S_3 (that the planet mentioned in S_2 has precisely the orbit O)† is the solution to the problem with which we started. In this case we started with a problem of the Schema II-type: we introduced the assumption S_2 as a simplifying assumption in the hope of solving the original problem thereby more easily; and we had the good luck to be able to deduce S_3 – the solution to the original problem – from UG together with S_1, S_2, and the more important good luck that S_3 turned out to be true when the Berlin Observatory looked. Problems of the Schema II-type are sometimes mentioned by philosophers of science when the missing AS are *laws*; but the case just examined, in which the

† I use 'orbit' in the sense of space–time trajectory, not just spatial path.

263

missing AS was just a further contingent fact about the particular system is almost never discussed. I want to suggest that Schema II exhibits the logical form of what Kuhn calls a 'puzzle'.

If we examine Schema II, we can see why the term 'puzzle' is so appropriate. When one has a problem of this sort one is looking for something to fill a 'hole' – often a thing of rather under-specified sort – and that *is* a sort of *puzzle*. Moreover, this sort of problem is extremely widespread in science. Suppose one wants to explain the fact that water is a liquid (under the standard conditions), and one is given the laws of physics; the fact is that the problem is extremely hard. In fact, quantum mechanical laws are needed. But that does not mean that from classical physics one can deduce that water is *not* a liquid; rather the classical physicist would give up this problem at a certain point as 'too hard' – i.e. he would conclude that he could not find the right AS.

The fact that Schema II is the logical form of the 'puzzles' of normal science explains a number of facts. When one is tackling a Schema II-type problem there is no question of deriving a prediction from UG plus given AS, the whole problem is to find the AS. The theory – UG, or whichever – is *unfalsifiable in the context*. It is also not up for 'confirmation' any more than for 'falsification'; *it is not functioning in a hypothetical role*. Failures do not falsify a theory, because the failure is not a false prediction from a theory together with known and trusted facts, but a failure to *find* something – in fact, a failure to find an AS. Theories, during their tenure of office, are highly immune to falsification; that tenure of office is ended by the appearance on the scene of a better theory (or a whole new explanatory technique), not by a basic sentence. And successes do not 'confirm' a theory, once it has become paradigmatic, because the theory is not a 'hypothesis' in need of confirmation, but the basis of a whole explanatory and predictive technique, and possibly of a technology as well.

To sum up: I have suggested that standard philosophy of science, both 'Popperian' and non-Popperian, has fixated on the situation in which we derive predictions from a theory, and test those predictions in order to falsify or confirm the theory – i.e. on the situation represented by Schema I. I have suggested that, by way of contrast, we see the 'puzzles' of 'normal science' as exhibiting the pattern represented by Schema II – the pattern in which we take a theory as fixed, take the fact to be explained as fixed, and seek further facts – frequently contingent† facts about the particular system – which will enable us to fill out the explanation of the particular fact on the basis of the theory. I

---

† By 'contingent' I mean *not physically necessary*.

suggest that adopting this point of view will enable us better to appreciate both the relative unfalsifiability of theories which have attained paradigm status, and the fact that the 'predictions' of physical theory are frequently facts which were known beforehand, and not things which are surprising relative to background knowledge.

To take Schema II as describing everything that goes on between the introduction of a paradigm and its eventual replacement by a better paradigm would be a gross error in the opposite direction, however. The fact is that normal science exhibits a dialectic between two conflicting (at any rate, potentially conflicting) but interdependent tendencies, and that it is the conflict of these tendencies that drives normal science forward. The desire to solve a Schema II-type problem – explain the orbit of Uranus – led to a new hypothesis (albeit a very low-level one): namely, S_2. Testing S_2 involved deriving S_3 from it, and testing S_3 – a Schema I-type situation. S_3 in turn served as the solution to the original problem. This illustrates the two tendencies, and also the way in which they are interdependent and the way in which their interaction drives science forward.

The tendency represented by Schema I is the *critical* tendency. Popper is right to emphasize the importance of this tendency, and doing this is certainly a contribution on his part – one that has influenced many philosophers. Scientists do want to know if their ideas are wrong, and they try to find out if their ideas are wrong by deriving predictions from them, and testing those predictions – that is, they do this *when they can*. The tendency represented by Schema II is the *explanatory* tendency. The element of conflict arises because in a Schema II-type situation one tends to regard the given theory as something *known*, whereas in a Schema-I type situation one tends to regard it as *problematic*. The interdependence is obvious: the theory which serves as the major premise in Schema II *may* itself have been the survivor of a Popperian test (although it need not have been – UG was accepted on the basis of its explanatory successes, not on the basis of its surviving attempted falsifications). And the solution to a Schema II-type problem must itself be confirmed, frequently by a Schema I-type test. If the solution is a general law, rather than a singular statement, that law may itself become a paradigm, leading to new Schema II-type problems. In short, attempted falsifications do 'corroborate' theories – not just in Popper's sense, in which this is a tautology, but in the sense he denies, of showing that they are true, or partly true – and explanations on the basis of laws which are regarded as *known* frequently require the introduction of *hypotheses*. In this way, the tension between the attitudes of explanation and criticism drives science to progress.

11. Kuhn versus Popper

As might be expected, there are substantial differences between Kuhn and Popper on the issue of the falsifiability of scientific theories. Kuhn stresses the way in which a scientific theory may be immune from falsification, whereas Popper stresses falsifiability as the *sine qua non* of a scientific theory. Popper's answers to Kuhn depend upon two notions which must now be examined: the notion of an auxiliary hypothesis and the notion of a *conventionalist stratagem*.

Popper recognizes that the derivation of a prediction from a theory may require the use of auxiliary hypotheses (though the term 'hypothesis' is perhaps misleading, in suggesting something like putative laws, rather than assumptions about, say, boundary conditions). But he regards these as part of the total 'system' under test. A 'conventionalist stratagem' is to save a theory from a contrary experimental result by making an *ad hoc* change in the auxiliary hypotheses. And Popper takes it as a fundamental methodological rule of the empirical method to avoid conventionalist stratagems.

Does this do as a reply to Kuhn's objections? Does it contravene our own objections, in the first part of this paper? It does not. In the first place, the 'auxiliary hypotheses' AS are not fixed, in the case of UG, but depend upon the context. One simply cannot think of UG as part of a fixed 'system' whose other part is a fixed set of auxiliary hypotheses whose function is to render UG 'highly testable'.

In the second place, an alteration in one's beliefs, may be *ad hoc* without being unreasonable. '*Ad hoc*' merely means 'to this specific purpose'. Of course, '*ad hoc*' has acquired the connotation of 'unreasonable' – but that is a different thing. The assumption that certain stars have dark companions is *ad hoc* in the literal sense: the assumption is made for the specific purpose of accounting for the fact that no companion is visible. It is also highly reasonable.

It has already been pointed out that the AS are not only context-dependent but highly uncertain, in the case of UG and in many other cases. So, changing the AS, or even saying in a particular context 'we don't know what the right AS are' may be *ad hoc* in the literal sense just noted, but is not '*ad hoc*' in the extended sense of 'unreasonable'.

12. Paradigm change

How does a paradigm come to be accepted in the first place? Popper's view is that a theory becomes corroborated by passing severe tests: a prediction (whose truth value is not antecedently known) must be derived from the theory and the truth or falsity of that prediction must be ascertained. The severity of the test depends upon the set of basic sentences excluded by the theory, and also upon the improbability of the prediction relative to background knowledge. The ideal case is one in which a theory which rules out a great many basic sentences implies a prediction which is very improbable relative to background knowledge.

Popper points out that the notion of the number of basic sentences ruled out by a theory cannot be understood in the sense of cardinality; he proposes rather to measure it by means of concepts of *improbability* or *content*. It does not appear true to me that improbability (in the sense of logical ([im] probability) measures falsifiability, in Popper's sense: UG excludes *no* basic sentences, for example, but has logical probability *zero*, on any standard metric. And it certainly is not true that the scientist always selects 'the most improbable of the surviving hypotheses' on *any* measure of probability, except in the trivial sense that all strictly universal laws have probability zero. But my concern here is not with the technical details of Popper's scheme, but with the leading idea.

To appraise this idea, let us see how UG came to be accepted. Newton first derived Kepler's Laws from UG and the AS we mentioned at the outset: this was not a 'test', in Popper's sense, because Kepler's Laws were already known to be true. Then he showed that UG would account for the tides on the basis of the gravitational pull of the moon: this also was not a 'test', in Popper's sense, because the tides were already known. Then he spent many years showing that small perturbations (which were already known) in the orbits of the planets could be accounted for by UG. By this time the whole civilized world had accepted – and, indeed, acclaimed – UG; but it had not been 'corroborated' at all in Popper's sense!

If we look for a Popperian 'test' of UG – a derivation of a new prediction, one risky relative to background knowledge – we do not get one until the Cavendish experiment of 1781† – roughly a hundred years after the theory had been introduced! The prediction of S_3 (the orbit of Neptune) from UG and the auxiliary statements S_1 and S_2 can also be regarded as a confirmation of UG (in 1846!); although it is difficult to regard it as a severe test of UG in view of the fact that the assumption S_2 had a more tentative status than UG.

It is easy to see what has gone wrong. A theory is not accepted unless it has real explanatory successes. Although a theory may legitimately be preserved by changes in the AS which are, in a sense, '*ad hoc*' (although not *unreasonable*), its *successes* must not be ad hoc. Popper requires that the predictions of a theory must not be antecedently known to be true in order to rule out ad hoc 'successes'; but the condition is too strong.

† One might also mention Clairault's prediction of the perihelion of Halley's comet in 1759.

Popper is right in thinking that a theory runs a risk during the period of its establishment. In the case of UG, the risk was not a risk of definite falsification; it was the risk that Newton would not find reasonable AS with the aid of which he could obtain real (non-*ad hoc*) explanatory successes for UG. A failure to explain the tides by the gravitational pull of the moon alone would not, for example, have falsified UG; but the success did strongly support UG.

In sum, a theory is only accepted if the theory has substantial, non-*ad hoc*, explanatory successes. This is in accordance with Popper; unfortunately, it is in even better accordance with the 'inductivist' accounts that Popper rejects, since these stress *support* rather than *falsification*.

13. On practice

Popper's mistake here is no small isolated failing. What Popper consistently fails to see is that *practice is primary*: ideas are not just an end in themselves (although they are *partly* an end in themselves), nor is the selection of ideas to 'criticize' just an end in itself. The primary importance of ideas is that they guide practice, that they structure whole forms of life. Scientific ideas guide practice in science, in technology, and sometimes in public and private life. We are concerned in science with trying to discover correct ideas: Popper to the contrary, this is not *obscurantism* but *responsibility*. We obtain our ideas – our correct ones, and many of our incorrect ones – by close study of the world. Popper denies that the accumulation of perceptual experience leads to theories: he is right that it does not lead to theories in a mechanical or algorithmic sense; but it does lead to theories in the sense that it is a regularity of methodological significance that (1) lack of experience with phenomena and with previous knowledge about phenomena decreases the probability of correct ideas in a marked fashion; and (2) extensive experience increases the probability of correct, or partially correct, ideas in a marked fashion. 'There is no logic of discovery' – in that sense, there is no logic of *testing*, either; all the formal algorithms proposed for testing, by Carnap, by Popper, by Chomsky, etc., are, to speak impolitely, *ridiculous*: if you don't believe this, program a computer to employ one of these algorithms and see how well it does at testing theories! There are *maxims* for discovery and maxims for testing: the idea that correct ideas just come from the sky, while the methods for testing them are highly rigid and predetermined, is one of the worst legacies of the Vienna Circle.

But the correctness of an idea is not certified by the fact that it came from close and concrete study of the relevant aspects of the world; in this sense, Popper is right. We judge the correctness of our ideas by applying them and seeing if they succeed; in general, and in the long run, correct ideas lead to success, and ideas lead to failures where and insofar as they are incorrect. Failure to see the importance of practice leads directly to failure to see the importance of success.

Failure to see the primacy of practice also leads Popper to the idea of a sharp 'demarcation' between science, on the one hand, and political, philosophical, and ethical ideas, on the other. This 'demarcation' is pernicious, in my view; fundamentally, it corresponds to Popper's separation of theory from practice, and his related separation of the critical tendency in science from the explanatory tendency in science. Finally, the failure to see the primacy of practice leads Popper to some rather reactionary political conclusions. Marxists believe that there are laws of society; that these laws can be known; and that men can and should act on this knowledge. It is not my intention to argue that this Marxist view is correct; but surely any view that rules this out *a priori* is reactionary. Yet this is precisely what Popper does – and in the name of an *anti-a priori* philosophy of knowledge!

In general, and in the long run, true ideas are the ones that succeed – how do we know this? This statement too is a statement about the world; a statement we have come to from experience of the world; and we believe in the practice to which this idea corresponds, and in the idea as informing that kind of practice, on the basis that we believe in any good idea – it has proved successful! In this sense 'induction is circular'. But of course it is! Induction has no deductive justification; induction is not deduction. Circular justifications need not be totally self-protecting nor need they be totally uninformative:† the past success of 'induction' increases our confidence in it, and its past failure tempers that confidence. The fact that a justification is circular only means that that justification has no power to serve as a *reason*, unless the person to whom it is given as a reason already has some propensity to accept the conclusion. We do have a propensity – an *a priori* propensity, if you like – to reason 'inductively', and the past success of 'induction' increases that propensity.

The method of testing ideas in practice and relying on the ones that prove successful (for that is what 'induction' is) is not unjustified. That is an *empirical* statement. The method does not have a 'justification' – if by a justification is meant a proof from eternal and formal principles that justifies reliance on the method. But then, nothing does – not even, in my opinion, pure mathematics and formal logic.

† This has been emphasized by Professor Max Black in a number of papers.

17

'Degree of confirmation' and inductive logic*

I

Carnap's attempt to construct a *symbolic inductive logic*, fits into two major concerns of empiricist philosophy. On the one hand, there is the traditional concern with the formulation of Canons of Induction; on the other hand, there is the distinctively Carnapian concern with providing a formal reconstruction of the language of science as a whole, and with providing precise meanings for the basic terms used in methodology.

Of the importance of continuing to search for a more precise statement of the inductive techniques used in science, I do not need to be convinced; this is a problem which today occupies mathematical statisticians at least as much as philosophers.

But this general search need not be identified with the particular project of defining a *quantitative* concept of 'degree of confirmation'. I shall argue that this last project is misguided.

Such a negative conclusion needs more to support it than 'intuition'; or even than plausible arguments based on the methodology of the developed sciences (as the major features of that method may appear evident to one). Intuitive considerations and plausible argument might lead one to the conclusion that it would not be a good investment to spend ones *own* time trying to 'extend the definition of degree of confirmation'; it could hardly justify trying to, say, convince Carnap that this particular project should be abandoned. But that is what I shall try to do here: I shall argue that one can *show* that no definition of degree of confirmation can be adequate or can attain what any reasonably good inductive judge might attain *without* using such a concept. To do this it will be necessary (a) to state precisely the condition of adequacy that will be in question; (b) to show that no inductive method based on a 'measure function'† can satisfy it; and (c) to show that *some* methods (which can be precisely stated) *can* satisfy it.

---

* First published in P. A. Schilpp (ed.), *The Philosophy of Rudolf Carnap* (La Salle, Ill., The Open Court Publishing Co., 1963).

† This is Carnap's term for an *a priori* probability distribution. Cf. Carnap, 1950; and for an excellent explanation of leading ideas, *vide* also the paper by Kemeny in this volume.

From this we have a significant corollary: not every (reasonable) inductive method can be represented by a 'measure function'. Thus, we might also state what is to be proved here in the following form: the actual inductive procedure of science has features which are incompatible with being represented by a 'measure function' (or, what is the same thing, a quantitative concept of 'degree of confirmation').

II

Let us begin with the statement of the condition of adequacy. The first problem is the *kind of language* we have in mind.

What we are going to suppose is a language rich enough to take account of the *space–time arrangement* of the individuals. Languages for which DC (degree of confirmation) has so far been defined are not this rich: we can express the hypothesis that five individuals are black and five red, but not the hypothesis that ten *successive* individuals are *alternately* black and red. Extension of DC to such a language is evidently one of the next steps on the agenda; it would still be far short of the final goal (definition of DC for a language rich enough for the formalization of empirical science as a whole).

In addition to supposing that our language, L, is rich enough to describe spatial relations, we shall suppose that it possesses a second sort of richness; we shall suppose that L contains elementary number theory. The problem of defining DC for a language which is rich enough for elementary number theory (or more broadly, for classical mathematics) might seem an insuperable one, or, at any rate, much more difficult than defining DC for a language in which the individuals have an 'order'. But such is not the case. I have shown elsewhere (Putnam, 1956), that any measure function defined for an (applied) first order functional calculus can be extended to a language rich enough for Cantorian set theory; hence certainly rich enough for number theory, and indeed far richer than needful for the purposes of empirical science. The difficult (I claim: *impossible*) task is not the 'extension to richer languages' in the *formal* sense (i.e. to languages adequate for larger parts of logic and mathematics) but the extension to languages richer in a *physical* sense (i.e. adequate for taking account of the fact of *order*).

In short, we consider a language rich enough for

(a) the description of space–time order.
(b) elementary number theory.

The purpose of the argument is to show that DC *cannot* be adequately defined for such a language. This is independent of whether or not the particular method of 'extending to richer languages' used in the paper mentioned is employed. But by combining the argument of that paper with the present argument we could get a stronger result: it is not possible to define DC adequately in a language satisfying just (a).

To state our condition of adequacy, we will also need the notion of an *effective hypothesis* (a deterministic law).

Informally, an effective hypothesis is one which says of each individual whether or not it has a certain molecular property M; and which does so effectively in the sense that it is possible to *deduce* from the hypothesis what the character of any individual will be. Thus an effective hypothesis is one that we can *confront* with the data: one can deduce what the character of the individuals will be, and then see whether our prediction agrees with the facts as more and more individuals are observed. Formally, an hypothesis h will be called an effective hypothesis if it has the following properties:

(i) h is expressible in L.

(ii) if it is a consequence of h that $M(x_i)$ is true† (where M is a molecular predicate of L and x_i is an individual constant), then $h \supset M(x_i)$ is *provable* in L.

(iii) h is equivalent to a set of sentences of the forms $M(x_i)$ and $\sim M(x_i)$; where M is some molecular predicate of L, and x_i runs through the names of all the individuals.

The notion of an effective hypothesis is designed to include the hypotheses normally regarded as expressing putative universal laws. For example, if a hypothesis implies that each individual satisfies the molecular predicate‡ $P_1(x) \supset P_2(x)$, we require that (for each i) $(P_1(x_i) \supset P_2(x_i))$ should be deducible from h in L, for h to count as effective.

We can now state our condition of adequacy:

I. If h is an effective hypothesis and h is true, then the *instance confirmation* of h (as more and more successive individuals are examined) approaches 1 as limit.

We may also consider *weaker* conditions as follows:

I′. If h is an effective hypothesis and h is true, then the *instance confirmation* of h eventually becomes and remains greater than 0·9 (as more and more successive individuals are examined).

I″. (Same as I′, with '0·5' in place of '0·9'.)

† Logical formulas are used here only as names of themselves; never in their object-language use.

‡ I.e. the predicate $P_1 (\cdots) \supset P_2 (\cdots)$; we use the corresponding open sentence to represent it.

Even the weakest of these conditions is violated – *must* be violated – by every measure function of the kind considered by Carnap.

III

In I and its variants we have used the term 'instance confirmation' (Carnap, 1950) introduced by Carnap. The instance confirmation of a universal hypothesis is, roughly speaking, the degree of confirmation that the next individual to be examined will conform to the hypothesis.

It would be more natural to have 'degree of confirmation' in place of 'instance confirmation' in I, I', and I". However, on Carnap's theory, the degree of confirmation of a universal statement is always zero. Carnap does not regard this as a defect; he argues (ibid.) that when we refer to a universal statement as amply confirmed all we really mean is that the instance confirmation is very high. I shall make two remarks about this contention:

(1) This proposal is substantially the same as one first advanced by Reichenbach (1947) and criticized by Nagel (1939). The criticism is simply that a very high confirmation in this sense (instance confirmation) is *compatible with any number of exceptions*.

(2) The whole project is to define a concept of degree of confirmation which underlies the scientist's 'qualitative' judgments of 'confirmed', 'disconfirmed', 'accepted', 'rejected', etc. much in the way that the quantitative magnitude of *temperature* may be said to underlie the qualitative distinctions between 'hot' and 'cold', 'warm' and 'cool', etc. But a universal statement *may* be highly confirmed (or even 'accepted') as those terms are actually used in science. Therefore it must have a high degree of confirmation, if the relation of 'degree of confirmation' to 'confirmed' is as just described. To say that it only has a high *instance* confirmation is to abandon the analogy 'degree of confirmation is to confirmed as temperature is to hot'. But this analogy explains what it *is* to try to 'define degree of confirmation'.

(Carnap's reply is to maintain the analogy, and deny that a universal statement is ever really confirmed; what is really confirmed, in his view, is that no exceptions will be found in, say, our lifetime, or the lifetime of the human race, or *some* specifiable space–time region (which must be finite).)

IV

Before we proceed to the main argument, let us consider the possibility of obviating the entire discussion by *rejecting* I (and its weaker versions). To do this is to be willing to occupy the following position: (a) one

273

accepts a certain system of inductive logic, based on a function c for 'degree of confirmation', as *wholly adequate*; (b) one simultaneously admits that a certain effective hypothesis h is such that if it be true, we will never discover this fact by our system.

Such a position might indeed be defended by maintaining that certain effective hypotheses are *unconfirmable in principle*. For instance, suppose that we have an ordered series of individuals of the same order-type as the positive integers. Let h be the hypothesis that every individual in the series with a prime-numbered position is red and every individual with a composite position is black (count x_1 as 'composite').

In other words, x_1, x_4, x_6, x_8, x_9, x_{10}, etc. are all black: x_2, x_3, x_5, x_7, x_{11}, etc. are all red.

Someone might reason as follows:

The arithmetic predicates 'prime' and 'composite' do not appear in a single known scientific law; therefore such a 'hypothesis' is not a legitimate scientific theory, and it is only these that we require to be confirmable. In short, it is not a defect of the system if the hypothesis h just described cannot be confirmed (if its instance confirmation does not eventually exceed, and remain greater than o·9 or even o·5).

But this reasoning does not appear particularly plausible; one has only to say: 'Of course the situation described by h has not so far occurred in our experience (as far as we know); but *could we find it out* if it did occur'?

I think the answer is clearly 'yes'; existing inductive methods are capable of establishing the correctness of such a hypothesis (provided someone is bright enough to suggest it), and so must be any adequate 'reconstruction' of those methods.

Thus, suppose McBannister says: 'You know, I think this is the rule: the prime numbers are occupied by red!'

We would first check the data already available for consistency with McBannister's hypothesis. If McBannister's hypothesis fits the first thousand or so objects, we might be impressed, though perhaps not enough to 'accept'. But if we examined another thousand, and then a million, and then ten million objects and McBannister's suggestion 'held-up' – does anyone want to suggest that a reasonable man would *never* accept it?

A similar argument may be advanced if instead of the predicate 'prime' we have any recursive predicate of positive integers. It may take a genius, an Einstein or a Newton, to *suggest* such a hypothesis (to 'guess the rule', as one says); but once it has been suggested any reasonably good inductive judge can verify that it is true. One simply has to keep examining new individuals until any other, antecedently more plausible, hypotheses that may have been suggested have all been ruled out.

In short, if someone rejects I (and its several versions) he must be prepared to offer one of the following 'defenses':

(a) I know that if h is true I won't find it out; but I am 'gambling' that h is false.

(b) If h turns out to be true, I will *change my inductive method*.

Against the first 'defense' I reply that such a 'gamble' would be justifiable only if we could show that *no* inductive method will find it out if h is true (or at least, the *standard* inductive methods will not enable me to accomplish this). But in the case of McBannister's hypothesis about the prime-numbered objects and similar hypotheses, this cannot be urged. Against the second 'defense' I reply that this defense *presupposes that one can find out* if h 'turns out to be true'. But, from the nature of h, the only way to find out would be *inductively*. And if one has an inductive method that will accomplish this, then one's definition of degree of confirmation is evidently not an adequate reconstruction of that inductive method.

V

To simplify the further discussion, we shall suppose that there is only *one* dimension, and not four, and that the series of positions is discrete and has a beginning. Thus we may name the positions $x_1, x_2, x_3, \ldots$ etc. (Following a suggestion of Carnap's we will identify the positions and the individuals. Thus 'x_1 is red' will mean 'the position x_1 is occupied by something red' or 'red occurs at x_1'.) The modification of our argument for the actual case (of a four-dimensional space–time continuum) is simple.†

Next we suppose a function c for degree of confirmation to be given. Technically, c is a function whose arguments are sentences h and e, and whose values are real numbers, $0 \leqslant c(h, e) \leqslant 1$. The numerical value of $c(h, e)$ is supposed to measure the extent to which the statement expressed by h is confirmed by the statement expressed by e; thus $c(h, e)$ may conveniently be read 'the degree of confirmation of h on evidence e'.

Admissible functions c for degree of confirmation are required by Carnap to fulfill several conditions. One of these conditions is that the degree of confirmation of $M(x_i)$ should converge to the relative frequency

---

† Thus we may suppose that $x_1, x_2, \ldots$ are a subsequence of observed positions from the whole four-dimensional continuum; and that the hypotheses under consideration differ only with respect to these.

of M in the sample, as more and more individuals other than x_i are examined. This requirement can no longer be maintained in this form in the case of an *ordered* set of individuals; but the following version must still be required:

II. For every n (and every molecular property M), it must be possible to find an m such that, if the next m individuals (the individuals $x_{n+1}, x_{n+2}, \ldots, x_{n+m}$) are all M, then the DC of the hypothesis $M(x_{n+m+1})$† is greater than 0·5, regardless of the character of the first n individuals.

If n is 10, this means that there must be an m, say 10000000 such that we can say: if the individuals $x_{11}, x_{12}, \ldots, x_{10\,000\,000}$ are all red, then the probability is more than one-half that $x_{10\,000\,001}$ will be red (whether or not some $x_1, x_2, \ldots, x_{10}$ are non-red).

What is the justification of II? Let us suppose that II were violated. Then there must be an n (say, 10) and a property M (say, 'red') such that, for some assignment of 'red' and 'non-red' to $x_1, x_2, \ldots, x_{10}$ (say, x_1, x_2, x_3 are red; $x_4, x_5, \ldots, x_{10}$ are non-red) it holds that no matter how many of $x_{11}, x_{12}, \ldots$, are red, the DC that the *next* individual will be red does not exceed 0·5. Therefore the hypothesis h: x_1, x_2, x_3 are red; $x_4, x_5, \ldots, x_{10}$ are non-red; x_{11} and all subsequent individuals are red – violates I (and in fact, even I"). For no matter how many successive individuals are examined, it is not the case that the instance confirmation of h (this is just the probability that the *next* individual will be red) becomes and remains greater than 0·5.

Thus I entails II. But II is independently justifiable: if II were violated, then there would be a hypothesis of an exceptionally simple kind such that we could never find it out if it were true; namely a hypothesis which says *all* the individuals (with a specified finite number of exceptions) are M. For we would know that h above is true if we knew that 'all the individuals with seven exceptions are red', once we had observed $x_1, x_2, \ldots, x_{10}$. Thus if we want hypotheses of the simple form 'all individuals, with just n exceptions, are M' to be confirmable (to have an instance confirmation which eventually exceeds 0·5), we must accept II.

One more point: c cannot be an *arbitrary* mathematical function. For example, if the value of c were *never* computable, it would be no use to anybody. All the c-functions so far considered by Carnap and other workers in this field have very strong properties of computability. For instance, the DC of a singular hypothesis relative to singular evidence is

† Relative to a complete description with respect to M of the individuals $x_1, x_2, \ldots,$ x_{n+m}. A similar 'evidence' will be understood in similar cases.

always computable. However, this will not be assumed here (although it would materially simplify the argument); all I will assume is the very weak condition: the 'it must be possible to find' in II means *by an effective process*. In other words,† for each n (say, 10) there is some m (say, 10 000 000) such that one can *prove* (in an appropriate metalanguage M_L) that if $x_{11}, x_{12}, \ldots, x_{10\,000\,000}$ are 'red', then the DC that the *next* individual will be 'red' is greater than one-half.

If this is not satisfied, then (by an argument parallel to the above) there is some hypothesis of the simple form 'all the individuals, with just n exceptions, are M' such that we *cannot prove* at any point (with a few exceptions 'at the beginning') that it is more likely than not that the next individual will conform.

E.g. even if we have seen only 'red' things for a very long time (except for the seven 'non-red' things 'at the beginning'), we cannot prove that the DC is more than 0·5 that the next individual will be red.

We can now state our result:

THEOREM: there is no definition of DC which satisfies II (with the effective interpretation of 'it is possible to find') and also satisfies I.

The following proof of this theorem proceeds *via* what mathematical logicians call a 'diagonal argument'.

Let C be an infinite class of integers $n_1, n_2, n_3, \ldots$ with the following property: the DC of $\text{Red}(x_{n_1})$ is greater than 0·5 if all the preceding individuals are red; the DC of $\text{Red}(x_{n_2})$ is greater than 0·5 if all the preceding individuals *after* x_{n_1} are red; and, in general, the DC of $\text{Red}(x_{n_j})$ is greater than 0·5 if all the preceding individuals *after* $x_{n_{j-1}}$ are red.

The existence of a class C with this property is a consequence of II. For (taking $n = 0$) there must be an n_1 such that if the first $n_1 - 1$ individuals are red, the DC is greater than one-half that x_{n_1} is red. Choose such an n_1; then there must be an m such that if the individuals $x_{n_1+1}, x_{n_1+2}, \ldots, x_{n_1+m}$ are all red, the DC is more than one-half that x_{n_1+m+1} is red; call $n_1 + m + 1$ 'n_2':... etc.

Moreover, if we assume the 'effective' interpretation of 'it must be possible to find' in II, there exists a *recursive* class C with this property. (A class is 'recursive' if there exists a mechanical procedure for determining whether or not an integer is in the class.) We shall therefore assume that our chosen class C is 'recursive'.

A predicate is called 'arithmetic' if it can be defined in terms of polynominals and quantifiers.‡ For instance, the predicate 'n is

---

† What follows 'in other words' entails the existence of such an effective process, because it is effectively possible to enumerate the *proofs* in M_L.
‡ This usage is due to Gödel.

square' can be defined by the formula $(\exists m)(n = m^2)$, and is therefore arithmetic.

Now, Gödel has shown that every recursive class is the extension of an arithmetic predicate. (Gödel, 1931, Cf. Kleene, 1952, Theorem X, p. 292 and Theorem I, p. 241). In particular, our class C is the extension of some arithmetic predicate P. So we may consider the following hypothesis:

(1) An individual x_n is red if and only if $\sim P(n)$.

Comparing this with McBannister's hypothesis:

(2) An individual x_n is red if and only if n is prime.

We see that (1) and (2) are of the same form. In fact, the predicate 'is prime' is merely a particular example of a recursive predicate of integers.

Thus the hypothesis (1) is *effective*. It is expressible in L, because P is arithmetic; it satisfies condition (ii) in the definition of 'effective' (see above), because P is recursive; and (iii) is satisfied, since (1) says for each x_i either that $\mathrm{Red}(x_i)$ or $\sim \mathrm{Red}(x_i)$.

But the hypothesis (1) violates I. In other words, a scientist who uses c would never discover that (1) is true, even if he were to live forever (and go on collecting data forever). This can be argued as follows: however we interpret 'discover that (1) is true', a scientist who has discovered that (1) is true should reflect this in his behavior to this extent: he should be willing to bet at even money that the next individual will be non-red whenever (1) *says* that the next individual will be non-red (the more inasmuch as the a priori probability of 'non-red' is greater than 'red'). But, by the definition of C, the scientist will bet at *more* than even money (when he has examined the preceding individuals) that each of the individuals $x_{n_1}, x_{n_2}, x_{n_3}, \ldots$, is *red*. Thus he will make infinitely many mistakes, and his mistakes will show that he has never learned that (1) is true.

Finally, it is no good replying that the scientist will be right more often than not. The aim of science is not merely to be right about particular events, but to discover general laws. And a method that will not *allow* the scientist to accept the law (1), even if someone *suggests* it, and even if no exception has been discovered in ten billion years, is unacceptable.

VI

One might suspect that things are not so black as they have just been painted; perhaps it is the case that *every* formalized system of inductive logic suffers from the difficulty just pointed out, much as every formalized system of arithmetic suffers from the incompleteness pointed out by Gödel. It is important to show that this is not so; and that other approaches to induction – e.g. that of Goodman (1955), or that of Kemeny (1953) are not necessarily subject to this drawback.

Many factors enter into the actual inductive technique of science. Let us consider a technique in which as few as possible of these factors play a part: to be specific, only the direct factual support† (agreement of the hypothesis with the data) and the previous acceptance of the hypothesis.‡ Because of the highly over-simplified character of this technique, it is easily formalized. The following rules define the resulting method (M).:

1. Let $P_{t,M}$ be the set of hypotheses considered at time t with respect to a molecular property M. I.e. $P_{t,M}$ is a finite set of effective hypotheses, each of which specifies, for each individual, whether or not it is M.
2. Let $h_{t,M}$ be the effective hypothesis on M *accepted* at time t (if any). I.e. we suppose that, at any given time, various incompatible hypotheses have been actually suggested with respect to a given M, and have not yet been ruled out (we require that these should be consistent with the data, and with accepted hypotheses concerning other predicates). In addition, one hypothesis may have been accepted at some time prior to t, and may not yet have been abandoned. This hypothesis is called the 'accepted hypothesis at the time t'. So designating it is not meant to suggest that the other hypotheses are not considered as serious candidates for the post of 'accepted hypotheses on M' at some later t.
3. (Rule I:) At certain times t_1, t_2, t_3 ..., initiate an *inductive test with respect to M*. This proceeds as follows: the hypotheses in P_{t_i}, M at this time t_i are called the *alternatives*. Calculate the character $(M$ or not-$M)$ of the next individual on the basis of each alternative. See which alternatives succeed in predicting this. Rule out those that fail. Continue until (a) all alternatives but one have failed; or (b) all alternatives have failed; (one or the other must eventually happen). In case (a) *accept* the alternative that does not fail. In case (b) reject all alternatives.
4. (Rule II:) hypotheses suggested in the course of the inductive test are taken as alternatives (unless they have become inconsistent with the data) in the *next* test. I.e. if h is proposed in the course of the test begun at t_3, then h belongs to $P_{t_4,M}$ and not to $P_{t_3,M}$.

---

† This term has been used in a related sense by Kemeny and Oppenheim (1952).
‡ This factor has been emphasized by Conant (1947).

5. (Rule III:) if $h_{t,M}$ is accepted at the conclusion of any inductive test, then $h_{t,M}$ continues to be accepted as long as it remains consistent with the data. (In particular, while an inductive test is still going on, the previously accepted hypothesis continues to be accepted, for all practical purposes.)

Ridiculously simple as this method M is, it has some good features which are not shared by any inductive method based on a 'measure function'. In particular:

III. If h is an effective hypothesis, and h is true; then, using method M, one will eventually accept h if h is ever proposed.

The method M differs from Carnap's methods, of course, in that the acceptance of a hypothesis depends on which hypotheses are actually proposed, and also on the *order* in which they are proposed. But this does not make the method informal. Given a certain sequence of sentences (representing the suggested hypotheses and the order in which they are suggested), and given the 'time' at which each hypothesis is put forward (i.e. given the *evidence* at that stage: this consisting, we may suppose, of a complete description of individuals $x_1, x_2, \ldots, x_t$ for some t); and given, finally, the 'points' (or evidential situations) at which inductive tests are begun; the 'accepted' hypothesis at any stage is well defined.

That the results a scientist gets, using method M, depend on (a) what hypotheses he considers at any given stage, and even (b) at what points he chooses to inaugurate observational sequences ('inductive tests') is far from being a *defect* of M: these are precisely features that M shares with ordinary experimental and statistical practice. (Carnap sometimes seems to say† that he is looking for something *better* than ordinary experimental practice in these respects. But this is undertaking a task far more ambitious, and far more doubtful, than 'reconstruction'.)

In comparing the method M with Carnap's methods, the problem arises of correlating the essentially qualitative notion of 'acceptance' with the purely quantitative notion of 'degree of confirmation'. One method is this (we have already used it): say that a hypothesis is *accepted* if the instance confirmation is greater than 0·5 (if one is willing to bet at more than even money that the next individual will conform). In these terms, we may say: using Carnap's methods one will, in general, accept an effective hypothesis sooner or later if it is true, and in fact one will accept it infinitely often. But one won't *stick* to it. Thus these methods lack *tenacity*.

† Carnap, 1960, pp. 515–20; see esp. the amazing paragraph at the bottom of p. 518!

Indeed, we might say that the two essential features of M are

(i) *corrigibility*: if h is inconsistent with the data, it is abandoned; and
(ii) *tenacity*: if h is once accepted, it is not subsequently abandoned *unless* it becomes inconsistent with the data.

It is the first feature that guarantees that any effective hypothesis will eventually be accepted if true; for the other alternatives in the set $P_{t_i,M}$ to which it belongs must all be false and, for this reason, they will all eventually be ruled out while the true hypothesis remains. And it is the second feature that guarantees that a true hypothesis, once accepted, is not subsequently rejected.†

It would, of course, be highly undesirable if, in a system based on 'degree of confirmation' one had 'tenacity' in quite the same sense. If we are willing to bet at more than even money that the next individual will conform to h, it does not follow that if it *does* conform we should *then* be willing to bet that the next individual in turn will conform. For instance, if we are willing to bet that the next individual will be red, this means that we are betting that it will conform to the hypothesis that all individuals are red; and also that it will conform to the hypothesis that all individuals up to and including it are red, and all those thereafter green.‡ If it *does* conform to both these hypotheses, we cannot go on to bet that the next individual in turn will conform to both, for this would involve betting that it will be both red and green.§ But we can say this: for any effective hypothesis h, there should come a point (if h continues to be consistent with the data) at which we shall be willing to bet that the next individual will conform; and if the next individual conforms, we shall be willing to bet that the next in turn will conform; and so on. To say this is merely to say again: if it is true we ought *eventually* to accept it. And it is to this simple principle that M conforms, while the Carnapian methods do not.

Moreover, that the method M has the desirable property III is closely connected with a feature which is in radical disagreement with the way of thinking embodied in the 'logical probability' concept: the acceptance of a hypothesis depends on *which* hypotheses are actually proposed. The reader can readily verify that it is this feature (which, I believe, M shares

---

† It is of interest to compare III with the 'pragmatic justification of induction' given by Feigl (1950).
‡ The difficulty occasioned by pairs of hypotheses related in this way was first pointed out by Goodman (1946).
§ This raises a difficulty for Reichenbach's 'Rule of Induction'; the use of the rule to estimate the relative frequency of 'green' and 'grue' (see below) is another case in which contradictory results are obtained.

with the actual procedure of scientists) that blocks a 'diagonal argument' of the kind we used in the preceding section. In short, M is *effective*, and M is able to discover any true law (of a certain simple kind); but this is because what we will predict 'next', using M, does not depend *just* on the evidence. On the other hand, it is easily seen that any method that shares with Carnap's the feature: what one will predict 'next' depends *only* on what has so far been observed, will also share the defect: either what one should predict will not in practice be *computable*,† or some law will elude the method altogether (one is *in principle* forbidden to accept it, no matter how long it has succeeded).

This completes the case for the statement made at the beginning of this paper: namely, that a good inductive judge can do things, provided he does *not* use 'degree of confirmation', that he could not *in principle* accomplish if he *did* use 'degree of confirmation'. As soon as a scientist announces that he is going to use a method based on a certain 'c-function', we can exhibit a hypothesis (in fact, one consistent with the data so far obtained, and hence possibly true) such that we can say: if this is true *we* shall find it out; but you (unless you abandon your method) will never find it out.

Also, we can now criticize the suggested analogy between the 'incompleteness' of Carnap's systems, and the Gödelian incompleteness of formal logics. A more correct analogy would be this: the process of *discovery* in induction is the process of suggesting the correct hypothesis (and, sometimes, a suitable language for its expression and a mathematical technique that facilitates the relevant computation).

But once it has been suggested, the inductive checking, leading to its eventual acceptance, is relatively straightforward. Thus the suggestion of a hypothesis in induction is analogous to the *discovery of a proof* in formal logic; the inductive verification (however protracted, and however many 'simpler' hypotheses must first be ruled out) is analogous to the *checking* of a formal proof (however tedious). Thus one might say: the incompleteness we have discovered in Carnap's system is analogous to the 'incompleteness' that would obtain if there were no mechanical way of *checking* a proof, once discovered, in a formal logic. (Most logicians (e.g. Quine, 1952, p. 245) would hesitate at applying the word 'proof' in such a case.) On the other hand, in the system M, it may take a genius to *suggest* the correct hypothesis; but if it *is* suggested, we can verify it.

† Even in the case of induction by simple enumeration; i.e. there will be hypotheses of the simple form 'all individuals from x_n on are red', such that one will not be able to prove that one should accept them, no matter how many 'red' things one sees.

VII

The oversimplified method M ignores a great many important factors in induction. Some of these, like the reliability of the evidence, are also ignored by Carnap's methods. In addition there is the simplicity of the hypothesis (e.g. the data may be consistent with Mcbannister's hypothesis, and also with the simpler hypothesis 'no individual is red'); the 'entrenchment' of the various predicates and laws in the language of science (Goodman, 1955); etc.

Also, the method M is only a method for selecting among deterministic hypotheses. But we are often interested in selecting from a set of statistical hypotheses, or in choosing between a deterministic hypothesis and a statistical hypothesis (the use of the 'null hypothesis'† is a case in point). This is, in fact, the normal case: a scientist who considers the hypothesis 'all crows are black' is not likely to have in mind an alternative deterministic hypothesis, though he might (i.e. all the crows in such-and-such regions are black; all those in such-and-such other regions are white, etc.); he is far more likely to choose between this hypothesis and a statistical hypothesis that differs reasonably from it (e.g. 'at most 90 per cent of all crows are black').

It is not difficult to adapt the method M to the consideration of statistical hypotheses. A statistical hypothesis is ruled out when it becomes statistically inconsistent with the data at a pre-assigned confidence level. (A statistical hypothesis, once ruled out, may later 'rule itself back in'; but a deterministic hypothesis, as before, is ruled out for good if it is ruled out at all.) This involves combining the method M with the standard method of 'confidence intervals'. If a statistical hypothesis is true, we cannot guarantee that we shall 'stick to it': this is the case because a statistical regularity is compatible with arbitrarily long finite stretches of any character whatsoever. But the *probability* that one will stick to the true hypothesis, once it has been accepted, converges to 1. And if a deterministic hypothesis is true, we will eventually accept it and 'stick to it' (if someone suggests it).‡

Another approach, with a feature very similar to III above, has been suggested by Kemeny (1953). This method rests on the following idea: the hypotheses under consideration are assigned a *simplicity order*. This may even be arbitrary; but of course we would like it to correspond as well as possible to our intuitive concept of simplicity. Then one selects

---

† (The hypothesis that the character in question is randomly distributed in the population).

‡ The above is only a sketch of the method employed in extending M to statistical hypotheses. For statistical hypotheses of the usual forms, this method can be fully elaborated.

the simplest hypothesis consistent with the data (at a pre-assigned confidence level).

Thus, if we have three incompatible hypotheses h_1, h_2, h_3, we have to wait until at most one remains consistent with the data, if we use the method M. And this may take a very long time. Using Kemeny's method, one will, in general, make a selection much more quickly.

On the other hand, Kemeny's method does not make it unnecessary to take into account *the hypotheses that have in fact been proposed*, as one might imagine. (E.g. one might be tempted to say: choose the simplest hypothesis *of all those in the language.*) For one cannot effectively enumerate all the effective hypotheses on a given M in the language.† However, we may suppose that a scientist who suggests a hypothesis shows that it is effective (that it does effectively predict the relevant characteristic); and shows that it does lead to different predications than the other hypotheses. Then with respect to the class $P_{t_i, M}$ of hypotheses belonging to the inductive test we may apply the Kemeny method; since every hypothesis in the class is effective, and no two are equivalent. For instance, one might simply take the hypothesis with the fewest symbols as the simplest (i.e. a 10-letter hypothesis is simpler than a 20-letter); but this would be somewhat crude. But even a *very* crude method such as this represents an improvement on the method M above, and a closer approximation to actual scientific practice.

It is instructive to consider the situation in connection with an oversimplified example. The following excellent example is due to Goodman (1955):

(1) All emeralds are green.
(2) All emeralds are green prior to time t; and blue subsequently.

We might object to (2) on the ground that it contains the name of a specific time-point (t). This does not appear to me to be a good objection. The hypothesis that the first 100 objects produced by a certain machine will be red; the next 200 green; the next 400 red; etc. mentions a particular individual (the machine) and a particular time-point (the point at which the machine starts producing objects). But a scientist who is forbidden to open the machine or investigate its internal construction might 'behavioristically' acquire a considerable inductive confidence in this hypothesis.

Moreover, Goodman has shown how to rephrase (2) so that this objection is avoided. Define 'grue' as applying to green objects prior to t; and to blue objects subsequently. Then (2) becomes:

(2') All emeralds are grue.

† This is a consequence of Gödel's theorem.

What interests us about the hypotheses (1) and (2) (or (1) and (2'))
is this: if time t is in the future and all emeralds so far observed are
green, both are consistent with the data. But in some sense (2) is less
simple than (1). Indeed, if the language does not contain 'grue', (1) is
simpler than (2) by the 'symbol count' criterion of simplicity proposed
above. How do these hypotheses fare under the inductive methods so
far discussed?

Under the method M, there are three relevant possibilities: (2) may
be suggested at a time when no one has thought of (1) (highly implausible); or (1) and (2) may be suggested at the same time (slightly more
plausible); or (1) may be advanced long before anyone even thinks of
(2) (much more plausible, and historically accurate). In the last (and
actual) case what happens is this: (1) is compared with, say

(3) All emeralds are red.

and (1) is accepted. Much later someone (Goodman, in fact) suggests (2).
Then (1) is *still* accepted, in accordance with the principle of 'tenacity',
until and unless at time t (2) turns out to be correct.

In the case that (2) is suggested first we would, of course, accept (2)
and refuse to abandon it in favor of the simpler hypothesis (1) until
experimental evidence is provided in favor of (1) over (2) at time t. As
Conant has pointed out (Conant, 1947, Chapter 3) this is an important
and essential part of the actual procedure of science: a hypothesis once
accepted is not easily abandoned, even if a 'better' hypothesis appears
to be on the market. When we appreciate the connection between
tenacity and the feature III of our inductive method, we may see one
reason for this being so.

In the remaining case, in which (1) and (2) are proposed at the same
time, *neither* would be accepted before time t. This is certainly a defect
of method M.

Now let us consider how these hypotheses fare under Kemeny's
method (as here combined with some features of method M). If (1) is
suggested first, everything proceeds as it did above, as we would wish.
If (2) is suggested first, there are two possibilities: we may have a rule
of tenacity, according to which a hypothesis once adopted should not be
abandoned until it proves inconsistent with the data. In this case things
will proceed as with the method M. Or, we may adopt the rule that we
shift to a simpler hypothesis if one is suggested, provided it is consistent
with the data. In this case we must be careful that only a finite number
of hypotheses are simpler than a given hypothesis under our simplicity-
ordering; otherwise we may sacrifice the advantages of the principle of
tenacity (i.e. one might go on 'shifting' forever). Then we would adopt

(1) when it is suggested, even if we have previously accepted (2) and (2) is still consistent with the data. Lastly, if (1) and (2) are suggested at the same time, we will accept (1) (as soon as the 'null hypothesis' is excluded at the chosen confidence level).†

Thus the method incorporating Kemeny's proposal has a considerable advantage over M; it permits us to accept (1) long before t even if (2) and (2') are also available. In general, this method places a premium on simplicity, as M does not.

Another suggestion has been made by Goodman. Goodman rejects (2) as an inductive hypothesis as *explicitly* mentioning a particular time-point. This leaves the version (2'), however. So the notion of *entrenchment* is introduced. A predicate is better entrenched the more often it (or any predicate coextensive with it) has been used in inductive inferences. Under this criterion it is clear that 'green' is a vastly better-entrenched predicate than the weird predicate 'grue'. So in any conflict of this kind, the data are regarded as confirming (1) and not (2').

Goodman's proposal might be regarded as a special case of Kemeny's, Namely, we might regard the ordering of hypotheses according to 'entrenchment' as but one of Kemeny's simplicity-orders. On the other hand, we may desire to have a measure of simplicity as distinct from entrenchment. (Under most conceptions, simplicity would be a *formal* characteristic of hypotheses, whereas entrenchment is a *factual* characteristic.) In this case we might order hypotheses according to some weighted combination of simplicity and entrenchment (assuming we can decide on appropriate 'weights' for each parameter).

What has been illustrated is that the aspects of simplicity and entrenchment emphasized by Kemeny and Goodman (and any number of further characteristics of scientific hypotheses) can be taken into consideration in an inductive method without sacrificing the essential characteristics of *corrigibility* and *tenacity* which make even the method M, bare skeleton of an inductive method though it may be, superior as an inductive instrument to any method based on an *a priori* probability distribution.

VIII

At the beginning of this paper I announced the intention to present a precise and formal argument of a kind that I hope may convince Carnap. I did this because I believe (and I am certain that Carnap believes as well) that one should never abandon a constructive logical venture because of *merely* philosophical arguments. Even if the philosophical

† It is desirable always to count the null hypothesis as simplest; i.e. not to accept another until this is ruled out.

arguments are well taken they are likely to prove *at most* that the scope or significance of the logical venture has been misunderstood. Once the logical venture has succeeded (if it does succeed), it may become important to examine it philosophically and eliminate conceptual confusions; but the analytical philosopher misconstrues his job when he advises the logician (or any scientist) to stop what he is doing.

On the other hand, it is not the part of wisdom to continue what one is doing no matter what relevant considerations may be advanced against it.

If the venture is logical, so must the considerations be. And in the foregoing sections we have had to provide strict proof that there are features of ordinary scientific method which cannot be captured by any 'measure function'. (Unless one wants to try the doubtful project of investigating measure functions which are not effectively computable *even for a finite universe*.† And then one sacrifices other aspects of the scientific method as represented by M; its *effectiveness* with respect to what hypothesis one should select, and hence what prediction one should make.)

In short, degree of confirmation is supposed to represent (quantitatively) the judgments an ideal inductive judge would make. But the judgments an ideal inductive judge would make would presumably have this character: if a deterministic law (i.e. an effective hypothesis) h is true, and someone suggests it, and our 'ideal judge' observes for a *very* long time that h yields only successful prediction, he will eventually base his predictions on it (and continue to do so, as long as it does not fail). But this very simple feature of the inductive judgments he makes is represented by no measure function whatsoever. Therefore, *the aim of representing the inductive policy of such a 'judge' by a measure function represents a formal impossibility*.

Now that the formal considerations have been advanced, however, it becomes of interest to see what can be said on less formal grounds about the various approaches to induction. In the present section, let us see what can be said about the *indispensability of theories* as instruments of prediction on the basis of the inductive methods we have considered. We shall find that the method M and the method incorporating Kemeny's idea 'make sense' of this; the Carnapian methods give a diametrically opposite result.

To fix our ideas, let us consider the following situation: prior to the first large scale nuclear explosion various directly and indirectly relevant observations had been made. Let all these be expressed in a single

---

† If a particular measure function is computable for finite universes, the DC of a singular prediction on singular evidence is computable for *any* universe.

sentence in the observation vocabulary, e. Let h be the prediction that, when the two subcritical masses of uranium 235 are 'slammed together' to produce a single super-critical mass, there will be an explosion. It may be formulated without the theoretical expression 'uranium 235', namely as a statement that when two particular 'rocks' are quickly 'slammed together' there will be 'a big bang'. Then h is also in the observation vocabulary. Clearly, good inductive judges, given e, did in fact expect h. And they were right. But let us ask the question: is there any *mechanical rule* whereby given e one could have found out that one should predict h?

The example cited is interesting because there was not (or, at any rate, we may imagine there was not) any *direct* inductive evidence from the standpoint of induction by simple enumeration, to support h. No rock of this kind had ever blown up (let us suppose). Nor had 'slamming' two such rocks together ever had any effect (critical mass had never been attained). Thus the direct inductive inference *a la* Mill would be: 'slamming two rocks of this kind (or any kind) together does not make them explode'. But a *theory* was present; the theory had been accepted on the basis of *other* experiments; and the theory *entailed* that the rocks would explode if critical mass were attained quickly enough (assuming a co-ordinating definition according to which 'these rocks' are U-235). Therefore the scientists were willing to make this prediction in the face of an utter lack of direct experiential confirmation.†

(Incidentally, this is also a refutation – if any were needed – of Bridgman's view of scientific method. According to Bridgman, a theory is a summary of experimental laws; these laws should be explicitly formulated, and should be accepted only insofar as they are directly confirmed (apparently, by simple enumerative induction). Only in this way shall we avoid unpleasant 'surprises'.‡)

But, if this view is accepted, then the scientists in the experiment described above were behaving most irrationally; they were willing to accept, at least tentatively (and advise the expenditure of billions of dollars on the basis of) an experimental law that had never been tested once, simply because it was deduced from a theory which entailed *other* experimental laws which had been verified.

I believe that we should all want to say that even the most 'ideal inductive judge' could not have predicted h on the basis of e unless someone had suggested the relevant theories. The theories (in particular, quantum mechanics) are what connect the various facts in e (e.g. the fact that one gets badly burned if he remains near one of the 'rocks') with h. Certainly it appears implausible to say that there is a *rule*

† The physics in this example is slightly falsified, of course; but not essentially so.
‡ This seems the only possible reading of a good many passages in Bridgman, 1927.

whereby one can go from the observational facts (if one only had them all written out) to the observational prediction without any 'detour' into the realm of theory. But this is a consequence of the supposition that degree of confirmation can be 'adequately defined'; i.e. defined in such a way as to agree with the actual inductive judgments of good and careful scientists.

Of course, I am not accusing Carnap of believing or stating that such a rule exists; the existence of such a rule is a *disguised* consequence of the assumption that DC can be 'adequately defined', and what I hope is that establishing this consequence will induce Carnap, as it has induced me, to seek other approaches to the problem of inductive logic.

Thus let *O* be the observational language of science, and let *T* be a formalization of the full-fledged language of science, including both observational and theoretical terms. *O* we may suppose to be an applied First Order Functional Calculus; and we may suppose it contains only (qualitative) predicates like 'Red' and no functors. *T*, on the other hand, must be very rich, both physically and mathematically. Then we state: *if DC can be adequately defined for the language O, then there exists a rule of the kind described.*

Incidentally, it is clear that the possibility of defining DC for *T* entails the existence of a rule which does what we have described (since all the relevant theories can be expressed in *T*). But this is not as disturbing, for the creative step is precisely the invention of the theoretical language *T*.† What one has to show is that the possibility of defining DC just for *O* has the same consequence.

Carnap divides all inductive methods into two kinds. For those of the first kind, *the DC of h on e must not depend on the presence or absence in the language of predicates not occurring in either h or e.* Since *h* and *e* do not mention any theoretical terms, the DC of *h* on *e* must be the same, in such a method, whether the computation is carried out in *T* or *O*! In short, if we have a definition of DC in *O*, what we have is nothing less than a definition of *the best possible prediction in any evidential situation*, regardless of what laws scientists of the future may discover. For if the degree of confirmation of *h* on *e* is, say, 0·9 in the complete language *T*, then it must be 0·9 in the sub-language *O*.

For inductive methods of the second kind, the DC of *h* on *e* depends, in general on *K* (the number of strongest factual properties). But, with respect to the actual universe, each method of the second kind coincides with some method of the first kind (as Carnap points out).‡ Thus, if

† This has been remarked by Kemeny (1963).
‡ Carnap, 1952, p. 48. For a lengthier discussion of the plausibility of making DC dependent on κ, see Carnap, 1948, pp. 133–48; particularly p. 144.

there is any adequate method of the second kind (for the complete language T) there is also some adequate method of the first kind.

If we recall that the degree of confirmation of a singular prediction is effectively computable relative to singular evidence, we get the further consequence that it is possible in principle to build an electronic computer such that, if it could somehow be given all the observational facts, it would always make the best prediction – i.e. the prediction that would be made by the best possible scientist if he had the best possible theories. *Science could in principle be done by a moron* (or an electronic computer).†

From the standpoint of method M, however, the situation is entirely different. The prediction one makes will depend on what laws one accepts. And what laws one accepts will depend on what laws are proposed. Thus M does not have the counter-intuitive consequence just described. If two 'ideally rational' scientists both use M, and one thinks of quantum mechanics and the other not, the first may predict h given e while the second does not. Thus theories play an indispensable role.

This feature is intrinsic to M. We cannot take the class $P_{t_i, M}$ to be infinite; for the proof that each inductive test will terminate depends on it being finite. Also there is no *effective* way to divide all hypotheses into successive finite classes $P_{t_1, M}, P_{t_2, M}, P_{t_3, M}, \ldots$ in such a way that (a) every class contains a finite number of mutually incompatible effective hypotheses, and (b) every effective hypothesis is in some class.‡ M cannot be transformed into an effective method for selecting the best hypothesis from the class of *all* hypotheses expressible in the language (as opposed to the hypotheses in a given finite class). Thus science *cannot* be done by a moron; or not if the moron relies on the method M, at any rate.

The situation is even more interesting if one uses the Kemeny method. For the simplicity of hypotheses with the same observational consequences may vary greatly (even by the 'symbol count' criterion). A way of putting it is this: call two hypotheses 'essentially the same' if they have the same observational consequences. Then the relative simplicity of hypotheses that are 'essentially the same' may vary greatly depending on the language in which they are couched. (For instance, Craig (1956), has shown that every hypothesis can be 'essentially' expressed in O, in this sense; but the axiomatization required is infinite if the original hypothesis contains theoretical terms, so there would be infinite complexity.) Thus the hypothesis a scientist will accept, using a method which includes a simplicity order, will depend not only on

† Readers familiar with Rosenbloom's *Elements of Mathematical Logic* will recognize the identification of the computer with the 'moron'.
‡ This is a consequence of Gödel's theorem, as remarked above, p. 284 note.

what hypotheses he has been able to think of, but on the theoretical language he has constructed for the expression of those hypotheses. Skill in constructing theories within a language and skill in constructing theoretical languages both make a difference in prediction.

IX

There are respects in which all the methods we have considered are radically oversimplified: for instance, none takes account of the reliability of the data. Thus, Rule I of method M is unreasonable unless we suppose that instrumental error can be neglected. It would be foolish, in actual practice, to reject a hypothesis because it leads to exactly one false prediction; we would rather be inclined to suppose that the prediction might not really have been false, and that our instruments may have deceived us. Again there is the problem of assigning a proper weight to *variety* of evidence, which has been emphasized by Nagel. But my purpose here has not been to consider all the problems which might be raised. Rather the intention has been to follow through one line of inquiry: namely, to see what features of the scientific method can be represented by the method M and related methods, and to show that crucial features cannot be represented by any 'measure function'.

Again, I have not attempted to do any philosophic 'therapy'; to say what, in my opinion, are the mistaken conceptions lying at the base of the attempt to resuscitate the 'logical probability' concept. But one such should be clear from the foregoing discussion. The assumption is made, in all work on 'degree of confirmation', that there is such a thing as a 'fair betting quotient', that is, the odds that an ideal judge would assign if he were asked to make a fair bet on a given prediction. More precisely, the assumption is that *fair odds* must exist in any evidential situation, and *depend only on the evidence*. That they must depend on the evidence is clear; the odds we should assign to the prediction 'the next thing will be red' would intuitively be quite different (in the absence of theory!) if 50 per cent of the individuals examined have been red, and if all have been. But, I do not believe that there exists an abstract 'fairness of odds' independent of *the theories available to the bettors*. To suppose that there does is to suppose that one can define the best bet *assuming that the bettors consider the best possible theory*; or (what amounts to the same thing) assuming they consider all possible theories.

Such a concept appears to be utterly fantastic from the standpoint of the actual inductive situation; hence it is not surprising that any definition would have to be so non-effective as not to be of any use to anybody.

291

Since this assumption underlies the work of De Finetti (1931), and the 'subjective probability' approach of Savage (1954), I am inclined to reject all of these approaches. Instead of considering science as a monstrous plan for 'making book', depending on what one experiences, I suggest that we should take the view that science is a method or possibly a collection of methods for *selecting a hypothesis*, assuming languages to be given and hypotheses to be proposed. Such a view seems better to accord with the importance of the hypothetico-deductive method in science, which all investigators have come to stress more and more in recent years.

18
Probability and confirmation*

The story of deductive logic is well known. Until the beginning of the nineteenth century, deductive logic as a subject was represented by a finite and rather short list of well known patterns of valid inference. The paucity of the subject did not discourage scholars, however – there were professorships in Logic, courses in Logic, and volumes – fat, fat volumes – in Logic. Indeed, if anyone wants to see just how much can be made out of how little subject matter, I suggest a glance at any nineteenth-century text in traditional logic. The revolution in the subject was brought about by the work of the English logician Boole.

Boole's full contribution to the mathematics of his time has still to be fully appreciated. In addition to creating single handed the subject of mathematical logic, he wrote what is still the classic text on the subject of the calculus of finite differences, and he pioneered what are today known as 'operator methods' in connection with differential equations. To each of the subjects that he touched – calculus of finite differences, differential equations, logic – he contributed powerful new ideas and an elegant symbolic technique. Since Boole, mathematical logic has never stopped advancing: Schröder and Pierce extended Boole's work to relations and added quantifiers; Behmann solved the decision problem for monadic logic; Löwenheim pioneered the subject that is today called 'model theory'; Russell extended the system to higher types; and by 1920, the modern period in the history of the subject was in full swing and the stage was set for the epochal results of Gödel and his followers in the 1930s.

The hope naturally arises, therefore, that there might some day be a comparable revolution in *inductive* logic. Inductive logic, too, has a traditional stock of recognized patterns of inference – again a not very long list is all that is involved, notwithstanding the big books on the subject – and the hope is understandable that the mathematical method might be as fruitful in enriching and enlarging this subject as it has been in enriching and enlarging the subject of deductive logic. So far this hope has

* First published in *The Voice of America, Forum Philosophy of Science*, 10 (U.S. Information Agency, 1963).

not been rewarded. However, a number of American philosophers and logicians have been extremely active both in prosecuting the attempt to create a subject of mathematical inductive logic and in criticizing it; and this paper will give both a brief account of the progress that has been made and of the difficulties that have arisen. In the process I hope to give some hint of the extraordinary range and vitality of the discussion that is now taking place in the foundation of scientific methodology.

Carnap's school

The most important and ambitious attempt to erect a foundation for mathematical inductive logic is due to Rudolf Carnap, who came to the United States as a refugee from Hitler Germany in the 1930s, and who turned to inductive logic in the late 1940s. Although Carnap is now retired from his duties as a professor at the University of California at Los Angeles, he continues to carry on his investigations in this field and has inspired a number of younger logicians to become interested in it. It may be said that Carnap's work has today the sort of importance for the whole field of inductive logic that Frege's work had for deductive logic in the first years of this century: it is obviously unsatisfactory as it stands, but there is no real alternative approach. Either the difficulties with Carnap's approach must be surmounted, or the whole project must be abandoned.

Inductive logic is concerned with the relation of *confirmation*. Just as in deductive logic we consider certain correct inferences which lead from a set of sentences P (called the *premises* of the inference) to a conclusion S, so in inductive logic we deal with certain inferences which lead from a set of sentences E (called the *evidence* for the inductive inference) to a conclusion S. The premises of a good deductive inference are said to *imply* the conclusion S; the evidence, in the case of a warranted inductive inference, is said to *confirm* the conclusion S. The difference in terminology reflects some of the fundamental differences between the two subjects. The most important of these differences is that in a warranted inductive inference the sentences E may be true and the conclusion S may still be false. It is for this reason that we do not say that E *implies* S but choose the weaker terminology E *confirms* S. This difference has one immediately annoying consequence: it makes it somewhat difficult to state with any degree of precision what inductive logic is *about*.

Since deductive inferences are truth-preserving (this means: if the premises are true, the conclusion *must* be true) we can characterize deductive logic by saying that we seek mathematical rules for deriving conclusions from sets of premises which will preserve truth: which will lead from true premises only to true conclusions. But in *inductive* logic we seek what? Mathematical rules which will lead from true premises to conclusions which are *probably* true, relative to those premises. But 'probably' in what sense? If all we seek are rules that will be generally successful in the real world as a matter of actual empirical fact, then since the criterion for success is an empirical one, it does not seem very likely that it can be made mathematically precise, or that a precise and powerful mathematical theory of the class of inference-rules with this characteristic can ever be attained. If we seek rules which will lead to conclusions which are probably true on the evidence given in some *a priori* sense of 'probably true', then what reason is there to think that there exists a defensible and scientifically interesting *a priori* notion of probability?

Let us bypass these difficulties for the moment by agreeing that, just as in deductive logic there are certain clear cases of valid reasoning, so in inductive logic there are certain clear cases of warranted inductive reasoning, and we can at least attempt to see if these have any mathematically interesting structural properties. The possible gains from a successful subject of mathematical inductive logic are so great that we should not refuse even to let the subject commence because the relation of confirmation cannot be delineated in advance in a wholly precise way. We shall start, then, with an unanalyzed relation of *confirmation* and see what we might do to make it precise.

Carnap's leading idea is to assume that a *quantitative* measure of confirmation, which he calls *degree* of confirmation, can be worked out. Degrees of confirmation are real numbers between zero and one. A degree of confirmation of, say, $7/8$ corresponds to a 'fair betting quotient' of $7:1$, i.e. if one were forced to bet on the truth or falsity of S at odds of $7:1$, the evidence E being such that the proposition S is confirmed to the degree $7/8$, then one should be indifferent as to whether one's money is placed on S being true or on S being false. If the odds were $6:1$ then one should ask that one's money be placed on S; while if the odds were $8:1$ then one should ask that one's money be placed on not-S, i.e. one should bet that S is false. Thus the odds at which an 'ideal inductive judge' would regard it as a matter of indifference whether his money is placed on S or on not-S are determined by and reciprocally determine the exact 'degree of confirmation' of S on any evidence E.

Since the idealized 'betting situation' just alluded to plays a major role in Carnap's thinking, it is necessary to examine it more closely. The situation is obviously unrealistic in numerous respects. For example, if the bet concerns the truth or falsity of a scientific *theory*, then how should the bet ever be decided? Even if we can agree to call the theory *false* in certain definite cases, under what circumstances would we decide that the theory was certainly *true* and that the money should be paid to the affirmative bettor?

Even if we assume that we have available an 'oracle' who decides all of our bets for us, and by whose decisions we are willing to abide, is it reasonable to assume that even an *ideal* inductive judge could have such a thing as a *total* betting strategy which would enable him to assign *exact* 'fair betting quotients' in all possible evidential situations and to all possible hypotheses? This question is connected with the reasonableness of the 'subjective probability' notion, to which we shall come later. Suppose, finally, that we assume that an 'ideal inductive judge' could always arrive at a set of odds which he regarded as 'fair' in connection with any S, and given any evidence E. Is it reasonable to suppose that this judgment could be independent of both his empirical knowledge and the hypotheses he is clever enough to frame?

These questions may be answered in the following way. What we seek, following Carnap, is a precise definition of a 'c-function' – a function $c(E, S)$ whose first argument is a sentence (or more generally a set of sentences), whose second argument is a sentence, and whose values are real numbers between zero and one which can be used to guide our betting, and, more generally, our beliefs and actions, in the following way: if $c(E, S)$ has the value p/q, we shall take odds on S of $p:q-p$ as 'fair', and adjust our expectations accordingly. If such a function c can be constructed, and the function c turns out to 'learn from experience' and to 'predict' more cleverly than any existing inductive judge (or as well as the best existing inductive judges), then we shall certainly be highly satisfied.

If the notion of 'fair odds' is indeed vague in many situations, in connection with many hypotheses S, we can then *make* it precise by defining the odds $p:q-p$ to be fair, where $p/q = c(E, S)$. Such a function c would then determine a *total* betting strategy in the sense just described. How reasonable it is to hope for such a thing we cannot determine in advance: who in 1850 could have anticipated the evolution of deductive logic in the next hundred years? If the judgments of an inductive judge are based not just on the evidence E but on collateral information E' available to him, we may insist that this evidence E' be explicitly formulated, i.e. that the degree of confirmation be computed not on the basis of E but on the basis of E and E' (this is called the *requirement of total evidence in* inductive logic). And the function c will, of course, 'consider' all hypotheses that can be framed in the language. The development of new theoretical languages may modify our inductive predictions, but within a fixed scientific language we hope for an adequate function c with respect to the hypotheses that can be framed in that language and the evidence that can be expressed in that language.

Space does not permit me to go further into Carnap's approach than these introductory remarks. The details may be found in Carnap's book, *Logical Foundations of Probability*, and in his monograph *The Continuum of Inductive Methods*.

The diagonal argument

Carnap's published work has so far considered only hypotheses S and evidence E of a very simple and highly restricted kind. In particular, Carnap does not consider problems involving *sequential orderings*. For example, the evidence might be that 200 balls have been drawn from an urn and of these 162 were red and the remainder not-red, but not that every *third* ball (in some sequence) is red, nor would the hypothesis be that every third ball will continue to be red.

The difficulty is not just a difficulty with Carnap's notation. Carnap's 'c-functions' are defined over languages which contain names for relations (e.g. 'to the left of') as well as properties ('red', 'square', 'hard'). Thus we can express the fact that every third ball of the balls so far drawn has been red in Carnap's notation by introducing a primitive relation symbol R which is to be interpreted in such a way that 'Rab' means 'ball a is drawn before ball b'. Using the symbol R and the usual logical symbols we *can* then say that so-and-so-many balls have been drawn and every third one has been red, and we can also express the prediction that, say, the next two balls to be drawn will be not-red and the third one will again be red. But it then turns out that Carnap's c-functions assign a probability of about one-third to the hypothesis that any future ball will be red independently of its position in the sequence of drawings. In other words, Carnap's inductive logics are clever enough, if we think of them as mechanical learning devices, to 'learn from experience' that approximately one-third of the balls drawn have a certain non-relational property, but they are not 'clever enough' to learn that position in the sequence is relevant.

Since this point is crucial to the whole discussion, let me emphasize it somewhat. As I have already indicated, we may think of a system of inductive logic as a design for a 'learning machine': that is to say, a design for a computing machine that can extrapolate certain kinds of empirical regularities from the data with which it is supplied. Then the criticism of the so-far-constructed 'c-functions' is that they correspond to 'learning machines' of very low power. They can extrapolate the simplest possible empirical generalizations, for example: 'approximately nine-tenths of the balls are red', but they cannot extrapolate so simple a regularity as 'every other ball is red'. Suppose now that we had a sequence with a regularity that required the exercise of a high order of intelligence to discover. For instance, all the balls up to a certain point might be red, and from that point on the second (after the point in question) might be black, also the third, also the fifth...and, in general, all the balls at a *prime* place in the sequence (counting from the point in question) might be black, while the balls at a composite place in the sequence might be red. Needless to say, a 'learning machine', or an inductive logic that could extrapolate *such* a regularity is not today in existence. The question is whether such a thing will ever exist.

One difficulty with Carnap's program is this: to say that a certain prediction, say, 'the next ball will be red and the one after that will be black' is highly confirmed on the evidence given implies that, considering all *possible* hypotheses as to the regularity governing the sequence, this is the *best* prediction one can make. If there is such a thing as a correct definition of 'degree of confirmation' which can be fixed once and for all, then a machine which predicted in accordance with the degree of confirmation would be an *optimal*, that is to say, a cleverest possible learning machine. Thus any argument against the existence of a cleverest possible learning machine must show either that Carnap's program cannot be completely fulfilled or that the correct *c*-function must be one that cannot be computed by a machine. Either conclusion would be of interest.

The first alternative says, in effect, that inductive logic cannot be formalized in Carnap's way – on the basis of a quantitative notion of degree of confirmation; while the second alternative says that inductive logic must be a highly non-constructive theory: that is, it may *define* in some way the concept 'best possible prediction on given evidence', but the definition cannot be applied mechanically; rather we will have to hire a mathematician to try to *prove* that a given prediction is 'best possible' on the basis of the definition, and, since we cannot give the mathematician a uniform rule for finding such proofs, we have no guarantee that he will succeed in any given case.

Of course, neither alternative would be wholly fatal to Carnap's program. If the first alternative is correct, we can still try to construct better and better learning machines, even if we cannot hope to build, or even give precise sense to the notion of a 'best possible' one. And if the second alternative is correct, we can always *say* that any precise and correct mathematical definition of 'degree of confirmation' would be of interest, even if degree of confirmation does not turn out to be a function that a computing machine can compute, even when S is as simple as 'the next ball will be red and the one after that will be black', and E is merely a record of past drawings from an urn.

Is there, then, an argument against the existence – that is, against the possible existence – of a 'cleverest possible' learning machine? The answer is that there is. The argument appears in detail in a paper of mine in Schilpp (1963) but the leading idea is very simple. Let T be any learning machine, and consider what T predicts if the first, second,... and so on balls are all red. Sooner or later (if T is not hopelessly weak as a learning device) there must come an n such that T predicts 'the nth ball will be red'. Call this number n_1. If we let n_1 be black, and then the next ball after that be red, and the next after that again be red, and so on, then two things can happen. T's confidence may have been so shaken by the failure of its prediction at n_1 that T refuses to ever again predict that a future ball will be red. In this case we make all the rest of the balls red. Then the regularity 'all the balls with finitely many exceptions are red' is a very simple regularity that T fails to extrapolate. So T is certainly not a 'cleverest possible' learning device.

On the other hand, it can happen (and in fact it will, if T has any power as a learning machine at all), that, in spite of its failure on n_1, T 'regains confidence', and, if all the balls after n_1 have been red for a long time, T finally predicts 'the next ball will be red'. Call this next place in the sequence n_2 and make that ball black. Continuing in this way, we can effectively calculate an infinite sequence of numbers $n_1, n_2, ...$ such that if we make all the balls red *except* the ones at $n_1, n_2, ...$ and make *those* black, then we will *defeat* the machine T. That is, we will have constructed a regularity, depending on T, which it is beyond the power of T to extrapolate.

However, one can prove that it is always possible to build another machine which can extrapolate every regularity that T can extrapolate and also extrapolate the one that T *can't* extrapolate. Thus, there cannot be a 'cleverest possible' learning machine: for, for every learning machine T, there exists a machine T' which can learn everything that T can learn and more besides. This sort of argument is called a *diagonal argument* by logicians. Applied to inductive logic it yields the conclusion I stated before: that either there are better and better c-functions, but no 'best possible', or else there is a 'best possible' but it is not computable by a machine, even when S and E are restricted to the sorts of situations we have been considering – that is, drawing balls from an urn, noting their color, and trying to predict the color of future balls.

Confirmation of universal laws

It is possible that some of my readers who have mathematical training may wish to explore this fascinating field of inductive logic themselves.

They can certainly do so. The field is still in its infancy, and no special knowledge is needed to enter at this point beyond what is contained in Carnap's book. However, I should like to make a suggestion or two as to what might be fruitful lines for future investigation.

Carnap's approach in his detailed investigations has been to consider languages over finite universes – say, a 'universe' consisting of 300 balls of various colors in an urn. Once degree of confirmati)n has been defined for a universe with N objects, for arbitrary N, Carnap extends the definition to languages over an infinite universe by a limiting process. The restriction is imposed: when S and E are statements about finitely many explicitly mentioned individuals, that is, when S and E do not contain the words 'all' and 'some', then the degree of confirmation of S on E must not depend on N (the number of individuals in the language). Carnap calls this requirement the *requirement of fitting together*, and so far it has played a rather unhappily restricting role in inductive logic. As already mentioned, the methods so far constructed have extremely low 'learning power', and I suspect that this is connected with this way of proceeding from such highly restricted cases.

A method that would appear more natural to a mathematician used to working with continuous functions and spaces would be this: start with a language over a universe with continuously many individuals (say, space–time points), and whose primitive notions are continuous functions. The possible worlds or 'state descriptions', as Carnap calls them, relative to this language, would then form a continuous space. Each c-function would be a (presumably continuous) function over this space with an infinitesimal value on each state description.

If, now, there is a reason for defining degree of confirmation on some universe with only N individuals, for some finite n, we can do this by identifying those N individuals with N arbitrarily chosen individuals in the infinite world. In other words, instead of extending our c-function from the finite languages to the infinite language as Carnap does, we extend it from the infinite language to the various finite languages. In this way the 'requirement of fitting together' will automatically be met. Also, another serious difficulty in Carnap's approach disappears. Carnap's methods, in the monograph *The Continuum of Inductive Methods* assign to a universal law the degree of confirmation *zero* in the limit, that is, as N approaches infinity.

In other words the probability (in an infinite universe) that a universal law holds without exception is strictly zero on any finite amount of evidence. In a certain sense, then, strictly universal laws cannot be extrapolated at all: we can only extrapolate laws of the form '99 per cent of all individuals conform to the law L'. However, if we start directly with the infinite universe, there is no difficulty in constructing a c-function which assigns a specified universal law S the probability 0·90 on evidence E. We simply have to assign our infinitesimal degree of confirmation to each state description in which S holds and E also holds in such a way that the degree of confirmation of S on E (integrated over all state descriptions of the kind just mentioned) is 0·90. Indeed, given any initial measure function c_0 which does *not* permit the extrapolation of the law S, and given evidence E from which we wish to extrapolate S, we can easily *modify* c_0 so that on evidence E the degree of confirmation of S will be 0·90, while on evidence incompatible with E all degrees of confirmation remained unaltered. This idea is used in the proof that given a function c_0 we can construct a function c_1 which will extrapolate every regularity that c_0 will extrapolate and also some regularity that c_0 will not extrapolate.

Simplicity orderings

In a recent interesting book (Katz, 1962), Jerrold Katz reviews the notion of a *simplicity ordering*, first introduced into inductive logic by John Kemeny, who proved some interesting theorems about these orderings and their influence on inductive methods. A simplicity ordering is just what the name implies: an ordering $H_1, H_2, H_3, H_4, \ldots$ of hypotheses in order of decreasing simplicity or increasing complexity, by some measure of 'simplicity'. Katz emphasizes that any inductive method implicitly determines a simplicity ordering. For, on any given E, there will be infinitely many mutually incompatible hypotheses which are compatible with E. Suppose S_1, S_2 are two such hypotheses and we regard S_1 as highly confirmed. Since S_2 was also in agreement with E, why did we not regard S_2 as highly confirmed? We may say 'because S_1 was so much simpler than S_2'. But this is 'simplicity after the fact'. We *call* S_1 simpler than S_2 *because* it is the hypothesis to which we assign the high degree of confirmation.

More precisely: we may *either* suppose that the simplicity ordering is given first, and then it is determined that S_1 is to be preferred over S_2 for reasons of 'simplicity', *or* we may suppose that the c-function is given first, and then the simplicity ordering is determined by which we prefer. In general, evidence E cannot confirm a hypothesis S unless E is incompatible with every hypothesis S' such that S' is simpler than S. If a learning machine T is incapable of extrapolating a hypothesis S at all, then this means that S is not located in the simplicity ordering of hypotheses that can be confirmed by T at all. If T is based on a c-function, then to 'improve' T so that T will become able to confirm S what we have to do is modify the function c so that S becomes confirmed on suitable evidence E without destroying the ability of T to confirm other hypotheses when the evidence is incompatible with S. We have, so to speak, to 'insert' S into the simplicity ordering of the hypotheses confirmable by T.

A problem which, therefore, has considerable importance for inductive logic is this: given a simplicity ordering of some hypotheses, to construct a c-function which will be in agreement with that simplicity ordering, that is, which will permit one to extrapolate any one of those hypotheses, and which will give the preference always to the earliest hypothesis in the ordering which is compatible with the data. I believe that this problem should not be too difficult to solve, and that its function may provide an impetus for a new way of constructing c-functions in inductive logic.

One result which is of interest is this: if we require that any two hypotheses in a simplicity ordering must lead to different predictions, then it can be shown by techniques which are due to Gödel that *no* simplicity ordering can include all possible hypotheses and be such that we can effectively list the hypotheses in that order by using a computing machine. If we call an ordering 'effective' provided that a computing machine could be programmed to produce the hypotheses in question according to that ordering, then this says that no effective simplicity ordering can contain *all* hypotheses. Thus we can construct 'better and better' – that is, more and more inclusive – simplicity orderings, but there is no 'best' effective simplicity ordering. The program that I am suggesting, of starting with simplicity orderings and constructing c-functions, would therefore lead to a sequence of 'better and better' c-functions – that is, to inductive logics of greater and greater learning power – but not to an 'optimal' c-function. But this does not seem to be a serious drawback, to me anyway, since an 'optimal' c-function is hardly something to be hoped for.

Subjective probability and coherence

A c-function may also be interpreted as a measure of the degree to which a person with evidence E gives credence to a hypothesis S. On this interpretation, c-functions are just 'subjective probability metrics' of the kind studied by Good, Savage, De Finetti, and others of the 'subjective probability' school. The difference between Carnap's approach and that of, say, De Finetti is that De Finetti claimed to be studying *actual* belief whereas Carnap explicitly says that he is studying *rational* belief, that is, the belief or betting system of an idealized inductive judge (or a 'learning machine'). A c-function is called *coherent* if there does not exist a finite system of bets that events $e_1, \ldots, e_n$ will occur, with stakes $s_1, \ldots, s_n$, and odds which are fair according to the function c, such that if the amounts $s_1, \ldots, s_n$ are wagered on the events $e_1, \ldots, e_n$ respectively at the respective odds, then no matter which of the events occur and which do not occur, the bettor must lose money.

In other words, if your c-function is incoherent, then there exists a system of bets which I can invite you to make which is such that you *must* lose money if you make those bets, and yet the bets are 'fair' according to your own subjective probability system. It has been proved by Ramsey, De Finetti, and Kemeny that a c-function is coherent if and only if it is a normalized probability metric, that is, if and only if it obeys the usual axioms of the mathematical theory of probability. A stronger requirement of *strict coherence* has been introduced by Shimony, and has been shown to be equivalent to the requirement that the c-function must not assign the values 0 and 1 themselves to propositions which are neither logically false nor logically true.

In my opinion the requirements of coherence must be taken *cum grano salis*. Consider a total betting system which includes the rule: if it is ever shown that a hypothesis S is not included in the simplicity ordering corresponding to the betting system at time t, where t is the time in question, then modify the betting system so as to 'insert' the hypothesis S at a place n corresponding to one's intuitive judgment of the 'complexity' of the hypothesis S. This rule violates two principles imposed by Carnap. First of all, it violates the rule that if one changes one's degrees of confirmation in one's life, then this should be wholly accounted for by the change in E, that is the underlying c-function itself must not be changed. Secondly, it can easily be shown that even if one's bets at any one time are coherent, one's total betting strategy through time will not be coherent. But there is no doubt that this is a good rule nonetheless, and that one would be foolish not to improve one's c-function whenever one can. For we are not playing against a malevolent opponent but against nature, and nature does not exploit such 'incoherencies'; and even if we *were* playing against a malevolent opponent, there are many ways to lose one's 'bets' besides being 'incoherent'.

Conclusion

In this paper I have stressed the idea that the task of inductive logic is the construction of a 'universal learning machine'. Present-day inductive logics are learning devices of very low power. But this is not reason for alarm: rather we should be excited that such devices have been designed, even if they have very low power, and optimistic about the likelihood of their future improvement. In the future, the development of a powerful mathematical theory of inductive inference, of 'machines that learn', and of better mathematical models for human learning may all grow out of this enterprise.

Carnap was driven from Germany by Hitler, and his position has been condemned in the Soviet Union as 'subjective idealism'. Even in the United States, there have been a great many who could not understand this attempt to turn philosophy into a scientific discipline with substantive scientific results, and who have been led to extreme and absurd misinterpretations of the work I have been reporting to you here. Few, perhaps, would have expected traditional empiricism to lead to the development of a speculative theory of 'universal learning machines'; and yet, in retrospect, a concern with systematizing inductive logic has been the oldest concern of empiricist philosophers from Bacon on. No one can yet predict the outcome of this speculative scientific venture. But it is amply clear, whether this particular venture succeeds or fails, that the toleration of philosophical and scientific speculation brings rich rewards and that its suppression leads to sterility.

19
On properties*

It has been maintained by such philosophers as Quine and Goodman that purely 'extensional' language suffices for all the purposes of properly formalized scientific discourse. Those entities that were traditionally called 'universals' – properties, concepts, forms, etc. – are rejected by these extensionalist philosophers on the ground that 'the principle of individuation is not clear'. It is conceded that science requires that we allow something tantamount to quantification over non-particulars (or, anyway, over things that are not material objects, not space–time points, not physical fields, etc.), but, the extensionalists contend, quantification over *sets* serves the purposes nicely. The 'ontology' of modern science, at least as Quine formalizes it, comprises material objects (or, alternatively, space–time points), sets of material objects, sets of sets of material objects, . . ., but no *properties, concepts,* or *forms*. Let us thus examine the question: can the principle of individuation for properties ever be made clear?

I. Properties and reduction

It seems to me that there are at least two notions of 'property' that have become confused in our minds. There is a very old notion for which the word 'predicate' used to be employed (using 'predicate' as a term only for *expressions* and never for properties is a relatively recent mode of speech: 'Is existence a predicate?' was not a *syntactical* question) and there is the notion for which I shall use the terms 'physical property', 'physical magnitude', 'physical relation', etc., depending on whether the object in question is one-place, a functor, more than one-place, etc. Ignore, if possible the connotations of 'physical', which are rather misleading (I would be pleased if someone suggested a better terminology for the distinction that I wish to draw), and let me try to tell you what distinction it is that I mean to mark by the use of the terms 'predicate' (which I shall revive, in its classical sense) and 'physical property'.

* First published in N. Rescher *et al.* (eds.), *Essays in Honor of Carl G. Hempel*, (Dordrecht, Holland, D. Reidel, 1970).

305

The principle of individuation for predicates is well known: the property of being P (where 'property' is understood in the sense of 'predicate') is one and the same property as the property of being Q - i.e. to say of something that it is P and to say of something else that it is Q is to apply the *same predicate* to the two things - just in case 'x is P' is *synonymous* (in the wide sense of 'analytically equivalent to') 'x is Q'. Doubt about the clarity of the principle of individuation for predicates thus reduces to doubt about the notion of synonymy. While I share Quine's doubts about the existence of a clear notion of synonymy, I have more hope than he does that a satisfactory concept can be found, although that is not to be the subject of this paper.

Consider, however, the situation which arises when a scientist asserts that temperature *is* mean molecular kinetic energy. On the face of it, this is a statement of identity of properties. What is being asserted is that the *physical property* of having a particular temperature is *really* (in some sense of 'really') the *same property* as the property of having a certain molecular energy; or (more generally) that the *physical magnitude* temperature is one and the same physical magnitude as mean molecular kinetic energy. If this is right, then, since 'x has such-and-such a temperature' is not *synonymous* with 'x has bla-bla mean molecular kinetic energy', even when 'bla-bla' is the value of molecular energy that corresponds to the value 'such-and-such' of the temperature, it must be that what the physicist means by a 'physical magnitude' is something quite other than what philosophers have called a 'predicate' or a 'concept'.

To be specific, the difference is that, whereas synonymy of the expressions 'x is P' and 'x is Q' is required for the predicates P and Q to be the 'same', it is not required for the physical property P to be the same physical property as the physical property Q. Physical properties can be 'synthetically identical'.

This fact is closely connected with *reduction*. 'Temperature is mean molecular kinetic energy' is a classical example of a reduction of one physical magnitude to another; and the problem of stating a 'principle of individuation' for physical properties, magnitudes, etc., reduces, as we shall see, to the problem of describing the methodological constraints on reduction. Not all reductions involve properties or magnitudes; for example, 'Water is H_2O' asserts the identity of each body of water with a certain aggregation of H_2O molecules, give or take some impurities, not the identity of 'the property of being water' and 'the property of being H_2O' - although one might assert that those are the same physical property, too - but many reductions do; e.g. the reduction of gravitation to space-time curvature, of surface tension to molecular attraction, and so on.

I shall suppose, then, that there is a notion of property – for which I use the adjective 'physical', mainly because 'physical magnitude' already exists with a use similar to the use I wish to make of 'physical property', which satisfies the condition that the property P can be synthetically identical with the property Q, the criterion being that this is said to be true just in case P 'reduces' (in the sense of empirical reduction) to Q, or Q to P, or both P and Q 'reduce' to the same R.

II. Can one get an extensional criterion for the identity of properties?

The criterion for the identity of properties just given is not extensional, because the relation of reduction is not extensional. Water reduces to H_2O, and H_2O is coextensive with ($H_2O \vee$ Unicorn), but water does not reduce to ($H_2O \vee$ Unicorn). The difficulty is simply that

(x) (x is water $\equiv$ x is an aggregation of H_2O molecules)

is not merely true but nomological ('law-like'), while

(x) (x is water $\equiv$ x is an aggregation of H_2O molecules $\vee$ x is a unicorn)

is extensionally true (assuming there are no unicorns), but not law-like (unless the non-existence of unicorns is a law of nature, in which case things become still more complicated).

This raises the question: can one hope to get a criterion for the identity of properties (in the sense of 'physical property') expressible in an extensional language? The problem is related to such problems as the problem of getting a necessary and sufficient condition for 'nomological', and of getting one for causal statements, expressible in an extensional language, and what I shall say about this problem is closely related to the way in which I propose to treat those other problems.

III. Fundamental magnitudes

For reasons which will become clear later, I wish to begin by discussing the notion of a fundamental magnitude in physics. It seems clear that no *analytic* necessary and sufficient condition for something to be a fundamental magnitude can be given. At any rate, I shall not even try to give one. But just how serious is this? There do seem to be methodological principles, albeit vague ones, governing the physicist's decision to take certain terms as fundamental-magnitude-terms and not others. Relying on these principles, and on his scientific intuition, the physicist arrives

307

at a list of 'fundamental magnitudes'. At this point he *has* a necessary and sufficient condition for something to be a fundamental magnitude – his list. To be sure, this is an *empirical* necessary and sufficient condition, not an analytic one. But so what? If one has a confirmation procedure, even a vague one, for a term T, and by using that procedure one can arrive at a biconditional of the form (x) $(T(x) \equiv \cdots x \cdots)$ that one accepts as empirically true (and the condition $\cdots x \cdots$ occurring on the right side of the biconditional is precise), then what problem of 'explicating the notion of T-hood' remains? Such a term T may be regarded as a *programmatic* term: we introduce it not by a definition, but by a trial-and-error procedure (often an implicit one); and the program is (using the trial-and-error procedure) to find an empirically correct necessary-and-sufficient condition for 'T-hood' which is precise. If this is successful, then the notion of 'T-hood' is precise enough for all scientific purposes. Even if it is unsuccessful, one might succeed in discovering in each individual case whether T applies or not without ever obtaining any general necessary and sufficient condition: if even *this* is unsuccessful, someone is sure to propose that we drop the notion T altogether.

Even if it is not reasonable to ask for an analytic necessary and sufficient condition in the case of programmatic terms, it is surely reasonable to ask for some indication of the associated trial-and-error procedure, provided that we do not demand more precision in the reply than can reasonably be expected in descriptions of the scientific method at the present stage of knowledge. What is the associated 'trial-and-error procedure', or 'confirmation procedure', in the case of the term 'fundamental magnitude'?

One obvious condition is that fundamental magnitude terms must be 'projectible' in the sense of Goodman. Since this is a general requirement on all terms in empirical science, except complex-compound expressions, and since discussing it properly involves (as Goodman rightly stresses), attacking the whole problem of induction, I shall simply take it for granted. (Goodman's solution is, in effect, to say that a term is projectible if we do in fact project it sufficiently often. This leaves the whole problem of why we project some terms to begin with and not others up to psychology. I am inclined to believe that this, far from being a defect in Goodman's approach, is its chief virtue. It is hard to see, once one has passed Goodman's 'intelligence test for philosophers' (as Ullian has described Goodman's discussion of *green* and *grue*), how this question could be anything but a question for psychology. But anyone who feels that there is *some further* philosophical work to be done here is welcome to do it; my feeling is that what we have here is not so much an unsolved philosophical problem as an undefined one.)

A second condition is that these terms must characterize all things – i.e. all particles, in a particle formulation of physics, and all space–time points, in a field formulation of physics. (I believe that one will get different, though interdefinable, lists of fundamental magnitudes depending on which of these two types of formulation one chooses for physics.)

A third condition is that one of these terms must be 'distance', or a term with the aid of which 'distance' is definable, and that the positions of things must be predictable from the values of the fundamental magnitudes at a given time by means of the assumed laws. (This last requirement applies only before 'quantization'.)

A fourth condition is that the laws must assume an especially simple form – say, differential equations (and linear rather than nonlinear, first order rather than second order, etc., as far as possible), if these terms are taken as primitive.

Looking over these conditions, we see that what one has is not one trial-and-error procedure but two. For the laws (or, rather, putative laws) of physics are not fixed in advance, but are to be discovered at the same time as the fundamental magnitudes. If we assume, however, that the laws are to be expressible in a reasonably simple way as differential equations in the fundamental magnitudes, and that statistics enter (in fundamental particle physics) only through the procedure of passing from a deterministic theory to a corresponding quantum-mechanical theory (the so-called procedure of 'quantization'), then the double trial-and-error procedure is reasonably clear. What one does is to simultaneously look for laws expressible in the form specified (which will predict the position of particles), and to look for terms which are 'projectible' and by means of which such laws can be formulated.

To avoid misunderstandings, let me make it clear that I am *not* claiming that it is 'part of the concept' (as people say) of a fundamental law that it *must* be a differential equation, etc. I am saying that that is what we in fact look for *now*. If it turns out that we cannot *get* that, then we will look for the next best thing. We do not know now what the next best thing would be; partly this is a question of psychology, and partly it depends on what mathematical forms for the expressions of laws have actually been thought of at the time. I deny that the double trial-and-error procedure is fixed by *rules* (or, at least, it is a daring and so-far-unsupported guess that it *is* fixed by rules), unless one is willing to count 'look for laws in a form that seems simple and natural' as a *rule*. But the procedure *is* 'fixed' by the *de facto* agreement of scientists on what is a simple and natural form for the formulation of physical laws. It seems to me to be a great mistake in the philosophy of science to overplay the importance of *rules*, and to underestimate the importance of *regularities*. Regularities in what scientists take to be 'simple' and 'natural' may be a matter of psychology rather than methodology; but (a) the line between methodology and psychology is not at all that sharp; and (b) methodology may well *depend* on such psychological regularities.

IV. A criterion for the identity of 'physical$_2$' properties

H. Feigl has distinguished two notions of the 'physical'. In Feiglese, every scientific predicate is 'physical$_1$', i.e. 'physical' in the sense of having something to do with causality, space, and time; but only the predicates of *physics* are physical in the narrower sense, 'physical$_2$'. In this terminology, what I have been calling 'physical properties' should have been called 'physical$_1$ properties'. Our problem is to find a criterion for the identity of physical$_1$ properties. In this section I shall approach this problem by discussing the special problem of a criterion of identity for physical$_2$ properties. Assuming that the presently accepted list of fundamental magnitudes is complete, i.e. that there are no further fundamental magnitudes to be discovered, the natural procedure is to correlate physical$_2$ properties with equivalence classes of predicates definable with the aid of the fundamental magnitude terms. Each defined term in the vocabulary of physics (i.e. of elementary particle physics) corresponds to a physical$_2$ property and *vice versa*; two terms correspond to the same physical$_2$ property just in case they belong to the same equivalence class. But what should the equivalence relation be?

There are *two* natural proposals, I think, leading to two quite different notions of physical$_2$ property. One proposal, which I shall not investigate here, would be to take *nomological coextensiveness* as the equivalence relation; the other would be to take *logical equivalence*. I shall choose logical equivalence, because, although we want to allow 'synthetic identities' between physical$_2$ properties and, for example, observation properties (e.g. temperature is mean molecular kinetic energy), it does not seem natural or necessary to consider two terms as corresponding to the same physical$_2$ property when both are already 'reduced' (i.e. expressed in terms of the fundamental magnitude terms), and in their reduced form they are not logically equivalent.

How shall we understand 'logical equivalence', however? I propose to understand 'logical equivalence' as meaning logical equivalence in the narrower sense (not allowing considerations of 'synonymy'); so that P_1 and P_2 will be regarded as corresponding to the same physical$_2$ property only if: (a) P_1 and P_2 are built up out of fundamental magnitude terms alone with the aid of logical and mathematical vocabulary; and (b)

$(x)(P_1(x) \equiv P_2(x))$ is a truth of pure logic or mathematics. (The criterion as just given is for one-place predicates; it should be obvious how it is intended to be extended to relations and functors.)

The proposed criterion of identity implicity takes the stand that *no* relations among the fundamental magnitudes should be considered as 'analytic'. This seems reasonable to me in view of the strongly 'law-cluster' character of the fundamental magnitude terms, but a word of explanation may be in order. Consider, for the sake of an example or two, some of the relations among the fundamental magnitude terms that have seemed analytic in the past. For 'distance' (' $d(x, y)$ '), the following relation has often been considered to be 'part of the meaning'; ($d(x, y)$ is not equal to zero unless $x = y$. Yet just this relation is given up (for 'space-time distance', at least) by the Minkowskian metric for space-time. Similarly, that $d(x, y)$ has no upper bound is given up when we go over from Euclidean to Riemannian geometry. These examples indicate, to me at any rate, that, when fundamental magnitude terms are involved, it is foolish to regard *any* statement (outside of a logical or mathematical truth) as 'analytic'.

But is it safe to regard even logic and mathematics as analytic? The answer seems to depend on just what is being packed into the notion 'analytic'. If 'analytic' is a covert way of saying 'true by linguistic convention alone', then the view that logic and mathematics are 'analytic' is highly suspect. Certainly I do not presuppose this view (which I do not in any case accept) here. But if 'analytic' means 'true by virtue of linguistic convention *and* logic or mathematics', then trivially all truths of logic or mathematics are 'analytic'. But this thesis is compatible, for example, with the radical thesis that logic and mathematics are empirical, subject to revision for experimental reasons, etc. I do not wish to rule out this attitude towards logic and mathematics (which, in fact, I hold). Thus, when I say that *logical equivalence* is the criterion for the identity of physical$_2$ properties, I do not mean logical equivalence according to what we today take to be the laws of logic and mathematics; I simply mean equivalence according to whatever may in fact be the truths of logic. If we change our logic, then we may have to change our minds about what physical$_2$ properties are in fact identical; but the *criterion* of identity will not have changed; it will just be that we made a mistake in its application in some particular cases.

V. Basic terms of 'non-fundamental' disciplines

The issues involved in the reduction of theoretical terms in 'non-fundamental' disciplines to physical$_2$ terms are so well known by now, that

I shall be very brief. (I shall lean on the discussion by Kemeny and Oppenheim, which I regard as still being the best paper on the subject, (Kemeny, 1956), and on the subsequent paper by Oppenheim and myself (Oppenheim, 1958).) The basic requirement in every reduction, as enunciated by Kemeny and Oppenheim, is that all the observable phenomena explainable by means of the reduced theory should be explainable by means of the reducing theory. This means that the observation terms must be counted as part of the reducing theory – in the present case, physics – and that we must suppose that we have at least one true biconditional of the form $(x)(O(x) \equiv P(x))$, where P is a physical$_2$ term, for each undefined observation term O. (This requirement is not made by Kemeny and Oppenheim, but it seems the simplest way of ensuring that the maximum possible observational consequences will be derivable from the reducing theory.)

In the paper by Oppenheim and Putnam mentioned above, it is stressed that the reduction at issue need not be made *directly* to physics; if, for example, the laws of psychology are ever reduced to those of cell-biology (explanation in terms of reverberating circuits of neurons, etc.), while the laws of biology are reduced to laws of physics and chemistry, which is itself reduced to physics, then *the laws of psychology will have been reduced to those of physics* from the point of view of the logician of science, even if no one should ever care to write out the definition of a single psychological term directly in physical$_2$ language.

Once one has found a way of explaining the phenomena in physical$_2$ terms (in the sense just explained), then the next stop is to see if anything can be found (from the standpoint of the new explanation of the phenomena directly by means of the laws of physics) which answers to the old theoretical primitives. It is not necessary for this purpose that the old laws should be *exactly* derivable from the proposed identification. If we can find a relative interpretation of the old theories into the theory consisting of the laws of physics plus the 'bridge laws' connecting physical$_2$ terms with observation terms, which permits the deduction of a good approximation theory to the old theories, then we identify the things and properties referred to by the basic terms of the old theories with the things and properties referred to by the corresponding physical$_2$ terms (even if some conventional extension of meaning is involved, as in the case of 'water' and 'hot').

On the other hand, it may happen that some basic term of the old theories does not answer to *anything* (in the light of our new way of explaining the phenomena). In this case, we simply *drop* the old theories (or those laws involving the term in question, at any rate) and explain the phenomena by means of 'lower' level theories, including, in the last resort, direct explanation by means of physics-plus-bridge-laws.

The second case is classified by Kemeny and Oppenheim as *reduction by replacement*; the first case is classified as *reduction by means of biconditionals*. Both types of reduction are exemplified in science, and in some cases it is arguable whether, in view of the shifts of meaning involved, a given reduction should be classified as a reduction by replacement or by means of biconditionals. The important point is that *after* the reduction of a discipline, those basic terms that *remain*, that are still regarded as corresponding to 'physical properties' (in the sense of 'physical₁') at all, are reduced by means of biconditionals (or identity-signs, as in the case of 'Temperature *is* mean molecular kinetic energy'). For terms which are 'reduced by replacement' are *dropped*, so that the only basic terms that survive are the ones that we reduce by the other method.

VI. Psychological properties

What first led me to write a paper on the topic of 'properties' was the desire to study reduction in the case of *psychology*. I am inclined to hold the view that psychological properties would be reduced not to physical₂ properties in the usual sense (i.e. first-order combinations of fundamental magnitudes), but to *functional states*, where crude examples of the kinds of properties I call 'functional states' would be (a) the property of being a finite automaton with a certain machine table; and (b) the property of being a finite automaton with a certain machine table *and* being in the state described in a certain way in the table. To say that a finite automaton has a certain machine table is to say that *there are properties* (in the sense of physical₁ properties) which the object has (i.e. it always has one of them), and which succeed each other in accordance with a certain rule. Thus the property of having a certain machine table is a *property of having properties which...* – although a property of the first level (a property of things), it is of 'second order' in the old Russell–Whitehead sense, in that its definition involves a quantification over (first-order) physical₁ properties. This is a general characteristic of all 'functional' properties, as I use the term: although physical₁ properties in a wide sense, they are *second order* physical₁ properties. How then should a reduction to such properties be analyzed – e.g. *pain* to a certain functional state (as I proposed in an earlier paper)?

The answer is, that if we are willing to accept the hypothesis that all *first-order* physical₁ properties will turn out to be reducible to physical₂ properties, then all second-order physical₁ properties will automatically reduce to *second-order* physical₂ properties. If we succeed in reducing psychological properties to properties of the form: *the property (of second-order) of having (first-order) physical₁ properties which*..., then we make the further reduction to (second-order) physical₂ properties by simply making the theoretical identification of the foregoing physical₁ property with the corresponding physical₂ property, that is, with *the (second-order) physical₂ property of having (first-order) physical₂ properties which...*

It is likely, however, that this unusual type of reduction will have to be combined with the more familiar type if psychology is ever to be reduced. For, although a reduction of psychological states to properties of the kind just described would enable us to predict many of the aspects of the behavior of the corresponding species and to understand the functional organization of that behavior, there are undoubtedly aspects of human behavior whose explanation will require a reference not just to the functional organization of the human brain and nervous system, but to the details of the physical realization of that functional organization. An analogous complication has already appeared in the case of the reduction of chemistry to physics, and is beginning to appear in the case of molecular biology. Although many chemical phenomena can be explained 'at the chemical level', in some cases it is necessary to descend to the level of elementary particle physics, even to explain such familiar facts as the liquidity of water and the hardness of diamond; and although many cellular phenomena can be explained at the level of whole cells, nuclei, etc., in the most important cases it is necessary to 'descend' to explanation directly in physical-chemical terms.

It should be noted that if we accept the strict extensionalism which is urged by Quine, then all questions of reduction of properties trivialize upon the passing-over to corresponding questions about *sets*. *Temperature* as a physical magnitude which is not intrinsically quantified has no place in Quine's scheme: instead, we are urged to take as primitive 'temperature-in-degrees-centigrade', or some such. And the statement that temperature *is* mean molecular kinetic energy passes over into the harmless statement that 'temperature in degrees centigrade is directly proportional to mean molecular kinetic energy in c.g.s. units'. I have discussed this difficulty with Quine, and he has suggested meeting it by saying that 'temperature in degrees centigrade *is a quantification of* mean molecular kinetic energy'. (This would indicate the question '*Why* is temperature in degrees centigrade directly proportional to mean molecular kinetic energy in c.g.s. units?' is not a happy question.) Discussing this move would involve discussing: (a) whether it is really satisfactory to think of mean molecular kinetic energy as a class of equivalence-classes as Quine also suggests; and (b) whether the relation 'the function f is a quantification of S' does not, on the natural understanding of such phrases as 'a quantification of kinetic energy', turn out to be an *intensional* one. Of course, one *can* take the relation extensionally as meaning that temperature is a one-one function of the equivalence-classes, subject to a continuity condition; but then one will not have distinguished between the cases in which one magnitude is a *function* of another, and the cases in which one magnitude *reduces* to another, which is just our problem.

In the same way, there would be no sense, if Quine were right, in discussing whether pain is a brain-state, or a functional state, or yet another kind of state. 'Pain' is construed by Quine as a predicate whose arguments are an organism and a time; if the set of ordered pairs (O, t) such that O is in pain at t is identical with the set of ordered pairs (O, t) such that O satisfies some other condition at t, then *pain* (the relation) *is* (extensionally) the relation that holds between an organism and a time just in case the organism satisfies that other condition. Pain could be *both* a brain-state and a functional state. In some world, pain could even be 'identical' with pricking one's finger – if the organisms in that world experienced pain when and only when they pricked a finger.

Quine does not find this result counterintuitive, because he does not find intensional differences 'philosophically explanatory'. I believe that pointing to differences that are there *is* philosophically explanatory; and it seems to me that these particular differences are 'there'. But I do not expect that either of us will succeed in convincing the other.

VII. Prospect for an extensional criterion for the identity of properties

In the light of the foregoing discussion, I can give a brief answer to the question: can we get a criterion for the identity of properties (in the sense of physical$_1$ properties) which is expressible in extensional language? The answer is that we cannot *today*, as far as I am aware, but that prospects seem reasonably good for obtaining one eventually. The reduction of those observation terms that one might want to take as undefined terms in a reasonable formalization of science seems fully possible, notwithstanding some formidable complexities still to be unravelled. Also, it is assumed in present science that the number of fundamental magnitudes *is* finite (since there are assumed to be only four fundamental kinds of forces); and the assumption that the basic terms of the 'non-fundamental' disciplines will eventually be reduced is at least reasonable.

Of course, the present discussion is entirely empirical in spirit. Indeed, my whole purpose is to break away from two recent traditions ('recent tradition' is deliberate) which seem to me to be sterile and already exhausted: the tradition of 'explication' and the tradition of 'ordinary language analysis'. It may turn out that the number of fundamental magnitudes is infinite; or that some properties other than the ones studied in physics have also to be taken as 'fundamental' (although it is easy to see how the discussion should be modified in this case); or that there are *no* fundamental properties (e.g. there is a level still more 'fundamental' than the level of elementary particles, and a level still more 'fundamental' than that, etc.). If any one of these possibilities turns out to be real, then I am content to leave it to some philosopher of that future to reopen this discussion! The philosophical point that I wish to make is that at present, when we do *not* have a criterion for the identity of arbitrary physical$_1$ properties that is expressible in extensional language, we are still not all that badly off. We do have a criterion for the identity of physical$_2$ properties, as we presently conceive physical$_2$ properties, and this criterion can be extended to other physical$_1$ properties just as rapidly as we succeed in reducing the disciplines in which the corresponding physical$_1$ terms appear to physics. It does not appear unreasonable that we should be unable, in the case of physical$_1$ properties which have not been reduced, to answer the question of identity or non-identity with any certainty prior to the reduction. Of course, in some cases we *can* answer it; for example, properties which are not coextensive are certainly not identical.

VIII. Are properties dispensable?

That there are many assertions that scientists make that we do not know how to render in a formalized notation without something tantamount to quantification over properties is easily seen. First, consider the question we have mentioned several times: whether there are any fundamental magnitudes not yet discovered. Second, consider the scientist who utters a conjecture of the form 'I think that there is a single property, not yet discovered, which is responsible for such-and-such'. Thirdly, consider the assertion that two things have an unspecified observable property in common.

I believe that all of these cases really reduce to the second: the case of saying that something's having a property *P is responsible for* (or 'causes', etc.) such-and-such. Let us call a description of the form 'the property *P*, the presence of which (in such-and-such cases) is responsible for (or causes, etc.) such-and-such', a *causal description* of a property. Let us call a description of the form 'the property of being *P*' a *canonical description* of a property. Then the difficulty is basically this: that there are properties for which we know a causal description but no canonical description. And when we wish to speak of such properties, an existential quantifier over all properties seems unavoidable.

Consider the first case; the case of saying that there is a fundamental magnitude not yet discovered. This comes down to saying that there are phenomena (which itself involves a quantifier over observable properties!) for which some property P is responsible, such that the property P is not definable (in some specified way) in terms of the properties currently listed as 'fundamental'. Consider the third case: quantifying over observable properties. This might be handled in the case of humans by giving a list of all observable properties (although the impracticality of this proposal is obvious); but we also talk of properties that other species can observe and we cannot. But presumably this comes down to talking of those properties P that act as the stimuli for certain responses, and this could presumably be construed as a reference to the properties satisfying certain causal descriptions. Probably, then (although I do not feel absolutely sure of this), it is ultimately only in causal contexts that quantification over properties is indispensable.

One proposal which has been made for handling such references in an extensional way is this: the assertion that 'A's having the property P at t_0 is the cause of B's having the property Q at t_1,' for example, is handled by saying that '$P(A, t_0)$' is part of an *explanans*, whose corresponding *explanandum* is '$Q(B, t_1)$'. The '*explanans*' and the '*explanandum*' are respectively the premise and the conclusion in an argument which meets the conditions for an explanation as set forth in the familiar covering-law model. Does this obviate the need for property-talk?

I do not wish to discuss here the familiar objections to handling causal statements via the covering-law model (e.g. Bromberger's ingenious objection that this model would permit one to say that the period of a pendulum's being so-and-so *caused* the string to have such-and-such a length). But even without going into the adequacy of this model itself, two points need to be made.

First of all, the proposed analysis of causal statements only works when the properties in question are specified by canonical descriptions. When the property hypothesized to be the cause is mentioned only by a causal description – when part of the *explanans* is that there *exists* a property with certain causal efficacies – then this analysis does not apply. Of course, one could treat such explanations as programmatic: when the scientist says, 'I hypothesize that there is a property which is responsible for such-and-such, and which obeys some law of the following form', one could 'translate' this by 'I propose to introduce a new primitive P into the language of science, and I suggest that some theory containing the term P in such-and-such a way will turn out to be confirmed'; but this is clearly inadequate. (The theory might never be confirmed, because, for example, the money to perform the experiments was not forthcoming, and it might still be true that there *was* a property P which..., etc.) Or one might propose to substitute 'is true' for 'will turn out to be confirmed'; but then one runs into the difficulty that one is speaking of 'truth' in a language which contains a primitive one cannot translate into one's meta-language (and which has not, indeed, been given any precise meaning). Or one might say that the scientist has not made any *statement* at all; that he has just said, in effect '*Let's* look for a new theory of the following kind...'; but this seems just plain false.

Secondly, the covering-law theory of explanation uses the term 'nomological' ('law-like') which has admittedly never been explicated. What are the prospects for an explication of this term, in comparison with the prospects of the notion 'property'?

The following would be my program for arriving at a more precise description of the class of 'nomological statements': first, I think that we should try to formulate a hypothesis as to the *form* in which the fundamental laws can be written. This is a much weaker requirement than the requirement that we actually find the laws. The same mathematical form – for example, differential equations in the 'fundamental magnitudes' of classical physics – can be used for the laws of both classical physics and relativity physics. If one takes '$d(x, y)$' (the distance from x to y, where x and y are spatial points at one time) as primitive, then, indeed, this is context-dependent in relativity physics (i.e. $d(x, y)$, or even what x's and y's are spatial points *at one time*, is relative to the reference-system), but this is irrelevant to the statement of the *laws* since these are the same in all reference systems. The change in the geometry is just a change in the laws obeyed by $d(x, y)$; but laws are still expressible as differential equations valid at all points of space and time, and involving only the fundamental magnitudes. Conversely, it seems reasonable to say that any physical relation that can be expressed as a differential equation without boundary conditions, valid at *all* points in space and time, and in just the fundamental magnitudes, should count as a law. Once such a form has been found, the true statements of that form are defined to be the 'nomological' statements of physics. Secondly, as soon as one succeeds in reducing the basic terms of some 'non-fundamental' discipline to physics, one can define the concept 'nomological' for that discipline: namely, a statement in the vocabulary of that discipline is nomological if and only if it is equivalent, via the accepted reducing-biconditionals, to a nomological statement of physics.

It should not be supposed from the foregoing that I think that 'law of nature' *means* 'statement which is reducible to (or which itself is) a nomological statement of physics'. Rather, the situation seems to me to be as follows. Each of the scientific disciplines has pretty much its own list of 'fundamental magnitudes' and its own preferred form or forms for the expression of 'laws'. Thus the discussion in the section of this paper headed 'fundamental magnitudes' could be repeated, with suitable modifications, for each of the other disciplines. In each case there seem to be certain magnitudes which are 'dependent', in the sense that it is the business of the discipline (at least *prima facie* – a discipline may, of course, change its mind about what its 'business' is) to predict their time-course, and certain magnitudes which are independent, in the sense that they are introduced in order to help predict the values of the dependent magnitudes. In physics, for example, it was the position of particles that was, above all, dependent. In economics it would be prices and amounts of production. In each case the scientist looks for a set of properties including his 'dependent variables' which are 'projectible', and which will suffice for the statement of 'laws' – i.e. in order for properties to be accepted as the 'fundamental' ones, it must be possible to formulate (hopefully) true general statements in terms of those properties, which have one of the forms recognized as a form for the expression of laws at that stage in the development of science, and which will predict, at least statistically, the time-course of the dependent variables.

As we mentioned just before in the case of physics, it may be that one cannot *get* true general statements which will do what one wants in the form that one originally takes to be the preferred form for the expression of laws. For example, although the laws of relativity theory can be stated as differential equations in the classical 'fundamental magnitudes', just as the laws of classical physics were, the laws of quantum mechanics require additional 'fundamental magnitudes' (e.g. *spin*), and a more complex mathematical form – one has to introduce operators on Hilbert spaces. Similarly, it might be that the form preferred for the expression of laws in, say, economics at a particular time is too restrictive. When this turns out to be the case, one goes to the 'next best' form, where what is 'next best' is determined by the scientists in the field on the basis of their scientific intuition, in the light of the mathematical methods available at the time for their purposes.

The foregoing suggests that one might seek to explicate the notion of a fundamental law for each of the basic scientific disciplines, and then define a 'nomological statement' simply as a statement which is either itself a fundamental law of one of these disciplines, or which follows from the fundamental laws of one or more disciplines. However, this approach seems to overlook something.

What is overlooked is the enormous impact that the successes of reduction are having on our concept of a natural law. Thus, suppose the
Weber–Fechner Law is true without exception (which it in fact is not),
and that it is in terms of 'fundamental magnitudes' of psychology, and
of the right form to be a 'psychological law'. Then, if it were not for the
strong tendency in science in the direction of physicalistic reduction,
there would be no question but that it is a 'law of nature'. But let us
suppose that when we possess reduction-biconditionals for the concepts
involved, we find that the equivalent statement of physics is *not* 'necessary' at the level of physics, and that this is so because there is a perfectly
possible mutation such that, if it took place, then the 'Weber–Fechner
Law' would fail. In that case, it seems to me that we would not conclude
that the 'Weber–Fechner Law' was a natural law, albeit one with an
unstated scope limitation (although some ordinary language philosophers have urged such a course), but rather that it was not a law at all, but
a good approximation to a more complex statement which *is* a law. It
seems to me, in short, that a decisive condition for a statement's being
law *is* that the 'equivalent' physical$_2$ statement be a law of physics,
although this decisive condition is itself not part of the 'meaning' of the
word law, but rather a condition erected by science relatively recently.
(Actually, things are much more complicated than the foregoing suggests.
For the reductive definitions of the basic terms of the 'non-fundamental'
disciplines are themselves selected largely on the basis that they *enable us
to derive the laws* of those disciplines. On the other hand, once a reduction has been accepted, there is a great reluctance to change it.)

It is on the basis of the considerations just reviewed that I earlier
advanced the suggestion that the following be our program for gaining a
precise description of the class of nomological statement: to first try to
specify a form in which all the fundamental laws of physics (and only
laws, though not necessarily only fundamental ones) can be written; and
then to characterize the nomological statements as the statements which
follow from true statements of that form together with empirical reduction-biconditionals. (The remaining part of this program – finding a
suitable characterization of the law of physics – say, that they all be
differential equations in the fundamental magnitudes, valid at every point
of space and time, at least 'before quantization' – and finding the empirical reduction-biconditionals – is, of course, a task for science and not for
philosophy. In a sense, that is the whole point of this paper.)

It is evident that if this particular program for characterizing the
nomological statements ever succeeds, so must the program for characterizing 'identity of properties'. Indeed, the program for characterizing
the nomological statements is in one way harder of fulfillment than the

program for characterizing identity of properties, in that the latter program requires only that we know the reduction-biconditionals (and reduction-identities) and the fundamental magnitudes of physics, but not that we ever have a suitable characterization of the form of the physical laws.

What of the more conventional program for explicating 'nomological', as represented by the writing of Reichenbach, Goodman *et al.*? This program is to characterize fundamental laws (in all disciplines at once) as *true generalizations of a certain form*, where the specification of the form involves not only considerations of logical form in the usual sense, but such restrictions on the choice of the predicates as Goodman's requirement of projectability. It is, further, a part of this program to be independent of the sorts of empirical considerations that I have constantly been bringing up – one is apparently, to discover a form in which all and only laws of nature can be written (i.e. no *true* statement can have the form in question and not be a law of nature), and to do this by reflection on the meaning of 'law' (and, perhaps, on the methodology of science) alone. In short, what I hope scientists may be able to do empirically in the next couple of hundred years these philosophers would like to do *a priori*. Good luck to them!

It should be noted that if these philosophers ever succeed, then they will also succeed in providing us with one criterion for the identity of properties (though not the one I have suggested): for *nomological equivalence*, in spite of some counterintuitive consequences, is another criterion for the identity of physical₁ properties that I think would be workable (if one were willing to change one's intuitions a bit), and that deserves further investigation. Moreover, if one *could* 'explicate "nomological",' then one should also be able to explicate 'reduction law', and hence to explicate the criterion for the identity of physical₁ properties suggested in this paper.

In terms of the foregoing discussion, my answer to the question of whether quantification over properties is indispensable goes as follows: First, there are important locutions which are most naturally formalized as quantifications over properties, and for which there is no obvious formalization today in extensional language. Secondly, the concept of a property is intimately connected with the notions: *nomological, explanation, cause,* etc., and even comes close to being definable in terms of these notions. Yet these notions are generally admitted to be indispensable in science, even by those philosophers who reject *analytic, necessary, synonymy,* etc.(i.e. the notions most closely connected with the *other* concept of property mentioned at the beginning of this paper, the concept of a predicate, or the concept of a concept). The notion *is* indispensable, then, in the sense in which any notion is (i.e. we might use different *words*, but we would have to somehow be able to get the notion expressed); and, if the discussion of the prospects for a criterion of identity earlier in this paper was not unduly optimistic, science is well on its way to giving us as precise a criterion for the identity of properties as we could ask for. Let us, then, keep our properties, while not in any way despising the useful work performed for us by our classes!

20
Philosophy of Logic

I. What logic is*

Let us start by asking what logic *is*, and then try to see why there should be a philosophical problem about logic. We might try to find out what logic is by examining various definitions of 'logic', but this would be a bad idea. The various extant definitions of 'logic' manage to combine circularity with inexactness in one way or another. Instead let us look at logic itself.

If we look at logic itself, we first notice that logic, like every other science, undergoes changes – sometimes rapid changes. In different centuries logicians have had very different ideas concerning the scope of their subject, the methods proper to it, etc. Today the scope of logic is defined much more broadly than it ever was in the past, so that logic as some logicians conceive it, comes to include all of pure mathematics. Also, the methods used in logical research today are almost exclusively mathematical ones. However, certain aspects of logic seem to undergo little change. Logical results, once established, seem to remain forever accepted as correct – that is, logic changes not in the sense that we accept incompatible logical principles in different centuries, but in the sense that the style and notation that we employ in setting out logical principles varies enormously, and in the sense that the province marked out for logic tends to get larger and larger.

It seems wise, then, to begin by looking at a few of those principles that logicians have accepted virtually from the beginning. One such principle is the validity of the following inference:

(1)   All *S* are *M*
       All M are P
(*therefore*) All *S* are *P*

Another is the Law of Identity:

(2) *x* is identical with *x*.

---

* First published by Harper and Row in 1971. First published in Great Britain by Allen & Unwin in 1972.

Yet another is the *inconsistency* of the following:

(3) *p* and not-*p*.

And still another principle is the validity of this:

(4) *p* or not-*p*.

Let us now look at these principles one by one. Inference (1) is traditionally stated to be valid for all *terms S, M, P*. But what is a *term*? Texts of contemporary logic usually say that (1) is valid no matter what *classes* may be assigned to the letters *S, M*, and *P* as their extensions. Inference (1) then becomes just a way of saying that if a class *S* is a subclass of a class *M*, and *M* is in turn a subclass of a class *P*, then *S* is a subclass of *P*. In short, (1) on its modern interpretation just expresses the transitivity of the relation 'subclass of'. This is a far cry from what traditional logicians thought they were doing when they talked about Laws of Thought and 'terms'. Here we have one of the confusing things about the science of logic: that even where a principle may seem to have undergone no change in the course of the centuries – e.g. we still write:

All *S* are M

All M are P

(*therefore*) All *S* are P

– the *interpretation* of the 'unchanging' truth has, in fact, changed considerably. What is worse, there is still some controversy about what the 'correct' interpretation is.

Principle (2) is another example of a principle whose correct interpretation is in dispute. The interpretation favored by most logicians (including the present writer) is that (2) asserts that the relation of identity is reflexive: everything bears this relation (currently symbolized ' =') to itself. Some philosophers are very unhappy, however, at the very idea that ' =' is a relation. 'How can we make sense of a *relation*', they ask, 'except as something a thing can bear to *another* thing?' Since nothing can bear identity to a *different* thing, they conclude that whatever ' =' may stand for, it does not stand for a relation.

Finally, (3) and (4) raise the problem: what does *p* stand for? Some philosophers urge that in (4), for example, *p* stands for any *sentence* you like, whereas other philosophers, including the present writer, find something ridiculous in the theory that logic is about *sentences*.

Still, all this disagreement about fine points should not be allowed to obscure the existence of a substantial body of agreement among all logicians – even logicians in different centuries. All logicians agree, for example, that from the premises:

All men are mortal
All mortals are unsatisfied

one may validly infer:

All men are unsatisfied

even if they sometimes disagree about the proper *statement* of the general principle underlying this inference. Similarly, all logicians agree that if there is such a thing as the Eiffel Tower, then

The Eiffel Tower is identical with the Eiffel Tower

and all logicians agree that (if there is such a thing as 'the earth')

The earth is round or the earth is not round

even if they disagree about the statement of the relevant principle in these cases too. So there *is* a body of 'permanent doctrine' in logic; but it just doesn't carry one very far, at least when it comes to getting an exact and universally acceptable statement of the general principles.

II. The nominalism-realism issue

At this stage it is already clear that there are philosophical problems connected with logic, and at least one reason for this is also clear: namely, the difficulty of getting any universally acceptable statement of the general principles that all logicians somehow seem to recognize. If we explore this difficulty, further philosophical problems connected with logic will become clearer.

Philosophers and logicians who regard classes, numbers, and similar 'mathematical entities' as somehow make-believe are usually referred to as 'nominalists'. A nominalist is not likely to say:

(A) 'For all *classes* S, M, P: if all S are M and all M are P, then all S are P'.

He is more likely to write:

(B) 'The following turns into a true *sentence* no matter what *words* or *phrases* of the appropriate *kind* one may substitute for the letters S, M, P: "if all S are M and all M are P, then all S are P" '.

The reason is clear if not cogent: the nominalist does not really believe that classes exist; so he avoids formulation (A). In contrast to classes, 'sentences' and 'words' seem relatively 'concrete', so he employs formulation (B).

It is thus apparent that part of the disagreement over the 'correct' formulation of the most general logical principles is simply a reflection of philosophical disagreement over the existence or non-existence of 'mathematical entities' such as classes.

Independently of the merits of this or that position on the 'nominalism-realism' issue, it is clear, however, that (B) cannot really be preferable to (A). For what can be meant by a 'word or phrase of the appropriate kind' in (B)? Even if we waive the problem of just what constitutes the 'appropriate kind' of word or phrase, we must face the fact that what is meant is all *possible* words and phrases of some kind or other, and that *possible words and phrases* are no more 'concrete' than classes are.

This issue is sometimes dodged in various ways. One way is to say that the appropriate 'phrases' that one may substitute for S, M, P are all the 'one-place predicates' in a certain 'formalized language.' A formalized language is given by completely specifying a grammar together with the meanings of the basic expressions. Which expressions in such a language are one-place predicates (i.e. class-names, although a nominalist wouldn't be caught dead calling them that) is specified by a formal grammatical rule. In fact, given a formalized language L, the class of permissible substitutions for the dummy letters S, M, P in:

(5) If all S are M and all M are P, then all S are P

can be defined with great precision, so that the task of telling whether a certain string of letters is or is not a 'substitution-instance' of (1) (as the result of a permissible substitution is called) can even be performed purely mechanically, say, by a computing machine.

This comes close to satisfying nominalistic scruples, for then it seems that to assert the validity of (5) is not to talk about 'classes' at all, but merely to say that all substitution-instances of (5) (in some definite L) are true; i.e. that all the *strings of letters* that conform to a certain formal criterion (being a substitution-instance of (5) in the formalized language L) are true. And surely 'strings of letters' are perfectly concrete – or are they?

Unfortunately for the nominalist, difficulties come thick and fast. By a logical *schema* is meant an expression like (5) which is built up out of 'dummy letters', such as S, M, P, and the logical words if-then, all, some, or, not, identical, is (are), etc. Such schemata have been used by logicians, from Aristotle to the present day, for the purpose of setting out logical principles (although Aristotle confined his attention to a very restricted class of schemata, while modern logicians investigate all possible schemata of the kind just described). A schema may be, like (5), a 'valid' schema – that is, it may express a 'correct' logical principle

326

(what correctness or validity is, we still have to see), or it may be 'invalid'. For example:

If some S are P, then all S are P

is an example of an invalid schema – one that fails to express a correct logical principle. Ancient and medieval logicians already classified a great many schemata as valid or invalid.

Now, defining valid is obviously going to pose deep philosophical issues. But the definition of valid we attributed to the nominalist a moment ago, i.e. a schema S is valid just in case all substitution-instances of S *in some particular formalized language L* are true – is unsatisfactory on the face of it. For surely when I say that (5) is valid, I mean that it is correct *no matter what class-names* may be substituted for S, M, P. If some formalized language L contained names for all the classes of things that could be formed, then this might come to the same as saying 'all substitution-instances of (5) in L are true'. But it is a theorem of set theory that *no* language L *can* contain names for *all* the collections of things that could be formed, at least not if the number of things is infinite.

To put it another way, what we get, if we adopt the nominalist's suggestion, is not *one* notion of validity, but an infinite series of such notions: validity in L_1, validity in L_2, validity in L_3, . . . where each notion amounts simply to 'truth of all substitution-instances' in the appropriate L_i.

We might try to avoid this by saying that a schema S is valid just in case all of its substitution-instances in every L are true; but then we need the notion of *all possible formalized languages* – a notion which is, if anything, *less* 'concrete' than the notion of a 'class'.

Secondly, the proposed nominalistic definition of validity requires the notion of 'truth.' But this is a problematic notion for a nominalist. Normally we do not think of material objects – e.g. strings of actually inscribed letters (construed as little mounds of ink on paper) as 'true' or 'false'; it is rather *what the strings of letters express* that is true or false. But the *meaning* of a string of letters, or what the string of letters 'expresses', is just the sort of entity the nominalist wants to get rid of.

Thirdly, when we speak of *all* substitution-instances of (5), even in one particular language L, we mean all *possible* substitution-instances – not just the ones that happen to 'exist' in the nominalistic sense (as little mounds of ink on paper). To merely say that *those instances of* (5) *which happen to be written down are true* would not be to say that (5) is valid; for it might be that there is a false substitution-instance of (5) which just does not happen to have been written down. But *possible* substitution-instances of (5) – *possible* strings of letters – are not really physical objects any more than classes are.

One problem seems to be solved by the foregoing reflections. There is no reason in stating logical principles to be more puristic, or more compulsive about avoiding reference to 'non-physical entities', than in scientific discourse generally. Reference to classes of things, and not just things, is a commonplace and useful mode of speech. If the nomin-alist wishes us to give it up, he must provide us with an alternative mode of speech which works just as well, not just in pure logic, but also in such empirical sciences as physics (which is full of references to such 'non-physical' entities as state-vectors, Hamiltonians, Hilbert space, etc.). If he ever succeeds, this will affect how we formulate all scientific principles – not just logical ones. But in the meantime, there is no reason not to stick with such formulations as (A), in view of the serious problems with such formulations as (B). (And, as we have just seen, (B), in addition to being inadequate, is not even really nominalistic.)

To put it another way, the fact that (A) is 'objectionable' on nomin-alistic grounds is not really a difficulty with the science of logic, but a difficulty with the philosophy of nominalism. It is not up to logic, any more than any other science, to conform its mode of speech to the philosophic demands of nominalism; it is rather up to the nominalist to provide a satisfactory reinterpretation of such assertions as (5), and of any other statements that logicians (and physicists, biologists, and just plain men on the street) actually make.

Even if we reject nominalism as a demand that we here and now strip our scientific language of all reference to 'non-physical entities', we are not committed to rejecting nominalism as a philosophy, however. Those who believe that in truth there is nothing answering to such notions as class, number, possible string of letters, or that what does answer to such notions is some highly derived way of talking about ordinary material objects, are free to go on arguing for their view, and our unwillingness to conform our ordinary scientific language to their demands is in no way an unwillingness to discuss the philosophical issues raised by their view. And this we shall now proceed to do.

We may begin by considering the various difficulties we just raised with formulation (B), and by seeing what rejoinder the nominalist can make to these various difficulties.

First, one or two general comments. Nelson Goodman, who is the best-known nominalist philosopher, has never adopted the definition of validity as 'truth of all substitution-instances'. (It comes from Hugues Leblanc and Richard Martin.) However, Goodman has never tackled the problem of defining logical validity at all, so I have taken the liberty of discussing the one quasi-nominalistic attempt I have seen. Secondly, Goodman denies that nominalism is a restriction to 'physical' entities. However, while the view that only physical entities (or 'mental particulars', in an idealistic version of nominalism; or mental particulars and physical things in a dualistic system) alone are real may not be what Goodman intends to defend, it is the view that most people understand by 'nominalism', and there seems little motive for being a nominalist apart from some such view. (The distinction between a restriction to 'physical entities' and a restriction to 'mental particulars' or 'physical things and mental particulars' will not be discussed here, since it does not seriously affect the philosophy of logic.)

The first argument we employed against formulation (B) was that this formulation, in effect, replaces our intuitive notion of validity by as many notions of validity as there are possible formalized languages. Some logicians have tried to meet this difficulty by the following kind of move: Let L_0 be a formalized language rich enough to talk about the positive integers, and to express the notions 'x is the sum of y and z' and 'x is the product of y and z'. Let L_t be any other formalized language. Let S be a schema which has the property that all its substitution-instances in L_0 are true (call this property the property of being 'valid-in-L_0', and, analogously, let us call a schema 'valid-in-L_t' if all its substitution-instances in L_t are true). Then it is true – and the proof can be formalized in any language rich enough to contain both the notions of 'truth in L_0' and 'truth in L_t' – that S also has the property that all its substitution-instances in L_t are true. In other words, if a schema is valid-in-L_0, it is also valid-in-L_t. So, these logicians suggest, let us simply define 'validity' to mean valid-in-L_0. If S is valid, it will then follow – not by definition, but by virtue of the metamathematical theorem just mentioned – that all of its substitution-instances in L_t are true, no matter what language L_t may be. So 'validity' will justify asserting arbitrary substitution-instances of a schema (as it should, on the intuitive notion).

To this, one is tempted to reply that what I mean when I say 'S is valid' directly implies that every substitution-instance of S (in every formalized language) is true. On the proposed definition of valid all that I mean when I say that 'S is valid' is that S's substitution-instances in L_0 are true; it is only a mathematical fact, and not part of what I mean that then S's substitution-instances in any language are true. Thus the proposed definition of valid completely fails to capture the intuitive notion even if it is coextensive with the intuitive notion.

This reply, however, is not necessarily devastating. For the nominalistic logician may simply reply that he is not concerned to capture the 'intuitive' notion; it is enough if he can provide us with a notion that is philosophically acceptable (to him) and that will do the appropriate work.

Be this as it may, the fact remains that the language L_0 is one that itself requires talking about 'mathematical entities' (namely *numbers*), and that the proof of the statement that 'if S is valid-in-L_0, then S is valid-in-L_t' requires talking about arbitrary expressions of L_t (i.e. about all *possible* expressions of L_t). Thus neither the language L_0 nor the metamathematical theorem just mentioned is really available to a *strict* nominalist, i.e. to one who foreswears *all* talk about 'mathematical entities'.

The second argument we used was that the notion of 'truth' is not available to a nominalist. However, this claim is extremely debatable.

Our argument was, in a nutshell, that 'true' makes no sense when applied to a physical object, even if that physical object be an inscription of a sentence; it is not the physical sentence that is true or false, but *what the sentence says*. And the *things* sentences say, unlike the sentences (inscriptions) themselves, are not physical objects.

The natural response for a nominalist to make here would, I think, be to distinguish between

(6) S is true

and

(7) S is true as understood by Oscar at time t.

If S is a physical object (say, a sentence-inscription), then (6) makes little sense indeed, save as an elliptical formulation of some such fact as (7). But (7) represents a perfectly possible relation which may or may not obtain between a given inscription, an organism, and a time. (How reference to 'times' is to be handled by the nominalist, I shall not inquire; perhaps he must identify a 'time' with a suitable three-dimensional cross section of the whole four-dimensional space–time universe.) Why should it not be open to the nominalist to assert that some *sentences* are *true* in the sense of having the relation occurring in (7) to suitable organisms at suitable times? Granted that that relation is a complex one; the burden of proof is on the realist to show that that relation essentially presupposes the existence of non-physical entities such as propositions, meanings, or what have you.

Another form of the second argument takes the form of an 'appeal to ordinary language'. Thus it is contended that

(8) John made a true statement

is perfectly correct 'ordinary language' in certain easily imagined situations. Now there are two possibilities: (a) that (8) implies that statements exist (as non-physical entities); or (b) that (8) does not imply this. In case (b) there is no problem; we may as well go on talking about 'statements' (and, for that matter, about 'classes', 'numbers', etc.), since it is agreed that such talk does not imply that statements (or numbers, or classes) exist as non-physical entities. Then nominalism is futile, since the linguistic forms it wants to get rid of are philosophically harmless. In case (a), since (8) is true and (8) implies the existence of non-physical entities, it follows that these non-physical entities do exist! So nominalism is false! Thus nominalism must be either futile or false.

Against this the nominalist replies that what he wishes to do is to find a 'translation function' that will enable us to replace such sentences as (8) by sentences which do not even *appear* to imply the existence of non-physical entities. The effect of this will, he thinks, be to provide us with a terminology which is conceptually less confusing and more revealing of the nature of reality than the terminology we customarily employ. To be sure, such sentences as (8) are 'philosophically harmless' if correctly understood; but the problem is to make clear what this correct understanding is.

The nominalist can strengthen this somewhat by adding that it is not necessary, on his view, that the translation function preserve *synonymy*. It is enough that the proposal to understand such sentences as (8) on the model of their nominalistic translations should be a good one, in the sense of conducing to increased clarity.

Thus the fact that in 'ordinary language' the words 'true' and 'false' are normally applied to 'statements' does not convince the nominalist *either* that statements really exist as non-physical entities, *or* that a departure from ordinary language (in, say, the direction of (7)) is an intellectual sin.

Finally, there is the 'argument' that what (7) *means* is: there is a *statement* which S 'expresses' to Oscar at time *t*, and that *statement* is true. According to this argument, (7) involves a disguised reference to a non-physical entity (what S 'expresses'), and is thus not 'really' nominalistic.

This argument reduces either to the appeal to ordinary language just discussed or else to the mere claim that *really* only *statements* (construed as non-physical entities expressed by sentences) can be 'true' or 'false'. Since this claim is precisely what is at issue, this is not an argument at all, but a mere begging of the question.

All arguments that the notion of truth is unavailable to the nominalist seem thus to be bad arguments. On the other hand, it does not follow that the nominalist is simply *entitled* to the notion. Truth (or the triadic relation between inscriptions, organisms, and times that occurs in (7)) is hardly a primitive thing like 'yellow', so surely the nominalist owes us *some* account of what it is; an account consistently framed within the categories of his metaphysics. If he cannot provide such an account (and what nominalist has?), his right to use the notion does become suspect.

Before the reader (or the nominalist) replies too hastily with *tu quoque*, let him be reminded of the following facts: the 'intuitive' notion of truth seems to be inconsistent (cf. the well-known logical antinomies in connection with it); but given any formalized language L, there is a predicate 'true-in-L' which one can employ for all scientific purposes in place of the intuitive true (when the statements under discussion are couched in L), and *this* predicate admits of a *precise* definition using only the vocabulary of L itself plus set theory.† This is not wholly satisfactory – one would prefer a single predicate 'true' to an infinite collection of predicates 'true-in-L_1', 'true-in-L_2', etc. – but it is not unbearable, and the antinomies give strong reason to doubt that any notion of truth applicable to *all* languages and satisfying the intuitive requirements could be consistent. The realist is thus in a position not to explain the intuitive notion of truth, but to provide a battery of alternative notions that he can use in all scientific contexts as one would wish to use 'true', and that he can precisely define. But – today, at least – the nominalist cannot even do this much.

Our third argument was that reference to *all* the sentences of a formalized language (or even all the substitution-instances of a fixed schema) is not reference to 'inscriptions' (since it can hardly be supposed that all the infinitely many sentences of any formalized language are actually inscribed somewhere) but to abstract entities – 'possible inscriptions', or, according to some authors, to the 'types' or shape-properties which inscriptions exemplify. (These types are supposed to 'exist' independently of whether any inscriptions actually exemplify them or not; thus, they too are non-physical entities.) When we say 'all substitution-instances of (S) are true', we mean *even those substitution-instances that no one has actually written down*. Thus these 'substitution-instances '– especially the 'potential' ones – are no more 'physical' than classes are. To my knowledge, no rejoinder to this argument worth considering exists.

Our reconsideration of the three arguments has, thus, not altered our conclusion that (B) is not a nominalistic formulation. We see, however, that the deeper one goes into the first two of the three arguments the

† This was shown by Tarski. For a semipopular exposition see 'The semantic conception of truth' in H. Feigl and W. Sellars (eds.) *Readings in philosophical analysis*, New York, 52–84.

more complex (and also the more technical) the arguments are going to become.

We may summarize the upshot of this section by saying that at present reference to 'classes', or something equally 'non-physical' is indispensable to the science of logic. The notion of logical 'validity', on which the whole science rests, cannot be satisfactorily explained in purely nominalistic terms, at least today.

III. The Nominalism–Realism issue and logic

The issue of nominalism versus realism is an old one, and it is interesting to review the way in which it has become connected with the philosophy of logic. Elementary logic has enunciated such principles as (2), (4), (5), has listed such patterns of valid inference as (1), and has asserted the inconsistency of such forms as (3) since Aristotle's day. Modern 'quantification theory', as the corresponding branch of modern logic is called, or 'first order logic with identity', has an immensely larger scope than the logic of Aristotle, but the topic of concern is recognizably similar.

The basic symbols are:

(i) 'Px' for 'x is P', and similarly, 'Pxy' for 'x bears P to y', '$Pxyz$' for 'x, y, z stand in the relation P', etc.

(ii) '(x)' (read: 'for every x') to indicate that every entity x satisfies a condition; i.e. '$(x) Px$' means 'every entity x is P'.

(iii) '(Ex)' (read: 'there is an x such that') to indicate that some (at least one) entity x satisfies a condition; i.e. '$(Ex) Px$' means 'there is an entity x which is P'.

(iv) '$=$' (read: 'is identical with') for identity, i.e. '$x = y$' means 'x is identical with (is one and the same entity as) y'.

(v) '$\cdot$' for 'and', '$\lor$' for 'or', '$-$' for 'not', e.g. '$(Px \lor -Qx) \cdot Rx$' means 'either x is P or x is not Q; and x is R'.

In addition the symbols $\supset$ (read: 'if . . . then') and $\equiv$ (read: 'if and only if') are used with the definitions: '$Px \supset Qx$' ('If Px then Qx') is short for '$-(Px \cdot -Qx)$', and '$Px \equiv Qx$' is short for '$(Px \supset Qx) \cdot (Qx \supset Px)$'.

In this notation we can write down all the principles that Aristotle stated. For example, (5) becomes

(5') $((x) (Sx \supset Mx) \cdot (x) (Mx \supset Px)) \supset (x) (Sx \supset Px)$.

Also, by considering the full class of schemata that we can write with this notation, we are led to consider potential logical principles that Aristotle never considered because of his general practice of looking at inferences each of whose premisses involved exactly *two* class-names.

The most important thing, however, is that with modern notation we can analyze inferences which essentially involve two-or-more-term *relations;* it was essentially the failure to develop a logic of relations that trivialized the logic studied before the end of the nineteenth century, and that makes that logic – the traditional logic from Aristotle on, and even including the work of Boole, tremendously important as it was for later developments – utterly inadequate for the analysis of deductive reasoning in its more complicated forms.

In his many logical and philosophical writings, Quine has urged that quantification theory does not really assert, for example, formulation (A) of the preceding section. On Quine's view, when a logician builds a system one of whose theorems is, say, (5'), he does not thereby mean to assert (A). Rather, in (5) or (5'), *S, M, P* are mere 'dummy letters' standing for *any predicate you like;* and what the logician is telling you is that *all substitution-instances* of (5) or (5') are truths of logic.

On this view, the following is a 'truth of logic':

(9) If all crows are black and all black things absorb light, then all crows absorb light.

But the general principle (A):

For all classes *S, M, P:* if all *S* are *M* and all *M* are *P*, then all *S* are *P*

is not a truth of *logic*, on Quine's view, but a truth of *mathematics*.

Now, then, I do not much care just where one draws the line between logic and mathematics, but this particular proposal of Quine's seems to me hardly tenable.

My reasons are principally two. In the first place, logical tradition goes against Quine; for, from the beginning it has been the concern of logicians to state just such general principles as (A) and *not* to 'sort out' such truths as (9) from other truths. In the second place, I don't think all substitution-instances of a valid schema *are* 'true': some are obviously meaningless. For example:

(10) If all boojums are snarks and all snarks are eggelumphs, then all boojums are eggelumphs

does not appear to me to be a true statement – it has the *form* of a logically valid statement, but, I think, it is not a statement at all, and neither true nor false. Indeed, to call (10) true requires some revision of the usual logical rules. For it is a theorem of standard logic that if a statement of the form 'if *p* and *q*, then *r*' is true, then either *p* and *q* are both true and *r* is true, or *p* is true and *q* false and *r* true (or false), or *p* is false and q true and r true (or false), or p and q are both false and r is true (or false). But in the case of (10) the three components corresponding to p, q, and r are *neither* true *nor* false.

Of course, one could adopt the decision to extend the notion of truth and to call any statement that has the form of a logically valid statement true. But then:

(11) All boojums snark or not all boojums snark

(which has the form $p \vee -p$) would have to be counted as true, and this seems extremely misleading, since normally anyone who asserts (11) would be understood to be committed to:

(12) The statement that all boojums snark is either true or false.

In my view, logic, as such, does *not* tell us that (9) is true: to know that (9) is true I have to use my knowledge of the logical principle (A), *plus* my knowledge of the fact that the predicates 'x is a crow', 'x is black' and 'x absorbs light' are each true of just the things in a certain class, namely the class of crows, (respectively) the class of black things, (respectively) the class of things which absorb light. Even this 'knowledge' involves a certain idealization: namely, ignoring the fact that some of these predicates (especially black) are ill-defined (neither true nor false) in certain cases. However, even if we are willing to make this idealization, knowing that, say, 'x is a crow' is a predicate which is true (apart from possible borderline cases) of each thing in a certain class and false of each thing in the complement of that class is knowing a good bit about both language and the world. That 'x is a crow' is a pretty well-defined predicate, 'x is beautiful' is pretty ill-defined, and 'x is a snark' is meaningless, is not *logical* knowledge, whatever kind of knowledge it may be.

We have thus a disagreement between Quine and myself, since it is just such statements as (9) that Quine regards as 'truths of logic', while according to me each such statement reflects a complicated mixture of logical and extra-logical knowledge. But it is not important that the reader should agree with me here and not with Quine – all I insist on, for present purposes, is that the decision to call such statements as (A) 'principles of logic' is not ill-motivated, either historically or conceptually. There may be some choice here, to be sure; but it is important that one quite natural choice makes statements like (A), which refer explicitly to classes, part of *logic*.

The logical schemata so far considered have contained (x) [*for every individual x*] and (Ex) [*there exists an individual x such that*], but not (F) and (EF). Thus, given a 'universe of discourse' we can say, with the

335

notation as so far described, that some element of the universe is P by writing $(Ex)\ Px$, but we cannot say that there is a *set* or *class* of all the elements with the property P (in symbols: $(EF)\ (x)\ (Fx \equiv Px)$), because we don't have '(EF)'.

The decision of the great founders of modern logic – Frege, and, following him, Bertrand Russell – was unhesitatingly to count such expressions as (EF) as part of logic, and even to allow expressions such as (EF^2), with the meaning *for every class of classes*, (EF^3) with the meaning *for every class of classes of classes*, etc., as part of 'logic'.

My contention here is that there was no mistake in so doing. Their decision may not have been the only possible decision – indeed, in the introduction to the second edition of *Principia Mathematica*, Russell carefully refrains from claiming that it was – but it represented a perfectly natural choice. The question of where to 'draw the line' (if we want to draw a line at all) between logic and set theory (and hence between logic and mathematics) is one which has no non-arbitrary answer.

Suppose, however, we decide to draw the line at 'first order' logic ('quantification theory') and to count such expressions as (EF), (EF^2), etc., as belonging to 'mathematics'. Still, we are left with the problem: when a logician builds a system which contains such theorems as $(5')$, *what does he mean to be asserting?* He may, of course, not mean to be asserting anything; he may just be constructing an uninterpreted formal system, but then he is certainly not doing logic. The simple fact is that the great majority of logicians would understand the intention to be this: the theorems of the system are intended to be valid formulas. Implicitly, if not explicitly, the logician is concerned to make assertions of the form 'such-and-such is *valid*'; that is, assertions of the kind (A). Thus even first order logic would normally be understood as a 'metatheory'; insofar as he is making assertions at all in writing down such schemata as $(5')$, the logician is making assertions of validity, and that means he is implicitly making second order assertions: for to assert the validity of the first order schema $(5')$ is just to assert $(S)\ (M)\ (P)$ (schema $5'$) – and this is a second order assertion.

In sum, I believe that (a) it is rather arbitrary to say that 'second order' logic is not 'logic'; and (b) even if we do say this, the natural understanding of first order logic is that in writing down first order schemata we are implicitly asserting their validity, that is, making second order assertions. In view of this, it is easy to see why and how the traditional nominalism–realism problem came to concern intensely philosophers of logic: for, if we are right, the natural understanding of logic is such that all logic, even quantification theory, involves reference to classes, that is, to just the sort of entity that the nominalist wishes to banish.

IV. Logic versus mathematics

In view of the foregoing reflections, it is extremely difficult to draw a non-arbitrary line between logic and mathematics. Some feel that this line should be identified with the line between first order logic and second order logic; but, as we have just seen, this has the awkward consequence that the notions of validity and implication† turn out to belong to mathematics and not to logic. Frege, and also Russell and Whitehead, counted not only second order logic but even higher order logic (sets of sets of sets of . . . sets of individuals) as logic; this decision amounts to saying that there is no line 'between' mathematics and logic; mathematics is part of logic. If one wishes an 'in-between' view, perhaps we should take the one between second and third order logic to be the 'line' in question. However, we shall not trouble ourselves much with this matter. The philosophical questions we are discussing in this chapter affect the philosophy of mathematics as much as the philosophy of logic; and, indeed, we shall not trouble to distinguish the two subjects.

V. The inadequacy of nominalistic language

By a 'nominalistic language' is meant a formalized language whose variables range over individual things, in some suitable sense, and whose predicate letters stand for adjectives and verbs applied to individual things (such as 'hard', 'bigger' than, 'part of'). These adjectives and verbs need not correspond to observable properties and relations; e.g. the predicate 'is an electron' is perfectly admissible, but they must not presuppose the existence of such entities as classes or numbers.

It has been repeatedly pointed out that such a language is inadequate for the purposes of sciences; that to accept such a language as the only language we are philosophically entitled to employ would, for example, require us to give up virtually all of mathematics. In truth, the restrictions of nominalism are as devastating for empirical science as they are for formal science; it is not just 'mathematics' but physics as well that we would have to give up.

To see this, just consider the best-known example of a physical law: Newton's law of gravitation. (That this law is not strictly true is irrelevant to the present discussion; the far more complicated law that is actually true undoubtedly requires even more mathematics for its

† A is said to *imply* B, just in case the conditional $(A \supset B)$ with A as antecedent and B as consequent is *valid*. In short, 'implication is validity of the conditional'.

formulation.) Newton's law, as everyone knows, asserts that there is a force f_{ab} exerted by any body a on any other body b. The direction of the force f_{ab} is towards a, and its magnitude F is given by:

$$F = \frac{g M_a M_b}{d^2}$$

where g is a universal constant, M_a is the mass of a, M_b is the mass of b, and d is the distance which separates a and b.

I shall assume here a 'realistic' philosophy of physics; that is, I shall assume that one of our important purposes in doing physics is to try to state 'true or very nearly true' (the phrase is Newton's) laws, and not merely to build bridges or predict experiences. I shall also pretend the law given above is correct, even though we know today that it is only an approximation to a much more complicated law. Both of these assumptions should be acceptable to a nominalist. Nominalists must at heart be materialists, or so it seems to me: otherwise their scruples are unintelligible. And no materialist should boggle at the idea that matter obeys some objective laws, and that a purpose of science is to try to state them. And assuming that Newton's law is strictly true is something we do only to have a definite example of a physical law before us – one which has a mathematical structure (which is why it cannot be expressed in nominalistic language), and one which is intelligible to most people, as many more complicated physical laws unfortunately are not.

Now then, the point of the example is that Newton's law has a content which, although in one sense is perfectly clear (it says that gravitational 'pull' is directly proportional to the masses and obeys an inverse-square law), quite transcends what can be expressed in nominalistic language. Even if the world were simpler than it is, so that gravitation were the only force, and Newton's law held exactly, still it would be impossible to 'do' physics in nominalistic language.

But how can we be sure that this is so? Even if no nominalist has yet proposed a way to 'translate' such statements as Newton's law into nominalistic language, how can we be sure that no way exists?

Well, let us consider what is involved, and let us consider not only the law of gravitation itself, but also the obvious presuppositions of the law. The law presupposes, in the first place, the existence of forces, distances, and masses – not, perhaps, as real entities but as things that can somehow be measured by real numbers. We require, if we are to use the law, a language rich enough to state not just the law itself, but facts of the form 'the force f_{ab} is $r_1 \pm r_2$', 'the mass M_a is $r_1 \pm r_2$', 'the distance d is $r_1 \pm r_2$', where r_1, r_2 are arbitrary rationals. (It is not necessary, or indeed possible, to have a name for each real number, but we need to be able to express arbitrarily good rational estimates of physical magnitudes.) But no nominalist has ever proposed a device whereby one might translate arbitrary statements of the form 'the distance d is $r_1 \pm r_2$' into a nominalistic language. Moreover, unless we are willing to postulate the existence of an actual infinity of physical objects, no such 'translation scheme' can exist, by the following simple argument: If there are only finitely many individuals, then there are only finitely many pairwise non-equivalent statements in the formalized nominalistic language. In other words, there are finitely many statements S_1, S_2, . . ., S_n such that for an arbitrary statement S, either $S \equiv S_1$ or $S \equiv S_2$ or . . . or $S \equiv S_n$, and moreover (for the appropriate i) $S \equiv S_i$ follows logically from the statement 'the number of individuals is N'.† But if we have names for two different individuals in our 'language of physics', say, a and b, and we can express the statements 'the distance from a to b is one meter $\pm$ one centimeter', 'the distance from a to b is two meters $\pm$ one centimeter', etc., then it is clear that we must have an *infinite* series of pairwise non-equivalent statements. (Nor does the non-equivalence vanish given the premiss 'the number of individuals is N'; it does not follow logically from that premiss that any two of the above statements have the same truth value.) Thus any 'translation' of 'the language of physics' into 'nominalistic language' must disrupt logical relations: for any N, there will be two different integers n, m such that the false 'theorem':

If the number of individuals is N, then the distance from a to b is n meters $\pm$ one cm. $\equiv$ the distance from a to b is m meters $\pm$ one cm.

will turn into a true theorem of logic if we accept the translation

† Here is a sketch of the proof of this assertion: suppose for example, $N = 2$ and introduce (temporarily) symbols 'a' and 'b' for the two individuals assumed to exist. Rewrite each sentence $(x)\,Px$ as a conjunction $Pa \cdot Pb$ and each sentence $(\exists x)\,Px$ as a disjunction $Pa \lor Pb$. Thus every sentence S of the language is transformed into a sentence S' without quantifiers. There are only finitely many atomic sentences (assuming the number of primitive predicates in the language is finite). If the number of these atomic sentences is n, then the number of truth functions of them that can be written is 2^{2^n}. One can easily construct 2^{2^n} quantifier-free sentences which correspond to these 2^{2^n} truth functions; then *any* sentence built up out of the given n atomic sentences by means of truth functional connectives will be logically equivalent to one of these sentences T_1, T_2, . . . , $T_2{}^{2^n}$. Moreover, if $S' \equiv T_i$ is a theorem of propositional calculus, then it is easily seen that $S \equiv (\exists a, b)\,(a \neq b \cdot T_i)$ is true in every two-element universe, and hence 'the number of individuals is two' (this may be symbolized $(\exists a, b)$ $(a \neq b \cdot (x)\,(x = a \lor x = b)\,)\,)$ implies $S \equiv (\exists a, b)\,(a \neq b \cdot T_i)$. Thus, if we let $S_1 =$ '$(\exists a, b)\,(a \neq b \cdot T_1)$', $S_2 = $ '$(\exists a, b)\,(a \neq b \cdot T_2)$', . . . , then (1) if the number of individuals is *two*, then every sentence S is equivalent in truth-value to one of the sentences S_1, S_2, . . . , $S_2{}^{2^n}$; and (2) the sentence $S \equiv S_i$ (for the appropriate i) is itself *implied by* the statement that the number of individuals is two. The same idea works for any finite number of individuals.

scheme. Thus a nominalistic language is *in principle* inadequate for physics.

The inadequacy becomes even clearer if we look at the matter less formalistically. The concept 'distance in meters' is an extremely complex one. What is involved in supposing that such a physical magnitude as distance can somehow be coordinated with *real numbers*?

One account (which I believe to be correct) is the following: It is clear that physics commits us to recognizing the existence of such entities as 'spatial points' (or space–time points in relativistic physics), although the nature of these entities is far from clear. Physicists frequently say that space–time points are simply 'events', although this is obviously false. Carnap and Quine prefer to think of points as triples of real numbers (or quadruples of real numbers, in the case of space–time points); however, this seems highly unnatural, since the identity of a spatial point does not intuitively depend on any particular coordinate system. I prefer to think of them as properties of certain events (or of particles, if one has point-particles in one's physics); however, let us for the moment just take them as primitive entities, not further identified than by the name 'point'. On any view, there is a relation $C(x, y, z, w)$ which we may call the relation of congruence, which is a physically significant relation among points, and which is expressed in word language by saying that the interval $\overline{xy}$ is *congruent* to the interval $\overline{zw}$. (I say 'on any view', because there is a serious disagreement between those philosophers who think that this relation can be operationally defined and those who, like myself, hold that all so-called operational definitions are seriously inaccurate, and that the relation must be taken as primitive in physical theory.) Let us take two points (say, the end points of the standard meter in Paris at a particular instant) and call them a_1 and a_2. We shall take the distance from a_1 to a_2 to be *one*, by definition. Then we may define 'distance' as a numerical measure defined for any two points x, y, as follows:

'The distance from x to y is r' is defined to mean that $f(x, y) = r$, where f is any function satisfying the following five conditions:

(1) $f(w, v)$ is defined (and has a non-negative real value) on all points w, v.

(2) $f(w, v) = 0$ if and only if w is the same point as v.

(3) $f(w, v) = f(w', v')$ if and only if $C(w, v, w', v')$ holds (i.e. if and only if the interval $\overline{wv}$ is congruent to the interval $\overline{w'v'}$).

(4) If w, v, u are colinear points and v is between w and u, then $f(w, u) = f(w, v) + f(v, u)$. (Here 'colinear' and 'between' can either be defined in terms of the C-relation in known ways, or taken as further primitives from physical geometry.)

(5) $f(a_1, a_2) = 1$.

It can be shown that there is a unique function f satisfying conditions (1)–(5).† Thus the content of the definition given above may also be expressed by saying that distance is defined to be the value of the unique function satisfying (1)–(5).

Let us call the account, above, a description of the 'numericalization'‡ of the physical magnitude distance. The point of interest in the present context is this: that even if we take 'points' as individuals, and the relation 'C (x, y, z, w)' as primitive, still we cannot account for the numericalization of distance without quantifying over functions. (Of course, we might avoid the whole problem, by identifying points with triples of real numbers and using the Pythagorean theorem to provide a definition of distance; but then the relation 'object O is at point P' would either have to be analyzed, or we would have to leave numericalization an essentially mysterious and unexplained thing.)

In short, even the statement-form 'the distance from a to b is $r_1 \pm r_2$', where r_1 and r_2 are variables over rational numbers, cannot be explained without using the notion of a function from points to real numbers, or at least to rational numbers. For any constant r_1, r_2 an equivalent statement can be constructed quantifying only over 'points'; but to explain the meaning of the predicate as a predicate of variable r_1, r_2, one needs some such notion as function or set. And the natural way, as we have just seen, even involves functions from points to reals.

It is easy for one and the same person to express nominalistic convictions in one context, and in a different context, to speak of distance as something defined (and having a numerical value) in the case of arbitrary points x, y. Yet, as we have just seen, this is inconsistent. If the numericalization of physical magnitudes is to make sense, we must accept such notions as function and real number; and these are just the notions the nominalist rejects. Yet if nothing really answers to them, then what at all does the law of gravitation assert? For that law makes no sense at all unless we can explain variables ranging over arbitrary distances (and also forces and masses, of course).

† Strictly speaking, this is only true if we require that f be a *continuous* function from space-points to reals. However, this property of continuity can be expressed without assuming that we already have a metric available on the space-points. I have left this out in the text only to simplify the discussion.

‡ The term used in every text on the philosophy of science is not 'numericalization' but 'measurement'. I have coined this barbarous term in order to stress that the problem is *not* one of *measuring* anything, but of *defining* something – i.e. a correspondence between pairs of points and numbers. The term 'measurement' is a hangover from the days of operationalism, when it was supposed that measurement was prior to definition, rather than vice versa.

VI. Predicative versus impredicative conceptions of 'set'

The set $\{x, y\}$ with just the two elements x, y is called the unordered pair of x and y. In terms of unordered pairs one can define ordered pairs in various ways. Perhaps the most natural, though not the customary, way is this: pick two objects a and b to serve as 'markers'. Then identify the ordered pair of x and y with the set $\{\{x, a\}, \{y, b\}\}$ – i.e. with the unordered pair whose elements are the unordered pair $\{x, a\}$ and the unordered pair $\{y, b\}$. Let us adopt the notation $<x, y>$ for this ordered pair, i.e. $<x, y>$ is defined to be $\{\{x, a\}, \{y, b\}\}$. Then it is easily seen that for any x, y, u, v:

$$<x, y> \ = \ <u, v>$$

if and only if $x = u$ and $y = v$. Thus, two 'ordered pairs' are identical just in case their elements are the same and are also ordered the same – which is all that is required of any definition of ordered pair.

In mathematics, a two-place relation is simply a set of ordered pairs. Since 'ordered pair' has just been defined in terms of 'unordered pair', and 'unordered pairs' are simply sets, it follows that 'relation' can be defined in terms of the one primitive notion *set*. If R is a relation such that for all u, v, y

$$\text{if } <u, v> \ \epsilon \ R \text{ and } <u, y> \ \epsilon \ R, \text{ then } v = y$$

then the relation R is called a 'function'. Since function has just been defined in terms of relation (and the notion ' $=$ ' which we count as part of elementary logic), it follows that function has been defined in terms of *set*.

It is well known that the natural numbers 0, 1, 2, 3, . . . can be defined in terms of *set*, in various ways. For example, one can identify 0 with the empty set, 1 with $\{0\}$, then 2 with $\{0, 1\}$, then 3 with $\{0, 1, 2\}$, etc. Also the elementary operations 'plus', 'time', etc., can all be defined from the notion of *set*. Rational numbers are naturally identified with ordered pairs of natural numbers with no common divisor (and such that the second member of the ordered pair is not zero); and real numbers may be identified with series of rational numbers, for example, where a 'series' is just a function whose domain is the natural numbers. Thus all of the 'objects' of pure mathematics may be built up starting with the one notion *set;* indeed, this is the preferred style in contemporary mathematics.

Instead of saying, therefore, that physics essentially requires reference to functions and real numbers, as we did in the previous section, we could simply have said that physics requires some such notion as *set,* for the notions of number and function can be built up in terms of that notion. In the present section we shall make a cursory examination of the notion of a set.

The most famous difficulty with the notion of a set is this: suppose we assume

(1) Sets are entities in their own right (i.e. things we can quantify over†).
(2) If ϕ is any well-defined condition, then there is a set of all the entities which satisfy the condition ϕ.

Then (assuming also that the condition ' $\sim x \in x$' is well defined), it follows that there is a set of all those sets x such that x does not belong to x. If y is that set, then:

(3) $(x)(x \in y \equiv \sim x \in x)$.

But then, putting y for x,

(4) $y \in y \equiv \sim y \in y$

and this is a self-contradiction!

Obviously, then, one of our assumptions was false. Which could it have been? We could say that ' $\sim x \in x$' is not a well-defined condition. But if $x \in y$ is a well-defined relation, for arbitrary sets x, y, then it would seem that $x \in x$, and also $\sim x \in x$, would have to be well-defined (in the sense of having a definite truth value) on all sets x. To give up either the idea that $x \in y$ is a well-defined relation or the idea that sets are entities we can quantify over would be to give up set theory altogether. But the only alternative is then to give up (or at least restrict) (2), which is highly counter-intuitive.

One way out of the difficulty is the so-called theory of types. On this theory, ' $x \in y$' is well defined only if x and y are of the appropriate types, where individuals count as the zero type, sets of individuals as type one, sets of sets of individuals as type two, etc. On this theory, ' $\sim x \in x$' is not even grammatical since no set can be said either to be or not to be a member of itself. One can ask whether a set belongs to any set of the next higher type, but not whether it belongs to itself (or to any set which is not of the next higher type).

Let R be some relation among individuals. A set a such that for all x, if $x \in a$, then $y \in a$, for at least one y such that Rxy, will for the moment be called an R-chain. Suppose we want to say that there is some R-chain containing an individual U. Then we write:

(5) $(\exists a)(a$ is an R-chain. $U \in a)$

† To 'quantify over' sets means to employ such expressions as 'for every set x' and 'there is a set x such that'.

343

where 'a is an R-chain' is short for '(x) $(x \in a \supset (\exists y) (y \in a \cdot Rxy))$'.

Now, the set β of all such U – all U such that some R-chain contains U – is a perfectly good set according to the theory of types, and also according to most mathematicians. Some few mathematicians and philosophers object to the idea of such a set, however. They argue that to define a set β as the set of all U such that there is an R-chain containing U is 'vicious' because the 'totality in terms of which β is defined' – the totality of all R-chains a – could contain β itself. In general, these mathematicians and philosophers say that a set should never be defined in terms of a 'totality' unless that totality is incapable of containing that set, or any set defined in terms of that set. This is, of course, rather vague. But the intention underlying all this is rather interesting.

Suppose I do not understand the notion of a 'set' at all, and, indeed, suppose I employ only some nominalistic language N. Suppose, now, that one day I decide that I understand two notions which are not nominalistic, or, at any rate, whose nominalistic status is debatable: the notions of 'formula' and 'truth'. In terms of these notions I can introduce a very weak notion of set. Namely, suppose I identify sets with the formulas of my nominalistic language which have just one free variable x – e.g. the set of red things I identify with the formula 'Red(x)'. The notion of 'membership' I explain as follows: if y is an individual and a is a 'set' (i.e. a formula with one free variable 'x'), then '$y \in a$' is to mean that a is true of y, where a formula $\phi(x)$ is true of an individual y just in case the formula is true when x is interpreted as a name of y. Thus, if a is the formula 'Red(x)', we have:

$y \in a$ if and only if a is true of y
i.e. if and only if 'Red(x)' is true of y
i.e. if and only if y is red.

So 'Red(x)' turns out to be the 'set of all red things' – as it should be.

I call this a 'weak' notion of set, because it still makes no sense to speak of *all* sets of individuals, let alone sets of higher type than one – one can speak of all formulas, to be sure, but that is only to speak of all sets of individuals definable in N. If new primitives are added to N, then, in general, the totality of sets in the sense just explained, will be enlarged. However, one can iterate the above procedure. That is, let N' be the language one obtains from N by allowing quantification over all sets of individuals definable in N, N'' the language we obtain from N' by allowing quantification over all sets of individuals definable in N', etc. Then all of these sets of individuals – the ones definable in N, in N', in $N'', \ldots$ are examples of 'predicative' sets: each of these sets presupposes a 'totality' which is defined 'earlier' (starting with the totality of individuals) and which does not presuppose it. (One can also introduce predicative sets of higher type, in terms of formulas about formulas, but this will not be done here.) The point that concerns us now is this: this notion of set, the predicative notion of set, is one which can be explained, up to any given level in the series N, N', N'', ... in terms of quantifying only over sets definable earlier in the series, and this whole way of speaking – of 'sets definable in N', 'sets definable in N'', etc. – can itself be regarded, if one wishes, as a mere *façon de parler*, explainable in terms of the notions of formula and truth.

In contrast to the foregoing, if one ever speaks of all sets of individuals as a well-defined totality, and not just all sets definable in some one language in the series N, N', N'', ... then one is said to have an impredicative notion of set.

VII. How much set theory is really indispensable for science?

In the foregoing, we argued that the notion set (or some equivalent notion, e.g. function) was indispensable to science. But we must now ask: does science need the 'strong' (impredicative) notion of set, or only the 'weak' (predicative) notion? For if we are interested in the nominalism–realism issue at all, we must assume not that the only alternatives are (a) nominalism, and (b) acceptance of the full notion 'all sets' (or even, 'all sets of individuals'). If we are inclined to be nominalistic at all, we may wish to keep our non-nominalistic commitments as weak as possible; and limiting these to the two notions truth and formula might seem highly desirable. Truth is a notion that some nominalists think they are entitled to anyway; and if formulas (in the sense of formula types, whether exemplified by any actual inscriptions or not) are 'abstract entities', and hence non-nominalistic, still they are relatively clear ones.

In the case of pure mathematics, the answer seems to be that a certain part of mathematics can be developed using only predicative set theory, provided we allow predicative sets of objects other than physical objects. For example, if we consider the formulas of N to be themselves individuals of some other language M, and then build up a series of languages M, M', M'', ... as sketched before, we can develop at least the arithmetic of rational numbers, and a rudimentary theory of functions of rational numbers. (We need some infinite domain of individuals to 'get started', however, which is why we have to take non-concrete objects, e.g. formulas as individuals, unless we are willing to postulate the existence of an actual infinity of physical objects.) Unfortunately, no satisfactory theory of real numbers or of real functions can be obtained in this way, which is why most mathematicians reject the predicative standpoint.

Turning to logic, i.e. to the notion of 'validity', we saw early in this chapter that a notion of validity, namely 'truth of all substitution-instances' (in, say, M), could be defined in what are essentially the terms of predicative set theory (truth, and quantification over formulas). We also saw that a more satisfactory notion requires the use of the expression 'all sets' – i.e. the notions of impredicative set theory.

Turning lastly to physics, we find the following. At first blush, the law of gravitation (we shall pretend this is the only law of physics, in the present chapter) requires quantification over *real* numbers. However, the law is equivalent to the statement that for every rational ε, and all rational m_1, m_2, d, there is a rational δ such that

$$\text{if } M_a = m_1 \pm \delta, M_b = m_2 \pm \delta, d = d_1 \pm \delta,$$

then

$$F = \frac{g m_1 m_2}{d_1} \pm \varepsilon$$

and this statement quantifies only over rational numbers. (There is the problem that the gravitational constant g may not be rational, however! which I shall neglect here.) Thus a language which quantifies only over *rational* numbers and which measures distances, masses, forces, etc., only by rational approximations ('the mass of a is $m_1 \pm \delta$') *is*, in principle, strong enough to at least *state* the law of gravitation.

Given just predicative set theory, one can easily define the rational numbers. Also one has enough set theory to define 'the cardinal number of S', where S is any *definable* finite set of physical things. Handling the 'numericalization' of such physical magnitudes as distance, force, mass using rational approximations and predicative sets is quite complicated, but still perfectly possible. Thus it appears *possible* (though complicated and awkward) to do physics using just predicative set theory.

In summary, then, the set theoretic 'needs' of physics are surprisingly similar to the set theoretic needs of pure logic. Both disciplines need *some* set theory to function at all. Both disciplines can 'live' – but live badly – on the meager diet of only predicative sets. Both can live extremely happily on the rich diet of impredicative sets. Insofar, then, as the indispensability of quantification over sets is any argument for their existence – and we will discuss why it is in the next section – we may say that it is a strong argument for the existence of at least predicative sets, and a pretty strong, but not *as* strong, argument for the existence of impredicative sets. When we come to the higher reaches of set theory, however – sets of sets of sets of sets – we come to conceptions which are today not needed outside of pure mathematics itself. The case for 'realism' being developed in the present section is thus a qualified one: at least sets of things, real numbers, and functions from various kinds of things to real numbers should be accepted as part of the presently indispensable (or nearly indispensable) framework of both physical science and logic, and as part of that whose existence we are presently committed to. But sets of very high type or very high cardinality (higher than the continuum, for example), should today be investigated in an 'if-then' spirit. One day they may be as indispensable to the very *statement* of physical laws as, say, rational numbers are today; then doubt of their 'existence' will be as futile as extreme nominalism now is. But for the present we should regard them as what they are – speculative and daring extensions of the basic mathematical apparatus of science.

VIII. Indispensability arguments

So far I have been developing an argument for realism along roughly the following lines: quantification over mathematical entities is indispensable for science, both formal and physical; therefore we should accept such quantification; but this commits us to accepting the existence of the mathematical entities in question. This type of argument stems, of course, from Quine, who has for years stressed both the indispensability of quantification over mathematical entities and the intellectual dishonesty of denying the existence of what one daily presupposes. But indispensability arguments raise a number of questions, some of which I should like briefly to discuss here.

One question which may be raised, for example, concerns the very intelligibility of such sentences as 'numbers exist', 'sets exist', 'functions from space–time points to real numbers exist', etc. If these are not genuine assertions at all but only, so to speak, pseudo-assertions, then *no* argument can be a good argument for believing them, and *a fortiori* 'indispensability arguments' cannot be good arguments for believing them.

But what reason is there to say that 'numbers exist', 'sets exist', etc., are unintelligible? It may be suggested that *something* must be wrong with these 'assertions', since one comes across them only in *philosophy*. But there is something extremely dubious about this mode of argument, currently fashionable though it may be. It is one thing to *show* that the locutions upon which a particular philosophical problem depends are linguistically deviant. If, indeed, *no* way can be found of stating the

347

alleged 'problem' which does *not* involve doing violence to the language, then the suspicion may be justified that the 'problem' is no clear problem at all; though, even so, it would hardly amount to certainty, since linguistically deviant expressions need not always be literally *unintelligible*. But it is no argument at all against the genuineness of a putative philosophical problem or assertion that its key terms are linguistically deviant (or, more informally, 'odd', or 'queer', or whatever), if that 'deviancy' (or 'oddness', or 'queerness', or whatever) was only established in the first place by appealing to the dubious principle that terms and statements that occur only in philosophy are *ipso facto* deviant. For the difficulty (it appears to be more than 'difficulty', in fact) is that there is no *linguistic* evidence for this startling claim. Every discipline has terms and statements peculiar to it; and there is no reason whatsoever why the same should not be true of philosophy. If the statement 'material objects exist', for example, does not occur outside of philosophy, that is because only philosophers are concerned with what entitles us to believe such an obvious proposition, and only philosophers have the patience and professional training to pursue a question of justification that turns out to be so difficult; what other science is concerned with entitlement and justification as such? Although the claim is frequently heard that philosophical propositions are by their very nature linguistically (or logically, or 'conceptually') confused, not one shred of *linguistic* evidence exists to show that such sentences as 'numbers exist', 'sets exist', and 'material objects exist', for that matter, are *linguistically* deviant; i.e. that these sentences violate any norms of natural language that can be ascertained to *be* norms of natural language by appropriate scientific procedures.

To put it another way; it would be startling and important if we could honestly *show* that locutions which are peculiar to philosophical discourse have something linguistically wrong with them; but it is uninteresting to claim that this is so if the 'evidence' for the claim is merely that certain particular locutions which are peculiar to philosophy *must* have something wrong with them *because* they are peculiar to philosophy and *because locutions which occur only in philosophical discourse are 'odd'*. The form of the argument is a straightforward circle: a principle P (that there is something wrong with locutions which occur only in philosophical discourse) is advanced; many supporting examples are given for the principle P (i.e. of philosophical statements and questions which are allegedly 'odd', 'queer', etc.); but it turns out that these supporting examples *are* supporting examples only if the principle P is assumed. I do not deny that, historically, many philosophical statements and arguments have contained (and in some cases, essentially depended upon) locutions which are 'queer' by any standard. What I claim is that there is nothing linguistically 'queer' about general existence questions ('Do numbers exist?', 'Do material objects exist?') *per se*, nor about general questions of justification or entitlement ('What entitles us to believe that material objects exist?'), either. (Yet these latter questions are rejected, and by just the circular reasoning just described, in John L. Austin's book *Sense and Sensibilia*, for example; and I am sure many philosophers would similarly reject the former questions.)

So far I have argued that there is no reason to classify such assertions as 'numbers exist' and 'sets exist' as linguistically deviant, apart from a philosophical principle which appears completely misguided. Moreover, there is an easy way to bypass this question completely. Even if some philosophers would reject the sentence 'numbers exist' as somehow not in the normal language, still, 'numbers exist with the property——' is admitted to be non-deviant (and even true) for many values of '——'. For example, 'numbers exist with the property of being prime and greater than 10^{10}', is certainly non-deviant and true. Then, if it should indeed be the case that 'numbers exist' *simpliciter* is not in the language, we could always bring it into the language by simply introducing it as a new speech-form, with the accompanying stipulation that 'numbers exist' is to be true if and only if there is a condition '——' such that 'numbers exist with the property——' is true.

What this amounts to is this: if the sentence

(1) $(\exists x) (x$ is a number $\cdot$ x is prime $\cdot$ $x > 10^{10})$

(i.e. the sentence so symbolized) is in the language, while

(2) $(\exists x) (x$ is a number)

(i.e. 'numbers exist') is not in the language, then ordinary language is not 'deductively closed': for (2) is deducible from (1) in standard logic (by the theorem '$(\exists x) (Fx \cdot Gx \cdot Hx) \supset (\exists x) Fx$'). But if ordinary language is not deductively closed in this respect, then we can deductively close it by introducing (2) into the language, and, moreover, this can be done essentially just one way. So we may as well count (2) as part of the language to begin with.

We have now rejected the view that 'numbers exist', 'sets exist', etc., are linguistically deviant, do not possess a truth value, etc.

A second reason that certain philosophers might advance for rejecting indispensability arguments is the following: these philosophers claim that the truths of logic and mathematics are *true by convention*. If, in

349

particular, 'numbers exist' and 'sets exist' are true by convention, then considerations of dispensability or indispensability are *irrelevant*.

This 'conventionalist' position founders, however, as soon as the conventionalist is asked to become specific about details. *Exactly how* is the notion of truth, as applied to sentences which quantify over abstract entities, to be defined in terms of the notion of *convention*? Even assuming that *some* mathematical sentences *are* 'true by convention', in the sense of being *immediately* true by convention, and that these could be listed, the conventionalist still requires some notion of *implication* in order to handle those truths of mathematics which are not, on any view, immediately conventional – i.e. which require proof. But the notion of implication (validity of the conditional) is one which requires set theory to define, as we have seen; thus conventionalism, even if correct, presupposes quantification over abstract entities, as something intelligible apart from the notion of a convention; mathematical truth ends up being explained as truth by virtue of immediate convention *and mathematics* – an explanation which is trivially correct (apart from the important question of just how large the conventional element in mathematics really is). Moreover, if the conventionalist is not careful, his theory of mathematical truth may easily end up by being in conflict with results of mathematics itself – in particular, with Gödel's theorem. However, discussion of this topic would lead us too far afield; for now I shall simply dismiss conventionalism on the ground that no one has been able to *state* the alleged view in a form which is at all precise and which does not immediately collapse.

A third reason that philosophers might once have given for rejecting indispensability arguments is the following: around the turn of the century a number of philosophers claimed that various entities presupposed by scientific and common sense discourse – even, in the case of some of these philosophers, material objects themselves – were merely 'useful fictions', or that we can not, at any rate, possibly know that they are *more* than 'useful fictions' (and so we may as well say that that is what they are). This 'fictionalistic' philosophy seems presently to have disappeared; but it is necessary to consider it here for a moment, if only because it represents the most direct possible rejection of the probative force of indispensability arguments. For the fictionalist says, in substance: ' *Yes*, certain concepts – material object, number, set, etc. – are indispensable, but *no*, that has no tendency to show that entities corresponding to those concepts actually exist. It only shows that those "entities" are *useful fictions*.'

If fictionalism has been rejected by present-day philosophers of science and epistemologists, this appears to have been in part for bad reasons. The fictionalists regarded the following as a logical possibility: that there might not in fact be electrons (or whatever), but that our experiences might be *as if* there were actually electrons (or whatever). According to the verificationism popular since the late 1920s, this is *meaningless*: if p is a proposition which it would be logically impossible to verify, then p does not represent so much as a logical possibility. But on this issue the fictionalists were surely right and the verificationists wrong: it may be absurd, or crazy, or silly, or totally irrational to believe that, e.g. we are all disembodied spirits under the thought control of some powerful intelligence whose chief purpose is to deceive us into thinking that there is a material world; but it is not *logically impossible*. This is not an essay on verificationism; but it is appropriate to say in passing that all of the verificationist arguments were bad arguments. The chief argument was, of course, to contend that 'material objects exist' *means* something to the effect that under certain circumstances we tend to have certain experiences; but all attempts to carry out the program of actually supplying a reduction of material object language to 'sense-datum' language have failed utterly, and today it seems almost certainly the case that no such reduction can be carried out. Given a large body of theory T, containing both 'sense-datum' sentences and 'thing sentences' (assuming, for the sake of charity, that a 'sense-datum' language could really be constructed), one could say what 'sense-datum' sentences are logically implied by T, to be sure; but this does not mean that the thing-sentences in T (much less in 'the language' considered apart from any particular theory) must be individually equivalent to sense-datum sentences in any reasonable sense of 'equivalent'. Another argument was a species of open-question argument: 'What more does it mean to say that material objects exist, than that under such-and-such conditions we tend to have such-and-such experiences?' But the open-question argument presupposes the success of phenomenalistic reduction. If you have a translation S' of a thing sentence S into phenomenalistic language, then it is well and good to ask 'What more does S mean than S'?', but you must not ask this rhetorical question unless you have constructed S'. Another play was to say: 'Pseudo-hypotheses, like the one about the demon, have only *picture meaning*'. Besides representing an objectionable form of argument (namely, assuming the philosophical point at issue and explaining your opponents' propensity to error psychologically), this contention is just false. The 'demon hypothesis' is not just a *noise* that happens to evoke some 'pictures in the head'; it is a grammatical sentence in a language; it is one we can offer free translations of; it is subject to linguistic transformations; we can deduce other statements from it and also say what other statements imply it; we can say whether it is linguistically appropriate or inappropriate in a given context, and whether a discourse containing it is linguistically regular or deviant. The verificationists would retort: 'It doesn't follow it has *meaning*'. But they would just be wrong, for this is just what meaning is: being meaningful is being subject to certain kinds of recursive transformations, and to certain kinds of regularities; we may not know much more about the matter than that today, but we know enough to know that what the verificationists were propounding was not an analysis of meaning but a persuasive redefinition. The worst argument of all, however, was the one which ran as follows: 'If you *do* admit the demon hypothesis as a logical possibility, then you will be doomed to utter skepticism; for you will never be able to offer any reason to say that it is false'. In case anyone needs to hear a reply to this claim, that verificationism and verificationism alone can save us all from the bogey of skepticism here is one: If the demon hypothesis is so constructed as to lead to exactly the same testable consequences as the more plausible system of hypotheses that we actually believe (or to the same testable consequences as any system of hypotheses that rational men would find more plausible), then it is not logically false, but it is logically impossible that it should ever be rational to believe it. For rationality requires that when two hypotheses H_1, H_2 lead to the same testable predictions (either at all times, or at the present), and H_1 is *a priori* much more plausible than H_2, then H_1 should be given the preference over H_2. If, in particular, H_1 has been accepted, and all hypotheses *a priori* more plausible than H_1 have led to a false prediction, then we should not give up H_1 merely because someone confronts us with a *logical possibility* of its being false. (This is roughly Newton's 'rule 4' in *Principia*.)

But, it may be asked, 'Is there really such a thing as *a priori* plausibility?' The answer is that it is easily shown that all possible inductive logics depend implicitly or explicitly on this: An *a priori* ordering of hypotheses on the basis of 'simplicity', or on the basis of the kinds of predicates they contain, or of the form of the laws they propose, or some other basis. To refuse to make any *a priori decisions* as to which hypotheses are more or less plausible is just to commit oneself to never making any inductive extrapolation from past experience at all; for at any given time infinitely many mutually incompatible hypotheses are each compatible with any finite amount of data, so that if we ever declare that a hypothesis has been 'confirmed', it is not because *all* other hypotheses have been ruled out, but because all the remaining hypotheses are rejected as too implausible even though they agree with and even predict the evidence – i.e. at some point hypotheses must be rejected on *a priori* grounds if any hypothesis is ever to be accepted at all. Again, the skeptic will object, 'How do you know the demon hypothesis is less plausible than the normal hypothesis?' But the answer is that to accept a plausibility ordering is neither to make a judgment of empirical fact nor to state a theorem of deductive logic; it is to take a methodological stand. One can only say whether the demon hypothesis is 'crazy' or not if one has taken such a stand; I report the stand I have taken (and, speaking as one who has taken this stand, I add: and the stand all rational men take, implicitly or explicitly). In sum, we can 'rule out' the demon hypothesis without playing fast and loose with the notion of logical impossibility or with the notion of meaninglessness; we have only to admit that we have taken a stand according to which this hypothesis is *a priori* less probable than the normal hypothesis, and then observe the peculiar fact: it is a logical truth (because of the way the demon hypothesis was constructed) that if the demon (hypothesis) is true, it cannot be rational to believe it (assuming, of course, the following maxim of rationality: Do not believe H_1 if all the phenomena accounted for by H_1 are accounted for also by H_2, and H_2 is more plausible than H_1). But if it is a logical truth (relative to the above maxim of rationality) that it would always be irrational to believe the demon hypothesis, then that is enough; if we can justify rejecting it, we need not feel compelled to go further and try to show that it does not represent even a logical possibility.

Another fashionable way of rejecting fictionalism has its roots in instrumentalism rather than in verificationism. One encounters, for example, somewhat the following line of reasoning: to ask whether statements are 'true' cannot be separated from asking whether it is rational to accept those statements (so far, so good), since it is rational to accept *p is true* just in case it is rational to accept *p*. But the end purpose of our whole 'conceptual system' is just the prediction and control of experience (or that plus 'simplicity', whatever that is). The fictionalist concedes that the conceptual system of material objects (or whatever) leads to successful prediction (or as successful as we have been able to manage to date) and that it is as simple as we have been able to manage to date. But these are just the factors on which rational acceptance depends; so it is rational to accept our conceptual system, and rational to call the propositions that make it up 'true' (or 'as true as anything is', in Anthony Quinton's happy phrase, since we always reserve the right to change our minds).

Now, there is unquestionably some insight in this retort to fictionalism. Elementary as the point may be, it is correct to remind the fictionalist that we cannot separate the grounds which make it rational to accept a proposition *p* from the grounds which make it rational to accept *p is true*. I myself dislike talk of simplicity, because simplicity in any measurable sense (e.g. length of the expressions involved, or number of logical connectives, or number of argument places of the predicates involved) is only *one* of the factors affecting the judgments of relative plausibility that scientists and rational men actually make, and by no means the most important one. But this is not a crucial point; we have only to recognize that the instrumentalist is using the word simplicity to stand for a complicated matter depending on many factors, notwithstanding some misleading connotations the word may have. The fictionalist concedes that predicative power and 'simplicity' (i.e. overall plausibility in comparison with rival hypotheses, as scientists and rational men actually judge these matters) are the hallmarks of a good theory, and that they make it rational to accept a theory, at least 'for scientific purposes'. But then – and it is the good feature of the instrumentalist strategy to press this devastating question home to the fictionalist – what *further* reasons could one want before one regarded it as rational to *believe* a theory? If the very things that makes the fictionalist regard material objects, etc. as 'useful fictions' do not make it rational to believe the material object 'conceptual system', what could make it rational to believe anything?

Historically, fictionalists split into two camps in the face of this sort of question. A theological fictionalist like Duhem maintained that Thomistic metaphysics (and metaphysics alone) could establish propositions about reality as true; science could only show that certain propositions are useful for prediction and systematization of data. A skeptical fictionalist like Hans Vaihinger maintained, on the other hand, that nothing could establish that, e.g. material objects really exist; we can only know that they are useful fictions. But neither move is satisfactory. Inquirers not precommitted to the Catholic Church do not agree that Thomistic metaphysics is a superior road to truth than modern science; and skepticism only reduces to a futile and silly demand that a deductive (or some kind of *a priori*) justification be given for the basic standards of inductive inquiry, or else that they be abandoned. Moreover, there is something especially pathetic about the skeptical version of fictionalism; for Vaihinger and his followers in the philosophy of 'As-If' did not doubt that science will lead to (approximately) correct prediction, and thereby they did accept induction at one point (notwithstanding the lack of a deductive justification), although they refused to believe that science leads to *true* theories, and thereby rejected induction (or the hypothetico-deductive method, which Mill correctly regarded as the most powerful method of the inductive sciences) at another point. Why can we never know that scientific theories are true? Because, the fictionalist said, we can give no deductive proof that they are true, even assuming all possible observational knowledge. But neither can we give a deductive proof that the sun will rise tomorrow! The fictionalist was thus a halfhearted skeptic; he chose to accept induction partially (as leading to successful prediction of experience), but not totally (as leading to true belief about things).

While I agree so far with the instrumentalist strategy of argument, I am deeply disturbed by one point; the premiss that the purpose of science is prediction of experience (or that plus 'simplicity', where simplicity is some kind of a funny aim-in-itself and not a rubric for a large number of factors affecting our judgment of probable truth). This premiss makes it easy to refute the fictionalist: for if there is no difference between believing p and believing that p leads to successful prediction (at least when p is a whole conceptual system), then fictionalism immediately collapses. But this is just verificationism again, except that now 'the unit of meaning is the whole conceptual system'. It is hard to believe that there is such a thing as 'the aim of science' – there are many aims of many scientists; and it is just not the case that all scientists are primarily interested in prediction. Some scientists are primarily interested in, for example, discovering certain facts about radio stars, or genes, or mesons, or what have you. They want successful predictions in order to confirm their theories; they do not want theories in order to obtain the predictions, which are in some cases of not the slightest interest in themselves, but of interest only because they tend to establish the truth or falsity of some theory. Also, it is just not the case that simplicity is a thing that all scientists value as an end in itself; many scientists only care about simplicity because and when it is evidence of truth. At bottom the only relevant difference between the following two statements:

(3) The aim of science is successful prediction

and

(4) An aim of some scientists is to know whether or not it is true that mesons behave in such-and-such a way

besides the incredible pomposity of (3) ('the aim of science' indeed!), is that (3) is couched in observation language. But why should the aim of science, if there is such a thing, or even the aims of all scientists be statable in observation language any more than the content of science is expressible in observation language? Surely this is just a hangover from reductionism!

In sum, fictionalism has on the whole been rejected for a bad reason:

because verificationism has made the perfectly sound and elementary distinction between truth of scientific theory and truth of its observational consequences unpopular, and thereby dismissed just the point – the apparent gap between the two – that worried the fictionalists. But, as we have also seen, there is a rejoinder to fictionalism that does not depend on reductionist views about either the content or the 'aim' of science. The rejoinder is simply that the very factors that make it rational to accept a theory 'for scientific purposes' also make it rational to believe it, at least in the sense in which one ever 'believes' a scientific theory – as an approximation to the truth which can probably be bettered, and not as a final truth. Fictionalism fails because (Duhem to the contrary) it could not exhibit a better method for fixing our belief than the scientific method, and because (Vaihinger to the contrary) the absence of a deductive justification for the scientific method in no way shows that it is not rational to accept it.

At this point we have considered the rejoinder to indispensability arguments, i.e. that it might be indispensable to believe p but that p might nonetheless really be false, and we have rejected this rejoinder, not for the usual verificationist or instrumentalist reasons, which seem to rest on false doctrines, but because it is silly to agree that a reason for believing that p warrants accepting p in all scientific circumstances, and then to add 'but even so it is not *good enough*'. Such a judgment could only be made if one accepted a trans-scientific method as superior to the scientific method; but this philosopher, at least, has no interest in doing *that*.

IX. Unconsidered complications

In this chapter, I have chosen to go into detail on one group of questions – those having to do with the indispensability of quantification over abstract entities such as sets – at the cost of having to neglect many others. One group of questions which I might have considered has to do with the existence of what I might call 'equivalent constructions' in mathematics. For example, numbers can be constructed from *sets* in more than one way. Moreover, the notion of *set* is not the *only* notion which can be taken as basic; we have already indicated that predicative set theory, at least, is in some sense inter-translatable with talk of formulas and truth; and even the impredicative notion of set admits of various equivalents: for example, instead of identifying functions with certain *sets*, as I did, I might have identified *sets* with certain functions. My own view is that none of these approaches should be regarded as 'more true' than any other; the realm of mathematical fact admits of many 'equivalent descriptions': but clearly a whole essay could have been devoted to *this*.

Again, we discussed very briefly the interesting topic of conventionalism. Even if the conventionalist view has never been made very plausible (or even clear), it raises fascinating issues. The question of to what extent we might revise our basic logical principles, as we have had to revise some of our basic geometrical principles in mathematical physics, is an especially fascinating one. Today, the tendency among philosophers is to assume that in no sense does logic itself have an empirical foundation. I believe that this tendency is wrong; but this issue too has had to be avoided rather than discussed in the present section. My purpose has been to give some idea of the many-layered complexity which one encounters in attacking even one part of the philosophy of logic; but I hope I have not left the impression that the part discussed in this chapter is all there is.

Bibliography

Althusser, L. 1965. *Pour Marx* and *Lire le Capital*, Paris.

Bohr, N. 1951. 'Discussion with Einstein on epistemological problems in atomic physics', in P. A. Schilpp (ed.), *Albert Einstein Philosopher-Scientist*, New York, 199–242.

Bridgman, P. 1927. *The Logic of Modern Physics*, New York.

Carnap, R. 1939. *The Foundations of Logic and Mathematics*, 4:3 of the *International Encyclopedia of Unified Science*, Chicago.

1948. 'On the application of inductive logic', *Philosophy and Phenomenological Research*, VIII, 133–48.

1950. *Logical Foundations of Probability*, Chicago.

1952. *The Continuum of Inductive Methods*, Chicago.

1955. 'Testability and meaning' in H. Feigl and M. Brodbeck (eds.), *Readings in the Philosophy of Science*, New York, 47–92. Reprinted from *Philosophy of Science*, 3 (1936) and 4 (1937).

1956. 'The methodological character of theoretical concepts' in H. Feigl *et al.* (eds.), *Minnesota Studies in the Philosophy of Science*, Minneapolis, 1–74.

Chomsky, N. 1957. *Syntactic Structures*, The Hague.

Church, A. 1956. *Introduction to Mathematical Logic*, Princeton, N.J.

Conant, J. 1947. *On Understanding Science*, New Haven, Conn.

Craig, W. 1953. 'On axiomatizability within a system', *Journal of Symbolic Logic*, 18:1 (March), 30–2. See also 'Replacement of auxiliary expressions', *Philosophical Review*, LXV, 1 (January 1956), 38–55.

Davis, M. 1958. *Computability and Unsolvability*, New York.

De Finetti, B. 1931. 'Sul significato suggestivo della probabilità', *Fundamenta mathematicae*, XVII, 298–329.

Donnellan, K. 1962. 'Necessity and criteria', *The Journal of Philosophy*, LIX, 22 (25 October), 647–58.

Einstein, A., Podolsky, B. and Rosen, N. 1935. 'Can quantum mechanical description of reality be considered complete?', *Physical Review*, 47:2, 777–80.

Feigl, H. 1950. 'De principiis non disputandum...?' in M. Black (ed.), *Philosophical Analysis*, Ithaca, 119–56.

Feyerabend, P. 1957. 'The quantum theory of measurement', in S. Koerner (ed.) *Observation and Interpretation in the Philosophy of Physics*, New York, 121–30.

Finkelstein, D. 1964. 'Matter, space, and logic', in R. Cohen and

M. Wartofsky (eds.), *Boston Studies in the Philosophy of Science*, V, 199–215:

Fodor, J. A. 1961. 'Of words and uses', *Inquiry*, 4:3 (Autumn), 190–208.

Fodor, J. A. and Katz, J. J. 1963. 'The structure of a semantic theory', *Language*, 39, 170–210.

Gödel, K. 1931. Über formal unentscheidbare Sätze der Principia Mathematica und verwandter System I'. *Monatschefte für Mathematik und Physik*, XXXVIII, 173–98.

1951. 'A remark about the relationship between relativity theory and idealistic philosophy' in P. A. Schilpp (ed.), 555–62. *Albert Einstein Philosopher-Scientist*, New York.

Goodman, N. 1946. 'A query on confirmation', *Journal of Philosophy*, XLIII, 383–5.

Grünbaum, A. 1962. 'Geometry, chronometry, and empiricism', in H. Feigl and G. Maxwell (eds.), *Minnesota Studies in the Philosophy of Science*, Minneapolis, 405–526.

Hanson, N. 1958. *Patterns of Discovery*, Cambridge.

1963. *The Concept of the Positron*, Cambridge.

Hempel, C. 1963. 'Implications of Carnap's work for the philosophy of Science', in P. A. Schilpp (ed.), *The Philosophy of Rudolf Carnap*, La Salle, Ill., 685–707.

Katz, J. 1962. *The Problem of Induction and its Solution*, Chicago.

Kemeny, J. 1953. 'The use of simplicity in induction', *Philosophical Review*, LXII, 391–408.

1963. 'Carnap's theory of probability and induction', in P. A. Schilpp (ed.), *The Philosophy of Rudolf Carnap*, La Salle, Ill.

Kemeny, J. and Oppenheim, P. 1952. 'Degree of factual support', *Philosophy of Science*, XIX, 308–24.

1956. 'On reduction', *Philosophical Studies*, VII, 6–19.

Kleene, S. 1952. *Introduction to Metamathematics*, New York.

Kuhn, T. 1962. *The Structure of Scientific Revolutions*, 2:2 of the *International Encyclopedia of Unified Science*, Chicago.

London, F. and Bauer, E. 1939. *La Théorie de l'Observation en Mécanique Quantique*, Paris.

Ludwig, G. 1954. *Die Grundlagen der Quantenmechanik*, Berlin.

Margenau, H. and Wigner, E. 1962. 'Comments on Professor Putnam's comments', *Philosophy of Science*, XXIX, 3 (July), 292–3.

Maxwell, G. 1962. 'The necessary and the contingent' in H. Feigl and G. Maxwell (eds.), *Minnesota Studies in the Philosophy of Science*, III, Minneapolis, 405–526.

Mehlberg, H. 1958. 'The observational problem of quantum theory', read at the May 1958 meeting of the American Philosophical Association, Western Division.

Nagel, E. 1939. 'Principles of the theory of probability', *International Encyclopedia of Unified Science*, I, 6, Chicago, 63ff.

1961. *Structure of Science*, New York.

BIBLIOGRAPHY

Newton, I. 1947. *Principia*, F. Cajori (ed.) Berkeley, Calif.
Oppenheim, P. and Putnam, H. 1958. 'Unity of science as a working hypothesis', in H. Feigl, G. Maxwell and M. Scriven (eds.), *Minnesota Studies in the Philosophy of Science*, II, 3–36.
Peano, G. 1957. *Opera Scelte*, Rome.
Popper, K. 1959. *The Logic of Scientific Discovery*, London.
Putnam, H. 1956. 'A definition of degree of confirmation for very rich languages', *Philosophy of Science*, XXIII, 58–62.
1957. 'Mathematics and the existence of abstract entities', in *Philosophical Studies*, 7, 81–8.
1961. 'Comments on the paper of David Sharp', *Philosophy of Science*, XXVIII, 3, 234–9.
1971. *Philosophy of Logic*, New York.
Quine, W. 1952. *Methods of Logic*, London.
1957. 'The scope and language of science', *British Journal for the Philosophy of Science*, 8, 1–17.
1964. *Set Theory and Its Logic*, Harvard.
Reichenbach, H. 1944. *Philosophic Foundations of Quantum Mechanics*, California.
1947. *The Theory of Probability*, Berkeley.
1958. *The Philosophy of Space and Time*, New York. (An English translation of Reichenbach's *Philosophie der Raum-Zeit-Lehre*, Berlin 1928.)
Rosenbloom, P. 1950. *Elements of Mathematical Logic*, New York.
Savage, J. 1954. *The Foundations of Statistics*, New York.
Scheffler, I. 1960. 'Theoretical terms and a modest empiricism', in A. Danto and S. Morgenbesser (eds.), *Philosopher of Science*, Cleveland, Ohio, 159–73. See also 'Prospects of a modest empiricism', *Review of Metaphysics*, 10:3 (March, 1957), 383–400, and 10:4 (June, 1957), 602–25.
Schilpp, P. (ed.) 1963. *The Philosophy of Rudolf Carnap*, La Salle, Ill.
Sharp, D. H. 1961. 'The Einstein-Podolsky-Rosen paradox re-examined', *Philosophy of Science*, 28:3 (July), 225–33.
Shimony, A. 1965. 'Quantum physics and the philosophy of Whitehead', in R. Cohen and M. Wartofsky (eds.), *Boston Studies in the Philosophy of Science*, II, (New York), 307–30.
Tarski, A. 1949. 'The semantic conception of truth', in H. Feigl and W. Sellars (eds.), *Readings in philosophical analysis*, New York, 52–84.
van Heijenoort, J. (ed.) 1967. *From Frege to Gödel*, Cambridge, Mass.
Wheeler, J. A. 1962. 'Curved empty space-time as the building material of the physical world' in E. Nagel, P. Suppes and A. Tarski (eds.), *Logic, Methodology and the Philosophy of Science*, Stanford, 361–74.
Ziff, P. 1960. *Semantic Analysis*, Ithaca.

Index

Printed in the United States
58159LVS00007B/35